Mass spectrometry for chemists and biochemists

Cambridge Texts in Chemistry and Biochemistry

GENERAL EDITORS

S. J. Benkovic
Professor of Chemistry
Pennsylvania State University, Philadelphia

D. T. Elmore
Professor of Biochemistry
The Queen's University, Belfast

J. Lewis
Professor of Inorganic Chemistry
University of Cambridge

E. L. Muetterties
Professor of Inorganic Chemistry
University of California, Berkeley

K. Schofield
Professor of Organic Chemistry
University of Exeter

J. M. Thomas
Professor of Physical Chemistry
University of Cambridge

B. A. Thrush
Professor of Physical Chemistry
University of Cambridge

Mass spectrometry for chemists and biochemists

M. E. ROSE
Department of Biochemistry
University of Liverpool

R. A. W. JOHNSTONE
Department of Organic Chemistry
University of Liverpool

CAMBRIDGE UNIVERSITY PRESS
Cambridge
London   New York   New Rochelle
Melbourne   Sydney

CHEMISTRY

6654-4798

Published by the Press Syndicate of the University of Cambridge
The Pitt Building, Trumpington Street, Cambridge CB2 1RP
32 East 57th Street, New York, NY 10022, USA
296 Beaconsfield Parade, Middle Park, Melbourne 3206, Australia

First published 1982

Printed in Great Britain at
the University Press, Cambridge

Library of Congress catalogue card number: 81-10122

British Library cataloguing in publication data

Rose, M. E.
Mass spectrometry for chemists and biochemists.
1. Mass spectrometry
I. Title II. Johnstone, R. A. W.
545'.33   QD96.M3

ISBN 0 521 23729 7 hard covers
ISBN 0 521 28184 9 paperback

In memory of my late grandfathers   M. E. R.

Dedicated to my Mother and Father, and Christine,
    Steven and Fiona   R. A. W. J.

CONTENTS

INTRODUCTION

'Trout made into fish cakes is still trout'
Muggles in *The Minnipins*, Carol Kendall (1959)

The basis of mass spectrometry is the production of ions from neutral compounds and the examination of the subsequent decomposition of those ions. In mass spectrometry, a substance is characterized by investigating the *chemistry* of ions resulting from that substance. Because the technique involves a chemical reaction, the sample being investigated is not recoverable; however, only very small quantities of material are required for the analysis. Most other physical methods of analysis deal with a narrowly defined property of a molecule, but this is not so in mass spectrometry. As with any chemical reaction, the precise outcome is dependent on a considerable number of factors, such as temperature, concentration, effects of the medium and so on. It is this uncertainty which lends mass spectrometry its versatility, intricacy and charm.

Chemical reactions in solids, liquids and the gas phase are usually discussed in terms of isolated molecules whereas the actual behaviour is the result of 'group' effects arising from collisional activation of molecules. Most mass spectra are measured at low pressures when collisions between ions and molecules are rare so that interpretation of the mass spectra in terms of isolated species is more satisfactory. For practical purposes, these low concentrations of ions impose severe restraints on methods of investigating them. The concentrations of ions are so low that other common physical methods such as ultraviolet, infrared and nuclear magnetic resonance spectroscopy cannot be used to elucidate the structures and internal energies of the ions. At best, postulated ion structures can only be described as consistent with indirect evidence based on isotopic labelling, thermochemical arguments and reasoning by analogy. As its name implies, a mass spectrometer measures mass (or strictly mass-to-charge ratio) and gives no direct information on ion structures. Because of this lack of knowledge of structures, fundamental investigation of the mechanisms of ion decomposition has been severely hampered. It is stressed that even without any sound, basic understanding of ion structures and mechanisms of mass spectrometric fragmentations, a great deal of analytical information may be extracted by use of empirical rules and concepts. At this stage, it is worth reminding the reader that mass spectrometry alone seldom gives a unique solution to a problem and it is best used in conjunction with information obtained by other means.

Following the construction of the first mass spectrometers many years ago, there followed a period of modest interest by physicists and chemists until it was realized by the organic chemist that here was an instrument which could be of immense value to his work. When *Mass Spectrometry for Organic Chemists* first appeared (1972), the technique already held a prime position in chemistry, and particularly in organic chemistry, for elucidation of molecular structures. Both the applications and development of mass spectrometry were, and still are, increasing rapidly. In the intervening ten years, mass spectrometry has been applied with advantage in many other, diverse fields. Notably, its use has spread to biochemistry, medicine and toxicology. The National Gallery in London has taken receipt of a mass spectrometer to aid its investigations concerning conservation, history and authenticity of works of art. Mass spectrometers have travelled as far as, if not further than, any other analytical apparatus since several space probes have carried miniaturized mass spectrometers for chemical analysis of extra-terrestrial atmospheres and soils. Many of these wider applications are due to improved instrumentation. In particular, many recent advances can be attributed to three interconnected factors: (*a*) the linking of mass spectro-meters to computers, (*b*) improvements permitting ready combination of mass spectrometers with gas and liquid chromatographs and (*c*) development of sophisticated means of producing ions from neutral species. These three topics are accorded separate chapters in this book. Included in category (*c*) is the production of negative ions. In mass spectrometry, positive ions have been investigated more thoroughly than negative ones but studies of negative ions are increasing owing to improvements in the efficiency of their generation and to the complementary analytical information gained. Therefore, the majority of the text concerns positive ions but discussion of negative ions is included where appropriate.

The present text was conceived as a simple up-dating of *Mass Spectrometry for Organic Chemists*, but the increasing applications of the technique begged a wider scope for the book and the newer developments necessitated considerable rewriting and expansion. Therefore, it is appropriate that the title of the text be amended to accommodate these changes and *Mass Spectrometry for Chemists and Biochemists* was selected. To write a simple introductory text to this subject is a somewhat uncomfortable task since it becomes necessary to make statements which may be arguable, but the authors are of the opinion that simplicity and accuracy are not mutually exclusive in this context. After an introductory chapter, instrumentation for mass spectrometry is described briefly, then there follows a new chapter about the application of computers to mass spectrometry. Combined gas chromatography/mass spectrometry, being a routine and important tool in many different fields, is described in some detail in chapter 4. Technologically, combined liquid chromatography/mass spectrometry is less advanced but its use and worth are growing rapidly with developments in methodology. Hence, it is covered in the same chapter with emphasis on technique. A separate chapter was considered preferable to scattered references to the important topic of derivatization as an adjunct to mass spectrometric investigations. This subject is discussed in chapter 5, which contains a section on derivatization of inorganic compounds. One trend discernible in mass spectrometry is away from simple identification of unknown substances and towards a quantification of known compounds and more fundamental mechanistic studies. Chapter 6 is devoted to the first of these aspects and will be of most use to those interested in life sciences and analytical chemistry. The following three chapters are in part concerned with more fundamental topics such as elucidation of fragmentation mechanisms and ion structures, and measurement of thermochemical and kinetic parameters. In chapter 7, some additional methods of generating ions are discussed (with otherwise intractible substances in mind) whilst chapter 8 covers the special case of metastable ions. These ions are at the heart of several methods of analysis and will play an expanding role within mass spectrometry in the future because they can be used to probe not only the products of mass spectrometric reactions but also the reactions themselves. In this introductory text, it has been felt necessary to include some theoretical aspects of mass spectrometry as well as its empirical applications, because, by understanding fundamentals, an empiricist may make better use of the method. Such an approach is embodied in chapter 9. The next two chapters (10 and 11) deal with empirical application of mass spectrometry to structural elucidation, which forms a large part of routine mass spectrometry for the typical user. The examples of structural elucidation were chosen to illustrate many of the modern methods of mass spectrometric analysis. Throughout, the various aspects of mass spectrometry are introduced and explained briefly, but when more detailed explanations are desirable, these have been included in chapter 12 (Further discussion of selected topics). Miscellaneous topics which would not be in context elsewhere in the book are also included in this chapter.

The book is suitable for those without any prior knowledge of mass spectrometry and, whilst being an introductory text, does provide an approach to the more advanced topics for the interested reader. As part of this approach, an extensive bibliography is included, facilitating consultation of further, more specialized literature. For the convenience of the reader, a simple table of quantities (conversion factors) appears at the front of the book.

ACKNOWLEDGEMENTS

The authors thank Mrs Susan Beard for her excellent assistance, Mr M. A. Breen and Dr. L. J. Goad for permission to use some of their results, VG Analytical Limited for enthusiastic co-operation and Mrs B. A. Rose for her help and perseverance.

Table of quantities

1 kcal = 4.184 kilojoules (kJ)
1 electron volt (eV) = 1.602×10^{-22} kJ
1 torr = 133.3 newtons per square metre (Nm^{-2})
1 microgramme (μg) = 10^{-6} g
1 nanogramme (ng) = 10^{-9} g
1 picogramme (pg) = 10^{-12} g
1 femtogramme (fg) = 10^{-15} g

List of abbreviations

ADC, analogue-to-digital converter
CI, chemical ionization
DAC, digital-to-analogue converter
EI, electron impact
FD, field desorption
FI, field ionization
GC/MS, combined gas chromatography/mass spectrometry
IKES, ion kinetic energy spectroscopy
LC/MS, combined liquid chromatography/mass spectrometry
MIKES, mass-analysed ion kinetic energy spectroscopy
PFK, perfluorokerosene
QET, quasi-equilibrium theory
%RA, percentage relative abundance
%RIC, percentage reconstructed ion current
%TIC, percentage total ion current
TMS, trimethylsilyl

mass spectrum which may be recorded on a cathode ray tube, paper chart, photographic plate or via a computer. An example of part of a mass spectrum recorded as peaks on a paper chart is shown in figure 1.2. As a standard practice, it is usual to make a record of a mass spectrum in either a *normalized* (or *percentage relative abundance*, % RA) form or as a *percentage of total ion current* (% TIC). In a normalized record, the biggest peak in the spectrum is called the *base peak* and its height is put equal to 100 units; the relative heights of all other peaks are referred to this base peak and lie between 0 and 100 units (note that the base peak is not necessarily the molecular ion peak although it may be). The height of a peak represents the abundance of ions at that particular *m/z* value. Table 1.1 illustrates part of the mass spectrum of 1, 3-dimethylbenzene recorded in this way and figure 1.3 shows the same spectrum as a normalized line diagram. Very often not all the peaks in a mass spectrum appear in the normalized line diagram since peaks of less than one per cent of the size of the base peak are frequently arbitrarily omitted as unimportant. Care must be exercised in this respect because, even though ions may not be abundant, they can be important for elucidating structures from mass spectra. For example, the molecular ions of a compound undergoing extensive fragmentation will have low abundance but are still a very important feature of the spectrum. Fragment ions may decompose at about the same rate as they are formed when again their abundance would be low but they could be important for unravelling a fragmentation pathway. For compounds of high molecular weight, the low mass ions (e.g. below *m/z* 50) may be numerous and abundant but also may have little value for the interpretation of a spectrum; such low mass ions are sometimes omitted from the normalized line diagram.

To obtain a record of a mass spectrum by the percentage of total ion current method, abundances of all ions giving peaks of significant size, from the molecular ion down to a suitably chosen low mass (often around *m/z* 40, except in chemical ionization – see later, section 7.2), are added together as a

Table 1.1. *Normalized ion abundances^a in the mass spectrum of 1,3-dimethyl-benzene*

m/z	Abundance	*m/z*	Abundance	*m/z*	Abundance
38,	1.2	65	5.5	102	1.1
39	7.1	74	1.1	103	6.1
50	2.8	77	11.3	104	2.6
51	9.1	78	5.1	105	28.7
52	5.0	79	6.5	106	61.7
53	2.2	89	1.9	107	5.4
62	1.6	91	100.0		
63	4.2	92	7.8		

^{a}Ion abundances less than one per cent of the base peak have been omitted.

Figure 1.2. Part of a mass spectrum recorded on photographic paper (the three traces record the spectrum at increasing sensitivities from the lower to the upper trace).

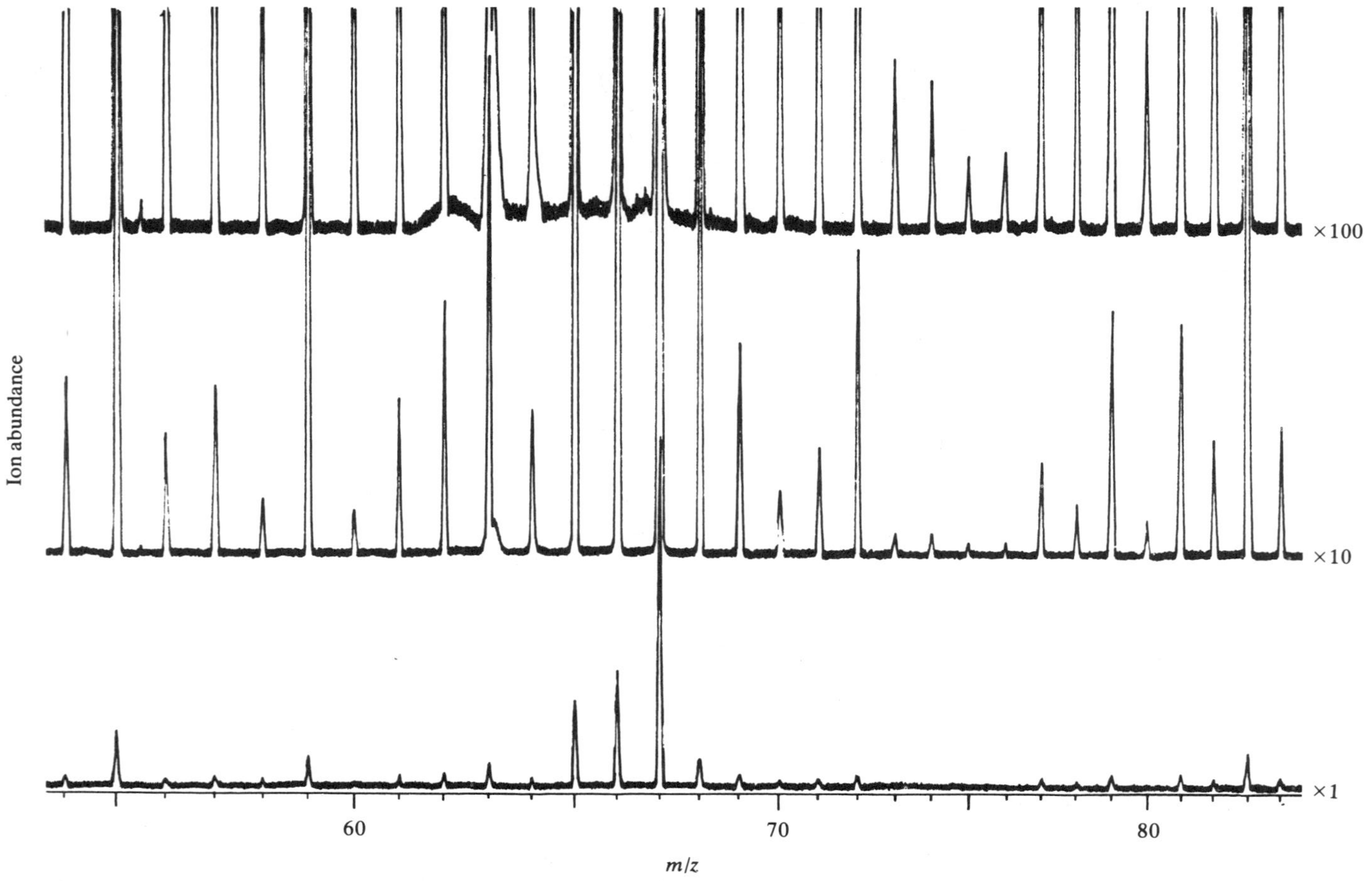

Table 1.2. *Ion abundances*[a] *in the mass spectrum of 1, 3-dimethylbenzene as % TIC*

m/z	Abundance	m/z	Abundance	m/z	Abundance
39	2.5	65	1.9	92	2.7
50	1.0	77	3.9	103	2.1
51	3.2	78	1.8	105	10.0
52	1.7	79	2.2	106	21.4
63	1.5	91	34.8	107	1.9

[a]Ion abundances of less than one per cent are omitted.

measure of the total ion current (TIC). The relative contribution of each ion to this total is then calculated as a percentage (% TIC). Table 1.2 records part of the mass spectrum of 1,3-dimethylbenzene from m/z 38 to the molecular ion with abundances of ions as % TIC. This last method of recording a mass spectrum is used less frequently than the normalization method but has some advantages in emphasizing the relative importance of an ion in the whole spectrum. It is important to know this relative importance when the mass spectrometer is used not to record a whole mass spectrum but to monitor the masses of only one ion or a small number of ions in a mass spectrum (see later, chapter 6). Also, the percentage ion current method finds some favour in comparing the spectra of isomers when changes in the relative

Figure 1.3. Line diagram showing normalized relative abundances of ions in the mass spectrum of 1,3-dimethylbenzene; m/z 91 = 100 units.

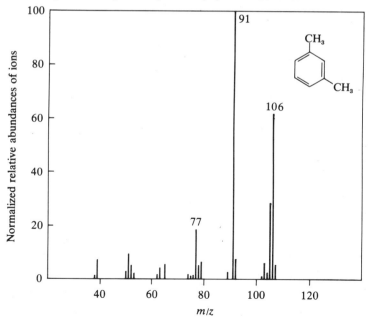

contributions of certain ions to the total ion current may be sufficient to distinguish between the isomers. The difference between the two ways of recording spectra in tabular form is more apparent than real since the relative abundance of a peak in a normalized spectrum is simply related numerically to its value as a percentage of total ion current. However, if the spectrum has been processed to remove 'background' ions of constant, unwanted impurities such as chemical ionization reactant gases (sections 2.3 and 7.2) or decomposition products of the stationary phases of gas chromatographic columns in combined gas chromatography/mass spectrometry systems (sections 2.2 and 4.2), this simple relationship may no longer exist. If such processing has been performed then the % TIC method is often replaced by an equivalent method: the percentage reconstructed ion current (% RIC). This latter sums the ion current from all consequential ions remaining after subtraction of 'background' ions. It should be noted that the normalization method is likely to be the less accurate because a small irregularity in recording or measuring the base peak will markedly affect all the other ion abundances normalized against it. The measurement of height of the base peak is prone to error, as caused for example by the recorder going off-scale. Usually, the percentage of total ion current method is virtually unaffected by such anomalies and is to be preferred, particularly if the mass spectral data are acquired and processed by a computer when the greater numeric difficulty of the method is unimportant. Most data systems allow either or both methods to be selected, with presentation of the spectrum in tabular or diagrammatic form. Table 1.3 shows a computer output listing of the mass spectrum of decahydronaphthalene in both % RA and % TIC forms.

Two main facets of mass spectrometry will be apparent from the above description. (i) In contrast with what happens in most other common methods of physicochemical analysis, some or all of the sample is consumed in mass spectrometry and is not recoverable. However, mass spectrometers are very sensitive instruments and spectra may be obtained from a few nanogrammes of material. (ii) Unlike other physical methods, mass spectrometry does not deal with a well-defined property of a molecule. The appearance of a mass spectrum depends not only on the compound itself but also upon the interval of time between ionization and detection of ions, upon the initial energy distribution in the molecular ions and hence on the method of ionization, and partly upon the physical characteristics of the instrument and therefore on its design and manufacture.

For these reasons it is not possible to guarantee that a mass spectrum is accurately reproducible from instrument to instrument -- even of the same manufacture when operating under apparently identical conditions. Normally, variations in ion abundance of a few per cent are acceptable in comparing spectra from different laboratories, but generally, good uniformity is found

Table 1.3. *Ion abundances[a] in the mass spectrum of decahydronaphthalene*

Mass	% RA	% TIC
41	64.02	5.25
42	9.79	0.80
43	7.19	0.59
50	0.67	0.06
51	3.02	0.25
52	1.82	0.15
53	12.99	1.07
54	21.23	1.74
55	46.84	3.84
56	25.85	2.12
57	1.94	0.16
63	0.46	0.04
65	3.73	0.31
66	8.24	0.68
67	100.00	8.20
68	59.97	4.92
69	35.17	2.89
70	3.19	0.26
71	0.45	0.04
77	5.02	0.41
78	1.08	0.09
79	9.43	0.77
80	5.99	0.49
81	86.55	7.10
82	78.53	6.44
83	15.70	1.29
84	17.16	1.41
85	0.82	0.07
91	2.14	0.18
93	2.87	0.24
94	3.50	0.29
95	60.05	4.93
96	99.19	8.14
97	11.35	0.93
108	0.81	0.07
109	30.31	2.49
110	9.07	0.74
111	0.42	0.03
123	0.89	0.07
137	0.91	0.07
138	99.35	8.15
139	10.37	0.85
140	0.31	0.03

[a] All peaks over m/z 40 of greater than 0.3 percentage of relative abundance are included.

and not too much difficulty is experienced from small inconsistencies. The popular method of identifying compounds by matching sample spectra against standard mass spectra stored in a computer library (section 3.4) is a testament to this.

1.2. Multiply charged ions

In mass spectrometry, ions are generally produced with single positive charges but may sometimes have two, three, or more charges so that the corresponding mass-to-charge ratios correspond to half, one-third, or lower fractional masses. A molecular ion with a single positive charge corresponds to a mass m, but with two positive charges it corresponds to $m/2$. Doubly charged ions are fairly common, especially in the mass spectra of aromatic compounds, but more highly charged ions are rarer. Figure 1.4 shows part of a mass spectrum containing some prominent doubly charged ions amongst the singly charged ones. Assignment of doubly charged ions is obvious when peaks occur at half-integer m/z values but doubly charged ions occur at integer values if their mass, m, is even. In the latter case, some integer peaks may consist of superimposed peaks from singly and doubly charged ions. This should be borne in mind particularly when examining the spectra of aromatic compounds and especially if a mass peak occurs at an m/z value corresponding to half the molecular weight of the compound under investigation. The doubly charged species may fragment either into two singly charged ions or into another doubly charged ion with ejection of a neutral particle.

Figure 1.4. Partial mass spectrum showing peaks due to doubly-charged ions marked with an asterisk on the middle trace.

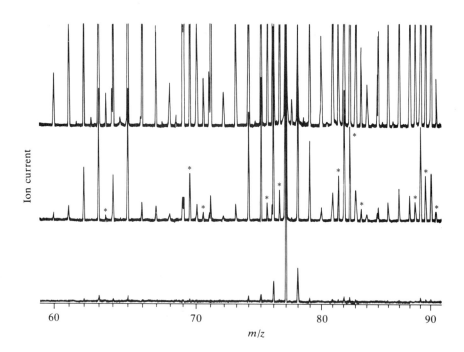

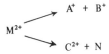

1.3.   Isotopes

Many elements in their natural states contain isotopes and, because a mass spectrometer measures mass-to-charge ratios, these isotopes appear in the mass spectrum. The most abundant isotope of carbon is ^{12}C, but natural carbon contains also ^{13}C and ^{14}C. Although as a beta-ray emitter the latter is extremely valuable for radio-tracer work, its natural abundance is so low as to make it almost inconsequential to mass spectrometry. Such is not the case for ^{13}C which occurs in a natural abundance of approximately 1.08 per cent in carbon. Hence, the mass spectrum of methane shows a molecular ion at m/z 16 ($^{12}CH_4$) together with an isotopic ion at m/z 17 ($^{13}CH_4$), the two ions having relative abundances of about 99 : 1. As the number of carbon atoms in a compound increases, so also do the chances of incorporating one ^{13}C atom into the molecule rather than a ^{12}C atom. A compound with ten carbon atoms would yield a molecular ion, $M^{+\cdot}$, and an isotopic ion one mass unit greater, $(M + 1)^{+\cdot}$, which would be about $10 \times 1.08 = 10.8$ per cent of the abundance of $M^{+\cdot}$ (see, for example, the case of decahydronaphthalene, $C_{10}H_{18}$; Table 1.3). The chances of finding two ^{13}C atoms in the same molecule increase with increasing numbers of carbon atoms so that $(M + 2)^{+\cdot}$ ions start to become more prominent. The chances of incorporating more than two ^{13}C atoms in a molecule are very small except with large numbers of carbon atoms, and $(M + 3)^{+\cdot}$ ions from this source may be safely ignored. For a compound with ten carbon atoms, the approximate relative heights of the $M^{+\cdot}$, $(M + 1)^{+\cdot}$, $(M + 2)^{+\cdot}$, and $(M + 3)^{+\cdot}$ peaks are 100 : 10 : 0.45 : 0.01 from which it can be seen how unimportant are peaks greater than $(M + 2)^{+\cdot}$. Tables are available giving the relative heights of $M^{+\cdot}$, $(M + 1)^{+\cdot}$, and $(M + 2)^{+\cdot}$ peaks for many elemental compositions (Beynon & Williams, 1963). All carbon compounds yield molecular and fragment ions accompanied by isotopic ions 1 and 2 mass units greater, making the appearance of the spectrum more complex. Figure 1.5 shows figure 1.1 modified so as to illustrate the appearance of the mass spectrum with ^{13}C isotopes included. Conversely, an estimate of the number of carbon atoms in an ion may be made by measuring the relative heights of the $M^{+\cdot}$ and $(M + 1)^{+\cdot}$ peaks, but the method does not yield good results when there are more than about ten or twelve carbon atoms.

Table 1.4 gives, for some elements commonly met in mass spectrometry, the approximate natural abundances of the more significant isotopes. It should be noted that sulphur, chlorine and bromine have particularly abundant isotopes separated by 2 mass units. For this reason, chlorine- and

Table 1.4. *The more important natural isotope abundances for elements commonly occurring in mass spectrometry*[a]

Element	Isotope (percentage of natural abundance)[b]		
Hydrogen	^{1}H (99.99)		
Boron	^{10}B (19.8)	^{11}B (80.2)	
Carbon	^{12}C (98.9)	^{13}C (1.1)	
Nitrogen	^{14}N (99.6)	^{15}N (0.4)	
Oxygen	^{16}O (99.8)	^{18}O (0.2)	
Fluorine	^{19}F (100.0)		
Silicon	^{28}Si (92.2)	^{29}Si (4.7)	^{30}Si (3.1)
Phosphorus	^{31}P (100.0)		
Sulphur	^{32}S (95.0)	^{33}S (0.7)	^{34}S (4.2)
Chlorine	^{35}Cl (75.5)	^{37}Cl (24.5)	
Bromine	^{79}Br (50.5)	^{81}Br (49.5)	
Iodine	^{127}I (100.0)		

[a] Metals frequently possess many abundant isotopes and, because of the diversity of organometallic compounds, a listing of metal isotopes has not been included here. Many reference texts contain listings of isotope abundances; see also Beynon (1960) and Kiser (1965).

[b] Percentages are given correct to the first decimal place. Trace isotopes are not included as they have little consequence in mass spectrometry at their natural abundance levels.

bromine-containing compounds especially are readily recognized in mass spectrometry and, by examining the isotope pattern in the molecular ion region, the numbers of chlorine and bromine atoms in the original molecule may be determined (Beynon, 1960; Biemann, 1962). Figure 1.6 shows the appearance of the molecular ion region for a dichlorobenzene; note the major peaks spaced 2 mass units apart due to the ^{35}Cl and ^{37}Cl isotopes of the chlorine atoms and the minor peaks caused by contributions from the ^{13}C isotope. A method of calculating simple isotopic patterns is presented in section 12.2.

Figure 1.5. The simple spectrum shown in figure 1.1 with the addition of ^{13}C-isotopic masses for ions M$^{+\cdot}$, A$^+$, B$^+$ and C$^+$.

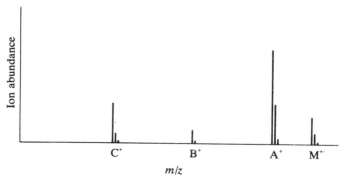

Normalized mass spectra are sometimes drawn with the isotopic contributions simplified. For example, for two peaks at m/z 253, 254 of relative heights as shown in figure 1.7a, the ^{13}C isotope contribution of m/z 253 contributes to the height of the peak at m/z 254; the ^{13}C isotope of m/z 254 occurs at m/z 255. If there were eighteen carbon atoms in the elemental composition of the ion at m/z 253, then approximately 18 × 1.1 = 19.8 per cent of its peak height is the ^{13}C contribution in m/z 254. By subtracting this ^{13}C contribution from m/z 254 the relative abundances of the ions appear as in figure 1.7b. Similarly the ^{13}C isotope peak of m/z 254 at m/z 255 disappears. This device of removing isotope contributions is extremely helpful for many organometallic compounds since metals frequently have several isotopes and so yield complex-looking spectra; for example, mercury has seven isotopes spread over 9 mass units. Ions containing metal atoms can therefore afford many isotope peaks and removing their separate contributions simplifies the interpretation of such a spectrum. Figure 1.8 shows the mass spectrum of diethylmercury both with and without the isotope contributions; the spectrum without isotope peaks is much easier to interpret.

Figure 1.6. Contributions of carbon and chlorine isotopes to the pattern of peaks in the molecular ion region of a dichlorobenzene. The molecular ion (M$^{+\cdot}$) is usually considered to be the peak having contributions from the isotopes of lowest mass but strictly all six ions are molecular ions.

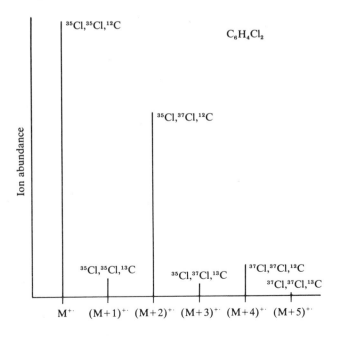

Figure 1.7. Appearance of part of the mass spectrum of a compound with 18 carbon atoms at m/z 253–255 (a) with ^{12}C- and ^{13}C-isotope contributions (b) with ^{13}C-isotope contribution removed and therefore showing only ^{12}C-isotopes.

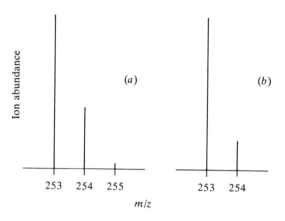

Figure 1.8. Partial mass spectrum of diethylmercury, (a) including all isotope peaks, and (b) mono-isotopic spectrum based on ^{202}Hg and ^{12}C.

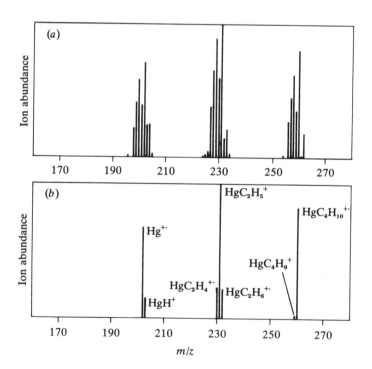

1.4.   Metastable ions

The sharp peaks of the normal ions in a mass spectrum are usually accompanied by some much broader, smaller peaks due to the so-called *metastable ions*. The maxima of these metastable ion peaks frequently occur at non-integral m/z values and these ions are easily distinguished from normal ions (figure 1.9). The peak shapes are approximately Gaussian but occasionally flat-topped or dish-shaped ones are found extending over several mass units. The origin of metastable ions is discussed later. These ions are very useful for determining fragmentation pathways; if an ion of mass m_1 decomposes to give an ion of mass m_2, then a metastable ion may be found at mass $m^* = m_2^2/m_1$. Therefore, observation of a metastable ion at this position

Figure 1.9. Metastable ion peaks indicated by arrows in a mass spectrum (normal ions give narrow peaks at integral mass; metastable ions usually give broad peaks at non-integral mass). Note also the fairly abundant doubly charged ions at m/z 57.5.

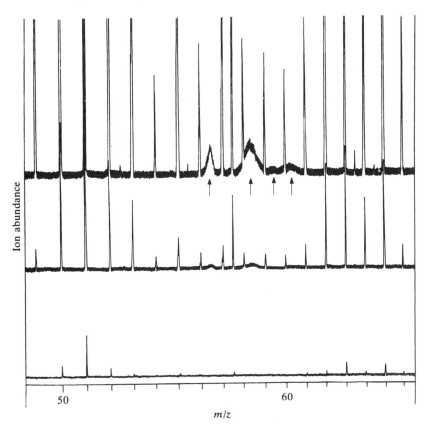

m/z

(m^*) confirms the fragmentation of a *precursor* ion (m_1) to a *product* ion (m_2). Thus, in the mass spectrum of toluene, abundant ions occur at m/z 91 ($C_7H_7^+$) and m/z 65 ($C_5H_5^+$) and the appearance of a metastable ion at m/z 46.4 ($= 65^2/91$) indicates that at least some of the ions at m/z 65 arise through ejection of C_2H_2 (26 mass units) from the ion at m/z 91.

1.5. Elemental compositions of ions

On the atomic scale, ^{12}C is given a mass of 12.0000 but other elements have fractional masses, e.g. ^1H has mass 1.0078 and ^{16}O has mass 15.9949 (table 1.5). Because of this, the masses of most ions in a mass spectrum fall near but not at integer values. For instance, the molecular ion of acetone at m/z 58 actually has a mass of 58.0418.

Table 1.5. *Masses of some elements commonly occurring in mass spectrometry[a]*

^1H	1.007 83	^{29}Si	28.976 49
^2H	2.014 10	^{30}Si	29.973 76
^{10}B	10.012 94	^{31}P	30.973 76
^{11}B	11.009 31	^{32}S	31.972 07
^{12}C	12.000 00	^{34}S	33.967 86
^{13}C	13.003 35	^{35}Cl	34.968 85
^{14}N	14.003 07	^{37}Cl	36.965 90
^{15}N	15.000 11	^{79}Br	78.918 39
^{16}O	15.994 91	^{81}Br	80.916 42
^{19}F	18.998 40	^{127}I	126.904 66
^{28}Si	27.976 93		

[a]Given correct to five decimal places. Notice how on this atomic scale, with ^{12}C taken as standard, the masses fall either just above or just below integral values.

The resolving power of a mass spectrometer is a measure of its ability to separate two ions of any defined mass difference. Basically, for two overlapping peaks M_1 and M_2 (figure 1.10), the resolution may be defined in terms of the mass difference (ΔM) between them such that the peaks are said to be resolved if $(h/H) \times 100 \leq 10$, where H is the height of the peaks and h measures the depth of the 'valley' between them. The resolution is then the value of $M_1/\Delta M$ when $(h/H) \times 100$ is equal to 10. For example, when two masses (100.000, 100.005) are separated by a 10 per cent valley, the resolution of the instrument is 100.000/0.005, i.e. 20 000. An instrument of even modest resolving power can distinguish readily between adjacent integral masses; to separate m/z 100 from m/z 101 a resolution of only 100 is required. A high-resolution instrument, however, can separate an ion at m/z 100.000 from one at m/z 100.005. At low resolution, a mass spectrum consists of a series of peaks at integer m/z values but at medium resolution any of these peaks may be split due to the presence of ions of different elemental compositions. An ion at m/z 28 might have the composition

N_2 or CO or C_2H_4; if all three ions were present, a low resolution spectrum would show only one peak at m/z 28 but a medium to high resolution instrument would reveal the three ions at m/z 28.0061, 27.9949, and 28.0313 respectively (figure 1.11). Thus at higher resolution small mass differences may be detected and this property is utilized for accurate mass measurement of ions as described in section 3.4.

Figure 1.10. Two overlapping ion peaks M_1 and M_2 of height H and overlap h. If $(h/H) \times 100 \leqslant 10$, these peaks are considered to be resolved.

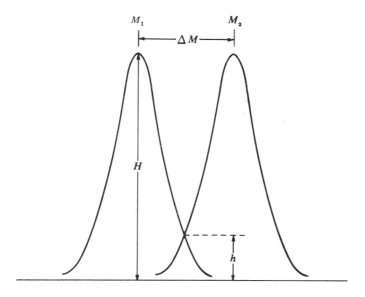

Figure 1.11. The appearance of the mass spectrum of a mixture of CO, N_2 and C_2H_4 at m/z 28 with (a) low resolving power showing no separation of masses and (b) medium resolving power (3000) showing good separation.

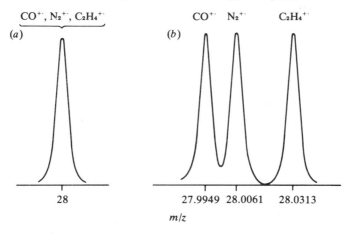

1.6.   Appearance of the mass spectrum

Beginning with the simple picture of ion decomposition, and introducing the complications of isotopes, multiply charged ions, metastable ions, and peak splitting due to differing elemental compositions of ions, an overall impression of what to expect and what to look for in a mass spectrum is gained. The mass spectrum of aniline has been chosen to illustrate these features (figure 1.12). There is a further, practical complication evident in this figure: because of the great range in heights of peaks from the small metastable ion peaks to the large base peak, it is necessary to be able to record simultaneously the spectrum at different sensitivities. Figure 1.12 shows three traces increasing in sensitivity from the bottom of the chart upwards. The lowest trace contains all the features of the *routine* or *low-resolution spectrum*, the middle trace is used to aid the measurement of small peaks in the lower trace, and the upper trace is used for both counting the spectrum and looking for metastable ion peaks. In counting a mass spectrum it is usual first to select some readily identifiable peaks of known m/z values. Air is usually present in or can be introduced into a mass spectrometer and provides a number of convenient 'marker' ions. Thus, peaks due to nitrogen (m/z 28), oxygen (m/z 32), and water (m/z 18) are observed, together with a doublet at m/z 40 due to argon and C_3H_4 from hydrocarbon 'background' ions. With very little experience, these ions are easily identified and used as starting points for counting up all the other ions in the spectrum. On a more sophisticated level, electromechanical mass markers may be purchased which put a mass count directly onto the spectrum as it is being recorded. If the spectrum is acquired and processed by a computer, the output will be of the form represented by figure 1.3 or table 1.3. Note that metastable ion peaks are lost in this processing because of the way a computer 'counts' a spectrum; this is discussed in chapter 3.

Figure 1.12. Mass spectrum of aniline. The molecular ion (M⁺·) and the ¹³C-isotope ion (M + 1)⁺· are shown. Metastable ions (m*) and doubly charged ions (m²⁺) are marked. Integral mass-to-charge ratios are given below the middle trace where the spectrum is counted; where a peak does not appear, as at m/z 59, a check on the top trace will often reveal ions of low abundance.

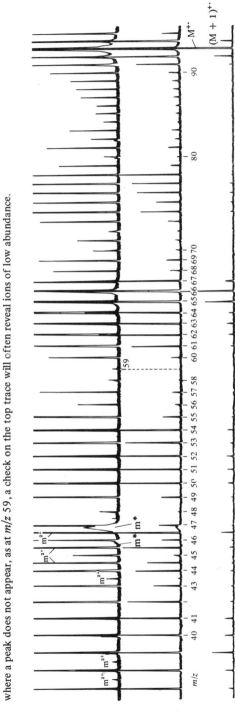

2   Instrument design

2.1.   Introduction

There are many different types of mass spectrometer and new designs are being developed all the time (McCormick, 1977, 1979). The discussion presented here is restricted to the types most usually encountered in chemical applications. Instruments for measuring mass spectra are extensively described in the literature and references to such information are included where appropriate. Mass spectrometer manufacturers frequently publish news releases which provide much detail of specific instruments and can keep the mass spectrometrist abreast of commercial developments. The essential features of a mass spectrometer are presented in block form in figure 2.1, and in figure 2.2 each of these features is clearly seen in photographs of a modern mass spectrometer capable of high resolving powers. A discussion of each unit in turn is necessary for a proper understanding of how a mass spectrum is produced.

2.2.   Inlet systems

There are several means of introducing samples into the mass spectrometer and the inlet system used will normally depend on the volatility and nature of the sample.

2.2.1.   Cold inlet

Gases or compounds that are very volatile at room temperature and a pressure of about 10^{-2} torr, are allowed to 'leak' into the mass spectrometer through a glass sinter and led along a glass tube to the ion source. The pressure inside the mass spectrometer is about 10^{-6} torr.

Figure 2.1. Essential features of a mass spectrometer in diagrammatic form. The system is maintained under vacuum.

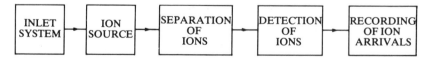

2.2.2.   Hot inlet systems

The first of these is very similar to the cold inlet except that it may be heated to about 300 °C to volatilize compounds which are then led along a heated line to the ion source. To prevent metal-catalysed decomposition or rearrangement of a sample, it is preferable to fabricate the whole system from glass, in which case the unit is commonly known as an AGHIS (all-glass heated inlet system). An alternative is the septum inlet which comprises a heated stainless-steel reservoir into which liquid samples are injected via a septum. At the low pressures and high temperatures employed, the vaporized sample diffuses through a system of valves into the ion source. A septum inlet is shown on the mass spectrometer in figure 2.2*a*. Sophisticated septum inlets exist in which the sample vapour does not come into contact with any metal surfaces or the septum. Both of the hot inlet systems described here and the cold inlet may require up to a milligramme of sample. The former inlets provide the most convenient methods of 'leaking' into the source small quantities of a reference compound, such as perfluorokerosene, required for mass calibration (see section 3.2).

2.2.3.   Direct probe inlet

Compounds that are not sufficiently volatile to be introduced through the hot or cold inlets may be inserted directly into the ion source by means of a probe passing through a vacuum lock. At the low pressures of about 10^{-7}–10^{-6} torr inside the ion source and with heating, very many compounds are sufficiently volatile to yield good mass spectra. The system is so easy to use and requires so little material (a few nanogrammes are quite sufficient) that it is frequently the method of choice even for relatively volatile substances. These direct insertion probes are often heated by radiation from or contact with the ion source itself, but probes are now available which may be heated or cooled independently of the source. Compounds of molecular weight up to about 2000 may be measured fairly readily with a direct insertion probe, assuming, of course, that the analyser part of the instrument is commensurate with such a mass. As a modification, if the probe is designed to project not just into the source, but right up to or slightly into the electron beam used for ionization (see next section), the technique is known as 'in-beam' or direct ionization. This topic will be discussed in greater detail in section 12.3 but it may be noted that the method is particularly useful for large, thermally unstable molecules of low volatility such as carbohydrates.

2.2.4.   Gas chromatographic inlets

The effluents from gas chromatographic columns consist of carrier gas admixed with the compounds being investigated. Carrier gas flow rates of 50 ml/min are common in gas chromatography with packed columns,

(a)

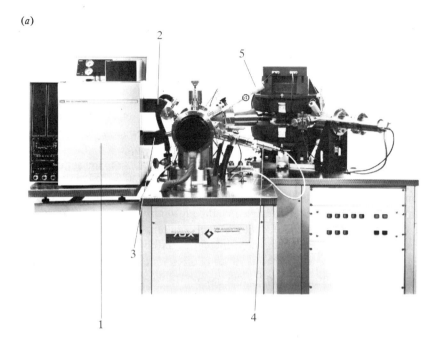

(b)

but such quantities of gas cannot be allowed directly into the ion source because large increases in pressure would result and lead to arcing of the high potentials inside many sources. Consequently, it is necessary to remove as much carrier gas as possible, ideally without diminishing at the same time the amount of sample going into the source. Differentiation between carrier gas and entrained sample can be effected by molecular separators (section 4.2). A 100 per cent efficient molecular separator would remove all the carrier gas and pass all the sample into the ion source, but most separators are far from this efficiency (McFadden, 1979). The sources of quadrupole mass spectrometers (see below) do not operate at high potentials and, when evacuated with highly efficient pumps, some are able to cope with high carrier gas flow rates without the need for a molecular separator. Another situation in which the need for a molecular separator is obviated occurs when the conventional packed gas chromatographic column is replaced by any one of the various types of capillary column. These higher resolution columns operate best with flow rates in the range of 0.5–5 ml/min. With modern pumping systems, even an ion source using high potentials will operate successfully when the whole of such gas chromatographic effluent is transferred to it through a glass (or glass-lined metal) tube. The pressure in an ion source under these conditions is usually no more than that obtaining when a packed column and molecular separator are employed. Finally, when the carrier gas of any type of gas chromatographic column is used also as the reactant gas for chemical ionization (see section 4.2), again all of the effluent is passed directly to the ion source.

2.2.5.   *Liquid chromatographic inlets*

Gas chromatography and mass spectrometry are both gas-phase techniques and hence, to some degree, compatible. The coupling of liquid chromatography and mass spectrometry involves a different dimension of technical challenge because the liquid effluent must, in most applications, be stripped of solvent before introduction of the solute into the ion source.

Figure 2.2(*a*) Front view of a modern, double-focussing magnetic sector mass spectrometer (the Micromass 70X) to show the inlet system. Key: 1, a linked gas chromatograph; 2, inlet line incorporating a molecular separator; 3 direct inlet line for gas chromatographic capillary column; 4, direct insertion probe; 5, septum inlet. See figure 2.2(*b*) for inlet line for chemical ionization reactant gas. Photograph by kind permission of VG Analytical Ltd, Altrincham, England. (*b*) Angled view of a modern, double-focussing magnetic sector mass spectrometer (the Micromass 7070) to show the ion source, analyser region and detector. 1, ion source; 2, electric sector; 3, magnetic sector; 4, detector (an electron multiplier). The inlet line for chemical ionization reactant gas is also labelled (5). Photograph by kind permission of VG Analytical Ltd, Altrincham, England.

If this solvent is not removed, unacceptably high pressures would develop in the ion source. It is mainly because of this difficulty that development of combined liquid chromatography/mass spectrometry (LC/MS) has lagged far behind that of combined gas chromatography/mass spectrometry (Arpino, 1975; McFadden, 1979). To date, the most successful LC/MS interface is based on a moving belt which differentiates between solvent and solute on the basis of their relative volatilities (section 4.3). The continuous moving belt has also proved to be an excellent method for rapid analysis of pure, solid samples. By depositing solutions of each sample successively onto the belt, they can be analysed at the approximate rate of one every 20 s. The fastest turnover time possible with the direct insertion probe is of the order of 5–10 min. The improved rate of analysis is used to best advantage with a computer which can process and output mass spectra at a similar rate.

2.2.6.   The total inlet system

A mass spectrometer may have more than one inlet. The mass spectrometer in figure 2.2 is typical in having four sample inlets: a direct probe, a septum inlet and two inlet lines from a gas chromatograph (one direct line for capillary columns, the other incorporating a molecular separator for packed columns). There is also an inlet line for the reactant gas when the instrument is operating in the chemical ionization mode. In many situations it is desirable to use more than one inlet simultaneously; this is possible as long as the source pressure remains within limits suitable for the type of spectrometer being used. For instance, the instrument in figure 2.2 will perform adequately when coupled to the gas chromatograph via the molecular separator, with a reactant gas for chemical ionization in the source and a reference compound 'leaking' into it from the septum inlet. Even at the relatively high pressure resulting from these simultaneous procedures, there is no serious arcing of the high (4 kV) potential in the source. For accurate mass measurement it is important to be able to ionize both the sample under investigation and a reference compound at the same time, necessitating at least two inlets on a high resolution mass spectrometer. The alternative to this is the double-beam instrument which instead has two sources (see section 2.4).

2.3.   Ion sources

An ion source may be defined simply as the region in which ionization occurs. As shown in figure 2.3, the region is usually enclosed in a small ion chamber in which the sample is ionized. The ions produced are pushed out of the chamber towards an exit slit by a low positive potential applied to a 'repeller' plate. On leaving the ion chamber exit, the ions are accelerated through a high potential of 1–8 kV and passed into the analyser of the mass spectrometer for separation according to mass-to-charge ratio. The source of a quadrupole mass spectrometer is essentially similar, but the ions produced are allowed to drift out of the source under the influence of a small electrostatic field. The lack of high potentials in the sources of these instruments makes them particularly suitable for chemical ionization and combined gas chromatography/mass spectrometry when higher pressures are required (Dawson, 1976). The neutral species (N) produced by fragmentation following initial ionization are, to all intents and purposes, unaffected by the fields inside the source and are disregarded by the mass spectrometer except in certain very specialized instruments, outside the scope of this book. There are several common means of ionizing compounds which are dealt with below (Milne & Lacey, 1974; Milberg & Cook, 1979).

2.3.1.   Electrons

The commonest method of ionization is by means of electron impact (EI) in which electrons, obtained from a heated filament *in vacuo*, are accelerated through some voltage (V) and directed across the ion chamber; the electrons thus have energy zV, where z is the electronic charge. The voltage (V) is continuously variable between about 5 and 100 V and, by convention, standard mass spectra are obtained at 70 V because maximum ion yield is obtained near this value. In practice, mass spectra may be obtained at any voltage down to the ionization energy of the compound under investigation, and it is often advantageous to reduce the energy of the impacting electrons much below 70 V. A volatilized sample molecule (M) and an electron of energy greater than the ionization energy of M will

Figure 2.3. Essential features of an ion source for electron impact or photoionization.

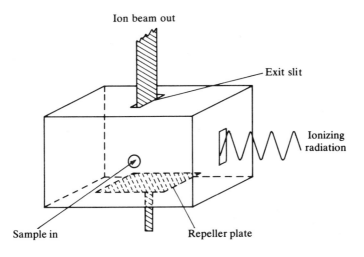

Ion beam out

Exit slit

Ionizing radiation

Sample in

Repeller plate

react if they pass close enough for the electron to impart sufficient of its energy to the neutral species. Not all of the energy of the electron is transferred.

$$M + e \rightarrow M^{+\cdot} + 2e$$

However, the reaction of an electron with a molecule is rather non-specific. Positive-ion formation is accompanied by other processes such as electron capture and electronic excitation (Cottrell, 1965).

$$M + e \rightarrow M^{-\cdot}$$

$$M + e \rightarrow M^* + e$$

Assuming that the mass spectrometer is tuned to analyse positive ions, any negative ions produced will be discharged at the positively charged repeller plate and, along with neutral species, will be pumped away. Since not all molecules (M) react with the electron beam and because those that do react can give a variety of products, these ion sources have low efficiencies. Probably less than 1 per cent of the sample molecules are converted into positively charged ions. To monitor negative ions, the repeller plate and acceleration potentials must be reversed in sign. Electron capture is less probable than electron removal by a factor of about 100, although this figure varies widely depending on the structure of the sample, and so negative-ion mass spectrometry by EI is inherently less sensitive than positive-ion mass spectrometry. In electron impact sources, the ions spend approximately 10^{-6} s after formation in the ion chamber before acceleration into the analyser of the mass spectrometer. Mass spectrometry investigates highly reactive, charged species which would be quenched rapidly or scattered at high pressures and therefore a high vacuum is necessary in the instrument.

2.3.2. *Chemical ionization*

The collision of an ion and a molecule can lead to a reaction giving a new charged species. The ion/molecule reaction between a methane ion and a methane molecule gives rise to the unusual but fairly stable CH_5^+ species.

$$CH_4^{+\cdot} + CH_4 \longrightarrow CH_5^+ + CH_3^{\cdot}$$

For effective ion/molecule reactions it is necessary to operate at source pressures of about 0.1–1.0 torr, much greater than for other ion sources, and this leads to some problems in preventing arcing of the high potentials in the source unless a quadrupole mass spectrometer is used. It is also necessary to construct the ion chamber in such a way that a pressure of 0.1–1.0 torr may be maintained within it whilst the pressure immediately outside it is 10^{-5}–10^{-4} torr. At a pressure of 0.1–1.0 torr, there are many ion/molecule collisions before the ions leave the ion chamber and collisional equilibration of energies may be expected. Because of this, the ions probably possess a distribution of internal energies similar to the thermal distributions found in solution chemistry. These so-called reactant gas ions (AH^+, such as CH_5^+ above) can ionize other neutral molecules (M), usually by proton transfer (Munson & Field, 1966; Mather & Todd, 1979).

$$AH^+ + M \rightarrow A + MH^+$$

The most popular reactant gases for chemical ionization (CI) are ammonia, isobutane and methane. At the higher pressures employed for CI mass spectrometry, electron capture becomes fairly efficient so that production of negative ions is just as likely as production of positive ions. Hence, negative-ion CI is a popular, developing technique (Mather & Todd, 1979) in which there are two basic ionization processes: electron capture,

$$M + e + B \rightarrow M^{-\cdot} + B$$

and reactant ion ionization,

$$R^- + MH \rightarrow M^- + RH$$

The neutral molecule (B) is a third body needed to remove excess of energy and R^- is a reactant gas ion such as OH^- (produced by electron bombardment of a mixture of methane and nitrous oxide). The sensitivity of positive ion CI is similar to that of positive-ion EI mass spectrometry, whilst negative-ion CI mass spectrometry has been claimed to be many times more sensitive than either of them, with femtogramme quantities of material giving good spectra. However, such sensitivities are recorded with atypical, highly electronegative molecules which favour negative-ion over positive-ion production. A more objective comparison suggests that negative-ion mass spectrometry is two or three times more sensitive than positive-ion mass spectrometry in the CI mode, but these figures are strongly dependent on the structure of the sample. Both negative- and positive-ion CI techniques, and instruments capable of performing both more or less simultaneously, will be discussed in chapter 7.

2.3.3.   Electric fields

By applying a high positive electric potential on an electrode in the shape of a point, sharp edge or fine wire, a very high potential gradient is produced around the regions of high curvature. The electric field (molecular orbitals) of a molecule experiencing such large potential gradients is distorted, and quantum tunnelling of an electron can occur from the molecule to the positively charged electrode; this is field ionization. The positive ion so formed is immediately repelled by the positive electrode into the mass spectrometer. Although this is an effective way of producing ions, there are difficulties in focussing them into a beam. Because the initially produced ions are repelled by the high positive potential on the electrode immediately after formation, the residence time of these ions in the source region is short (10^{-12} s) compared with the residence time of ions in an electron impact source. There are two different techniques employing this mode of ionization, termed field ionization (Beckey, 1977; Derrick, 1977) and field desorption (Beckey & Schulten, 1975; Beckey, 1977; Derrick, 1977). In the former method (section 7.5), the sample in the gas phase is passed close to the electric field to cause ionization. It is, like EI and CI techniques, suitable for combined gas chromatography/mass spectrometry although it is not so sensitive. In the field desorption (FD) technique, the sample is coated onto the electrode (called the emitter) so that ionization occurs in the condensed phase on or near the surface of the emitter. The precise mechanism of ionization by FD is complex and can give rise to controversy (Holland, Soltmann & Sweeley, 1976; Beckey & Röllgen, 1979; Holland, 1979; Beckey, 1979). FD mass spectrometry is particularly suitable for analysing high-molecular-weight, thermally unstable compounds (section 7.6).

2.3.4.   *Photons*

The ionization energies of most compounds lie below 13 eV so that high-intensity, short-wavelength radiation can be used for ionization. At energies greater than about 11 eV there are no materials with which to make windows to transmit the radiation into the source. Therefore, it is necessary to arrange for the photon source to emit radiation directly into the ion source. The helium discharge at 21.21 eV provides a convenient source of photons able to ionize all compounds. Unless a high intensity of radiation is attained inside the ion chamber, low ion yields will be obtained with consequent low sensitivity of the mass spectrometer. Ions formed in a photo-ionization source will reside there for about 10^{-6} s before acceleration and analysis. The utility of such sources for general mass spectrometry is limited for they offer little or no advantage over EI sources.

2.3.5.   *Combined sources*

It is often advantageous to be able to generate mass spectra of materials by two or even three ionization methods to yield complementary information. Combined electron impact/chemical ionization and combined electron impact/field ionization sources are most commonly used. It must be stressed that, because different source designs suit different ionization techniques, these combined sources are compromises. The EI/CI source will give adequate sensitivity in both EI and CI modes, but it will give ultimate sensitivity in neither. Mass spectrometers are expensive instruments, so there are advantages in possessing one unit which can perform in different modes. Other reported source combinations are electron impact/chemical ionization/ field desorption with or without the capability for field ionization, field

desorption/electron impact, field desorption/field ionization, field desorption/
field ionization/electron impact, and positive-ion chemical ionization/
negative-ion chemical ionization. The last instrument requires a quadrupole
mass spectrometer and further instrument modifications (section 7.4).
Undoubtedly the most popular and most advanced combined source is that
of EI/CI. This has reached a point where the switch from EI to CI mode
(and vice versa) can take less than a second, permitting alternate scans in
different ionization modes. This technique is put to best use in combined
gas chromatography/mass spectrometry in which it is possible to obtain
several electron impact and chemical ionization mass spectra in the time
it takes for a single compound to elute from the column. With many classes
of compound, for example derivatized peptides, complementary information
can be gleaned from the two types of mass spectrum leading to a surer and
far more thorough analysis than when using either alone.

2.3.6.   *Miscellany*

An introductory text cannot cover in any detail the many less
common sources in current use. Some of these are mentioned briefly here
and some discussed further in section 12.3. The literature cited can be
consulted for more information both on these more specialised sources
and on those discussed above (Wilson, 1971, 1973, 1975, 1977;
McCormick, 1977, 1979; Daves, 1979).

Electrohydrodynamic ionization (Simmons, Colby & Evans, 1974;
Stimpson & Evans, 1978) promises to benefit combined liquid
chromatography/mass spectrometry since the sample in a solvent is
ionized in an intense electric field without heating, making it attractive
for heat-sensitive materials which will not pass unchanged through a gas
chromatographic column.

Like electrohydrodynamic ionization, atmospheric pressure ionization
sources can sample entire vaporized liquid chromatographic effluents, or
gas chromatographic effluents without separation of the carrier gas. Ionization is brought about by the radioactive emitter ^{63}Ni (Horning, Carroll
et al., 1974) or by corona discharge (Carroll *et al.*, 1975). This last source
is very sensitive owing to high ionization efficiency. It is usually used with
quadrupole mass spectrometers.

A further method of ionization, particularly suited to liquid chroma-
tography/mass spectrometry, has been discovered (Blakley, Carmody &
Vestal, 1980). A solution of the sample is vaporized rapidly by heating and
most of the volatilized solvent removed by means of a molecular separator.
At the very high temperatures, the heavier particles, consisting of some of
the solvent and the involatile solute, are charged. These charged droplets
strike a heated metallic probe and break up, whereupon some molecules
are ionized by proton transfer. The mechanism of ionization is not yet

fully understood and it is most unusual in that heated metallic surfaces, which are usually avoided because they can catalyse decomposition of samples, form an integral part of the inlet. Even so, impressive mass spectra of intractible compounds like nucleosides are obtained.

For high-molecular-weight or thermally unstable compounds which do not give good mass spectra even with field desorption ionization, the unique method of radionuclide ionization (Macfarlane & Torgerson, 1976a, b) is likely to be useful. Spectra of neurotoxins and underivatized proteins have been recorded by this means in which fission fragments of the radionuclide ^{252}Cf interact with the sample, causing ionization (section 12.3). The radiation hazards of the instrument may restrict its general application and its challenge to instruments based on other ionization methods. It employs a time-of-flight analyser (see following section) and operates at very low resolution.

The techniques of 'in-beam' (direct) ionization, which also find best application to delicate compounds, require modifications to the direct probe inlet only and not to the source (sections 2.2 and 12.3).

Bombardment of solid surfaces, or solid samples coated on a surface, with ions (secondary-ion mass spectrometry), argon atoms (atom ionization) or laser beams (laser desorption) will also cause ionization (section 12.3). If the surface which holds the sample is a moving belt, the potential for these methods in combined liquid chromatography/mass spectrometry becomes clear.

For negative-ion formation, a low-pressure argon gas discharge will cause electron capture by a sample on which the discharge is focussed (von Ardenne, Steinfelder & Tümmler, 1971). The major application of spark source mass spectrometry is to qualitative and quantitative analysis of low levels of inorganic elements (Beynon, 1960, pp. 195-7).

2.4. Analysers (separation of ions)

The separation of ions according to their mass-to-charge ratios can be achieved with magnetic and electric fields alone or combined. Most of the basic differences between the various common types of mass spectrometer lie in the manner in which such fields are used to effect separation. It is not possible here to describe the many different ways in which the fields are combined, and so a description follows for a system typically used in mass spectrometry, the double-focussing instrument. Other systems are then briefly described as variants of this.

2.4.1. *Magnetic-sector mass spectrometers*

Double-focussing analysers. After the beam of ions has been accelerated through a potential of 4000–8000 V away from the ion source, it is divergent and slits are used to reduce the spread (figure 2.4). The ion beam then passes between a pair of smooth, curved metal plates called an electric sector where the electric potential maintained across the plates focusses the ion beam at a second slit, the monitor (figure 2.4). The electric sector is an energy focussing device. Thus, if V is the potential through which the ions are initially accelerated and E is the field in the electric sector, then equation (2.1) is obtained,

$$R = 2V/E \qquad (2.1)$$

where R is the radius of curvature of the ion path. The two equations governing the motion of the ions through the analyser are those relating their kinetic energy, $zV = \frac{1}{2}mv^2$, and centrifugal force, $zE = mv^2/R$ where v is the velocity of an ion of mass m and charge z; simple combination of these two equations gives equation (2.1). Ions accelerated through a potential (V) and passing through a uniform field (E) have the same radius of curvature (R) irrespective of mass-to-charge ratio. Therefore, if the field E is kept constant, the electric sector focusses the ions according to their translational energies.

The energy-focussed ion beam is passed through a magnetic field where mass separation is effected. As described above, after acceleration from the ion source, the ions possess a translational energy, zV ($= \frac{1}{2}mv^2$). On passing through the magnetic field, the ions experience a centrifugal force (mv^2/r where r is the radius of curvature) due to the constraint (Bzv) imparted by the magnetic field (B). From these quantities, equation (2.2) readily follows.

$$zV = \frac{1}{2}mv^2$$

$$BzV = mv^2/r$$

$$m/z = B^2r^2/2V \qquad (2.2)$$

Figure 2.4. Flight paths of ions in a double-focussing mass spectrometer of Nier–Johnson geometry. f_1, f_2 and f_3 are foci 1, 2 and 3.

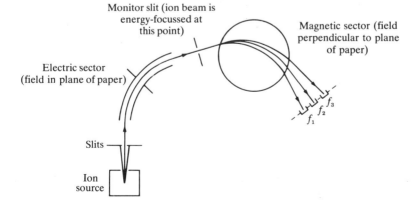

At constant values of the magnetic field (B) and the accelerating potential (V), each value of m/z will correspond to a radius of curvature r. Thus, after separation in the magnetic field, singly charged ions of mass $m_1, m_2, m_3, \ldots,$ m_n will be brought to foci $f_1, f_2, f_3, \ldots, f_n$ (see figure 2.4) corresponding to radii of curvature $r_1, r_2, r_3, \ldots, r_n$. If an electrical detector, for example an electron multiplier (see section 2.5), is placed at f_1 then arrival of ions (m_1/z) will be recorded and if other detectors were placed at $f_2, f_3, \ldots, f_n$ the ions $m_2/z, m_3/z, \ldots, m_n/z$ would also be recorded. Rather than have a series of detectors, it is simpler to vary the magnetic field (B) continuously so that each m/z species is brought to a focus at f_1 in turn, i.e. the radius of curvature (r) is kept constant; this is magnetic scanning of the spectrum. It is possible also to vary the accelerating voltage (V), with B constant, to bring each m/z species to the same focus (see equation 2.2). This method of voltage or electrical scanning is less desirable than magnetic scanning because the sensitivity of the instrument decreases with decrease in the accelerating voltage.

The combined use of electric and magnetic sectors gives a *double-focussing* mass spectrometer because the ion beam is focussed first for translational energy and then for mass-to-charge ratio. The layout of electric and magnetic fields described above (figure 2.4) is known as a *Nier–Johnson* geometry. A mass spectrometer employing this geometry is shown in figure 2.2. The energy-focussing electric sector allows this type of instrument to operate at high resolving powers and most accurate mass measurements are carried out on double-focussing spectrometers. According to equation (2.2), the higher the mass m, the higher must be the magnetic field strength. This places an upper limit on mass, defined by electromagnet technology but, in practice, a greater limitation almost always arises from the difficulty in production of tolerably stable high mass ions rather than their subsequent analysis. Modern analysers scan up to mass 2500 without difficulty.

Other layouts of electrical and magnetic fields are met in other types of mass spectrometer, but the basic aim remains the separation of ions according to m/z value with as great a resolution as is needed. Instruments using the *Mattauch–Herzog* geometry have their fields arranged as in figure 2.5. To detect the ions, a photographic plate or electron multiplier array may be placed in the plane of the ion foci $f_1, f_2, f_3, \ldots, f_n$ (figure 2.5) and the magnetic field kept constant. The ions then give rise to narrow bands on the photographic plate and the density of these bands is a measure of the number of ions arriving. The sensitivity range of a photographic plate is limited and it is usually necessary to make several exposures for increasing periods of time in order to record adequately all ions from the most to the least abundant. The photographic plate method of detection is clumsier than an electrical method since after each spectrum the plate must be developed. Also, the density of the bands must be found with a second instrument, a microdensitometer, which scans the bands and gives out optical density readings as electrical signals (section 2.5). The recently developed channel electron multiplier array (section 2.5), which consists of a series of electrical detectors, does not suffer from the sensitivity limitations of a photographic plate. One big advantage of the photographic plate or the multiplier array is that fluctuations in ion abundances during scans are immaterial since all ions are recorded simultaneously whereas, with a single electrical detector and magnetic scanning of the spectrum, fluctuations in ion abundances during the scan result in an alteration in the appearance of the spectrum.

Single-focussing analysers. If the electric sector is omitted from the instrument it becomes a *single-focussing* mass spectrometer. Since the divergent ion beam leaving the source is not focussed for kinetic energy, this type of instrument produces only low resolution spectra but it can usually distinguish integer m/z values up to masses of about 2000 without difficulty.

Reverse-geometry instruments. If the magnetic sector precedes the electric sector (the reverse of the situation in figures 2.2 and 2.4), the resulting mass spectrometer is still capable of performing in the same way as conventional instruments, but it has additional advantages. These are concerned with the detection and utilization of metastable ions and so discussion of this geometry (section 8.2) is postponed until the origin of such ions has been detailed (see chapter 8).

Figure 2.5. Flight paths of ions in a double-focussing mass spectrometer of Mattauch–Herzog geometry.

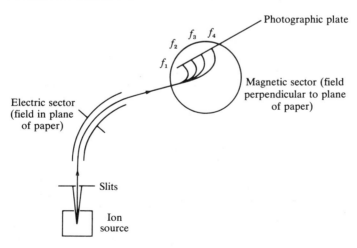

Triple analysers.

If a second electric sector is placed after the magnetic sector of a double-focussing mass spectrometer of conventional geometry the resulting triple focussing instrument, like the reverse geometry instrument, has special advantages for metastable ion studies (chapter 8).

Double-beam instruments.

A double-beam instrument would appear very like the mass spectrometer of figure 2.4, except that there would be two adjacent sources and two adjacent detectors. One source usually ionizes a reference material, the second source ionizes the sample. Rather than using two analysers, both ion beams are directed down the same flight tube (though they do not converge on each other) and hence experience the same electric and magnetic fields. They are detected and measured by separate detectors. Such an instrument offers some advantage for accurate mass measurements (section 3.4) because a reference compound and a sample can be ionized simultaneously without physical mixing.

2.4.2. Quadrupole mass spectrometers

All of the analysers discussed above are magnetic-sector instruments. The quadrupole, on the other hand, functions in an entirely different way. Four precisely parallel rods are arranged as in figure 2.6. Between each pair of opposite and electrically connected rods is applied a dc voltage and a superimposed radio-frequency (rf) potential. Under the influence of this combination of fields, ions in the analyser undergo complex trajectories (Lawson & Todd, 1972; Todd & Lawson, 1975; Campana, 1980). With the rf/dc voltage ratio constant, the voltages are varied to effect separation of ions according to mass. At any point in the scan, only one mass can pass through the system, all other ions being excluded. Because of this filtering action, the system is often called a quadrupole mass filter (Dawson, 1976).

Figure 2.6. Arrangement of the quadrupole mass filter, showing the complex flight path of a focussed ion. Ions that are not focussed collide with the rods.

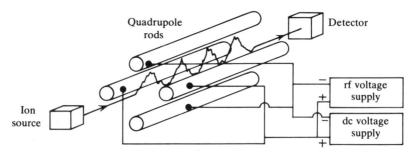

Its capabilities are about the same as for a single-focussing magnetic sector instrument, with maximum resolution near 2000 and highest detectable mass of about 1200, though it cannot provide information about metastable ions. It does have several advantages over the magnetic sector instrument in that it is more robust, somewhat cheaper and physically smaller. A typical commercial quadrupole mass spectrometer is shown in figure 2.7; note its relative simplicity compared with the magnetic sector instrument illustrated in figure 2.2. The usefulness of quadrupole instruments for combined gas chromatography/mass spectrometry (GC/MS) resulting from the source design has already been noted in the previous section. The analyser design enhances its capabilities for GC/MS since it can scan the entire mass range very rapidly (in a few milliseconds) in contrast to a magnetic sector instrument where eddy currents in the magnet become significant at scan rates faster than about 0.1 s/decade (a decade is a mass range covering a factor of ten, e.g. m/z 40 to 400). Fast scanning is very important if compounds elute in narrow bands from chromatographic columns. Computer control

Figure 2.7. A commercial quadrupole mass spectrometer (Micromass 12000 series) showing (1) ion source and attached inlets, (2) housing for quadrupole mass filter, (3) housing for electron multiplier detector and (4) linked gas chromatograph. Photograph by kind permission of VG Analytical Ltd, Altrincham, England.

of the rod voltages of a quadrupole can be effected more accurately than can corresponding control of the magnetic field of a magnetic sector instrument. In particular, the rod voltages can be switched very rapidly from one value to another in order to focus in turn a few selected ions, an advantage desirable for sensitive quantitative mass spectrometry (chapter 6). Generally, the quadrupole mass filter is automated more easily than a magnetic sector instrument.

The term quadrupole refers to the four poles of the mass filter. There are other mass spectrometers which operate on the same principle but which have different numbers of poles or different analyser design. Irrespective of this, these less common instruments have been called loosely quadrupole mass spectrometers. These analysers are referred to more correctly as monopoles, dipoles, hexapoles, dodecapoles, ring storage traps and so on (Todd & Lawson, 1975).

2.4.3. Time-of-flight mass spectrometers

These instruments also have no magnetic sector and separate ions of different masses by making use of their different velocities after acceleration through a potential (V). Since $zV = \frac{1}{2}mv^2$, the velocity (v) of an ion of mass (m) is $v = (2zV/m)^{1/2}$. The ion velocity is mass dependent so that, if a bunch of ions ($m_1, m_2, m_3, \ldots, m_n$) is accelerated and then allowed to pass along a field-free region, the ions will arrive at a detector at different time intervals depending on their velocities ($v_1, v_2, v_3, \ldots, v_n$). These time intervals are very short and it is necessary to display the mass spectrum on a cathode ray tube. Generally, the resolving power of these instruments is low but, because of their fast response times, they find considerable use in investigations of fast reactions. It is a time-of-flight analyser which is coupled to the Cf-radionuclide source (see section 2.3) because, unlike all the other analysers, it has no mass range limitations, thus allowing detection of ions of very high mass. For general mass spectrometry, the use of time-of-flight instruments is not widespread.

2.4.4. Ion cyclotron resonance mass spectrometers

These instruments employ crossed or alternately pulsed electric and magnetic fields. The latter development using Fourier transform methods has drastically reduced the time required to record a mass spectrum (it was of the order of 20 min) which had been the major factor limiting its use (Wilson, 1971, 1973, 1975, 1977). Their main application is the study of ion/molecule reactions and gas phase ionic equilibria. They are described more fully in section 12.4.

2.5. Detection and recording of spectra

There are four main ways of detecting ions: by photographic plate, Faraday cup, electron multiplier or scintillation counting. The photographic

plate is not very common but can be useful with mass spectrometers of
Mattauch–Herzog geometry for reasons described in the previous section.
In the simplest detector, ions impinge on a metal plate which is connected
to earth through a resistor. The resulting neutralization of the charge on
the ions leads to a flow of current through the resistor which can be
detected and affords a measure of ion abundance. The plate is not
infrequently replaced by a metal (Faraday) cup which is a more efficient
detector. The only real disadvantage of this detector is its lack of
sensitivity. The commonest detector is the electron multiplier which
consists of a series of about ten electrodes (dynodes). When an ion
impinges on the first electrode, it causes the release of a shower of electrons
to the second electrode and each of these electrons causes a further shower
of electrons at the third electrode. This cascading effect continues through
the whole series of electrodes and provides gains of the order of 10^6. The
channel electron multiplier array (McCormick, 1977), which can be
considered as a network of many such electron multipliers may, if its mass
range and resolution limitations are overcome, make redundant the photographic plate. The arrays are a very important development since all ions
can be detected simultaneously without scanning the magnetic sector of
the instrument. They should also attain very high sensitivity since they
record arrivals of all ions all the time whereas, during magnetic scanning,
the single electron multiplier spends much time recording the gaps between
ion arrivals and records sequentially ions of each mass. The fourth method is
the Daly detector, based on the scintillations produced by ions striking a
phosphor screen. The scintillations are monitored by a photomultiplier.
It is particularly useful for its capability for detecting metastable ions whilst,
in one mode of operation, ignoring normal ions (see section 8.1).

The amplified signals from the electron multiplier or Daly detector are
passed to a recorder. A pen recorder does not respond sufficiently rapidly to
be useful and instead, the signals are used to deflect mirror galvanometers
of different sensitivities. Ultraviolet light reflected from the mirrors is
focussed onto uv-sensitive paper. The resulting trace is illustrated in figure
1.12. At a more sophisticated level, these electrical signals are passed
indirectly, usually via a magnetic tape recording, or directly to a computer
which evaluates the incoming data and prints out the required information
as detailed in the following chapter.

The photographic plate method provides its own record of the spectrum,
but it is not directly amenable to computer processing. Where ions fall onto
the plate, the resulting bands are revealed by photographic development.
To achieve computer processing, these bands on the photographic plate must
be converted into an analogue electrical signal by using a microdensitometer.
A narrow beam of light is passed through the photographic plate (figure 2.8)
and is collected at a detector which puts out an electrical signal proportional

to the amount of light falling on it. The photographic plate is passed at constant speed through the light beam and the images on the plate cause fluctuations in the detected light level; the resulting electrical signal from the detector may be passed to a computer in the same way as that from an electrical or Daly detector. The method inevitably involves off-line working of the computer.

The signal from an electron multiplier array is made visible by accelerating the electron output onto a phosphor screen. The resulting image can be recorded by a Vidicon camera which stores the light impulses as charge. The signal from this camera is compatible with on-line data acquisition by computer.

2.6.   A total system

A modern, integrated gas chromatograph/magnetic sector mass spectrometer/computer system is illustrated in figure 2.9. In the figure, the main features are marked. The linking of mass spectrometers to computers is described in the next chapter.

Figure 2.8. Diagrammatic representation of the action of an optical micro-densitometer.

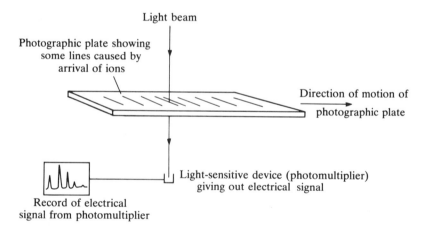

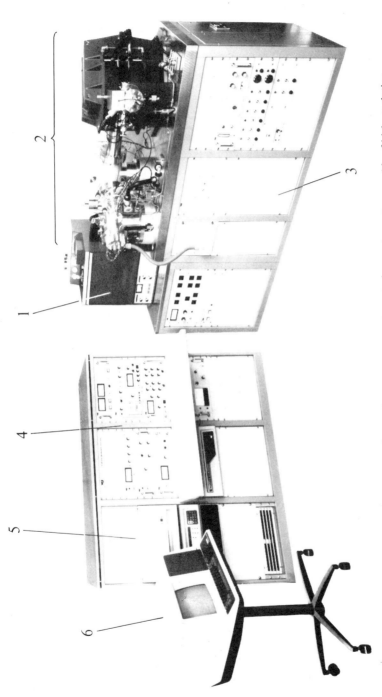

Figure 2.9. A typical integrated gas chromatograph/magnetic sector mass spectrometer/computer system capable of high resolution (the Micromass 7035). 1, gas chromatograph; 2, mass spectrometer; 3, housing for high vacuum pumps; 4, control panel; 5, computer; 6, visual display unit for observing results and keyboard for communicating with the computer. Photograph by kind permission of VG Analytical Ltd, Altrincham, England.

3   Data systems

3.1.   Introduction

The full potential of any mass spectrometer cannot be realized
without a data system. Whilst computers were introduced into mass
spectrometry to aid accurate mass measurement on double-focussing mass
spectrometers, they have proved useful also for less sophisticated mass
spectrometers. If the quantity of a sample to be examined is very small,
the human operator may not be able to react sufficiently rapidly to obtain
a representative mass spectrum during the brief period in which the compound evaporates in the ion source. A computer can obviate this difficulty
by controlling continuous repetitive scanning of the mass spectrometer
from the time that the sample is introduced into the source. The spectrum
or spectra of interest can be recalled from the data stored in the computer.
If a sample is contaminated with impurity it is often possible, but tedious
and time-consuming, to deconvolute manually the impurity peaks from the
sample peaks. A computer performs this operation in seconds. It is partly
due to the advent of computers that the technique of combined gas chroma-
tography/mass spectrometry (GC/MS) has proliferated in recent years. A data
system built around a computer is now a prerequisite for good quality GC/MS.

Mass spectrometers have been used in conjunction with a wide range of
data systems, from microcomputers and programmable desk calculators to
large, powerful computers. The larger and more complex the data system,
the less likely it is to be dedicated to the mass spectrometer. For purely
economic reasons, such a system usually serves several different instruments.
It may be used in the off-line mode in which the computer is not directly
coupled to the mass spectrometer or, if on-line, on a time-shared basis. The
commonest type of data system is of medium power, on-line and dedicated
to either one mass spectrometer or a number of mass spectrometers. It is
usually capable of three basic operations: (1) control of the scanning of the
mass spectrometer, (2) data acquisition from the mass spectrometer and
(3) data processing. Ideally it should perform all three essentially simultaneously. The subject of computers in mass spectrometry is reviewed

biennially (Ward, 1971, 1973; Mellon, 1975, 1977, 1979) and has been described in detail in a recent book (Chapman, 1978).

3.2. Data acquisition and mass calibration

3.2.1. On-line operation

The detection of ions was discussed in the previous chapter. It was stressed that the output from these detectors is, or can be converted to, an electrical current which is amplified as a continuously varying voltage (an analogue signal; figure 3.1*a*). This signal may be acquired directly by an analogue computer but, as most computers are digital devices, the signal must be converted into a digital form. This section will be concerned exclusively with digital data systems. The digitization step is accomplished by an analogue-to-digital converter (ADC) for which the mode of operation is represented in figure 3.1. The data system is equipped with a crystal oscillator clock which defines precise, regular time intervals at which the analogue signal is to be sampled by the ADC. This results in a regular series of voltage pulses (figure 3.1*b*), each of which is compared with a threshold value and is accepted for transmission to the data system only if it exceeds this pre-set threshold. This limitation is applied to remove unwanted (baseline) noise between mass peaks. The data system, freed from acquisition of unnecessary data, can process data in the time intervals between the mass peaks. The duration of these intervals is often greater than that of the mass peaks (see figure 1.12), particularly with high-resolution spectra in which the mass peaks are very narrow.

It is usual for there to be a further noise-rejection facility prior to transmission of the signal to the computer. Noise spikes or peaks are usually

Figure 3.1. Analogue signal (*a*) of two mass peaks and their computer compatible, digitized form (*b*).

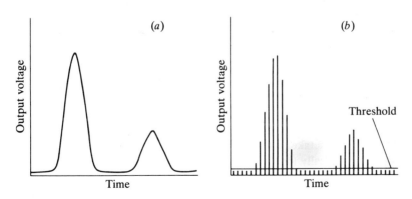

narrower than true mass peaks, so that rejection of spurious signals can be made on the basis of peak width.

A magnetic sector mass spectrometer is most often scanned from high to low mass exponentially with time. This gives rise to mass peaks of constant width throughout the mass range whilst the gaps between the peaks increase exponentially with decreasing mass (figure 1.12). For such a scan mode, the mass peak width in terms of time (t) is defined by equation (3.1):

$$t = 0.43 t_{10}/R \qquad (3.1)$$

where t_{10} is the time to scan a decade in mass and R is the resolution. At a resolving power of 10 000, when scanning the masses from m/z 400 to 40 in 10 s, the peak width is 0.43 ms. To define the shape of a mass peak accurately, the ADC must be capable of taking about 15 samples in this time, that is, a sampling rate of about 35 kHz (35 000 samples per s) is required. The same rate would be required by a 0.5–1 s/decade scan at low resolution. Typical ADC devices are capable of digitization at rates of 40–200 kHz. Conversely, the maximum scan speed, t_{max} (in seconds per decade), for a given sampling rate, S (in kHz), is given by equation (3.2):

$$t_{max} = 2.3 \times 10^{-3} \, NR/S \qquad (3.2)$$

which is derived simply from equation (3.1). The number of samples required across the peak is N (10–25 usually). Figure 3.2a shows the effect of insufficient sampling across a mass peak. Neither the position of the maximum nor its height are faithfully recorded at the lower sampling rate. With a peak which is an unresolved multiplet, the situation is even more critical, for the multiplicity may not even be recognized at low sampling rates (figure 3.2b).

Equations (3.1) and (3.2) are not relevant for quadrupole mass spectrometers which are scanned linearly with time, producing a linear mass scale, but simple calculations based on the same principles reveal that they have similar digitization requirements. Alternatively, with data-system control, it is possible to 'scan' a quadrupole mass filter in steps by rapid switching of the rod voltages. If the output of such a system is sampled at each step, the data are in digital form and directly amenable to acquisition by computer.

Once a mass peak has been adequately defined by sampling, it is passed directly to the computer where its area or height and time centroid are estimated. Ion abundance is obtained from the area or height of the peak (this area or height is often inappropriately termed 'intensity'). The time centroid (C) can be taken as the position of the largest digital sample or calculated numerically (figure 3.3) using equation (3.3):

$$C = \Sigma V_i t_i / \Sigma V_i \qquad (3.3)$$

where V_i is a digital voltage and t_i the time at which it is acquired.

The latter method is more demanding computationally but more reliable, particularly for high-resolution studies. The information stored by the computer may be in either of two forms. In the *profile* method, all of the digital samples are retained, the calculation of the centroids and sizes of the peaks being performed any time during or after acquisition. To acquire *centroid* data, these calculations must be carried out in real time, that is, as the data are acquired. Then, only the centroids and sizes of the peaks are stored by the data system and the original peak shapes are lost. Compared with the centroid method, the profile method requires less computer time but more storage space. It has the advantage that the shapes of the acquired peaks

Figure 3.2. Effect of sampling rate on peak definition. The analogue signals of (*a*) a normal peak and (*b*) a doublet peak from a mass spectrometer (top) are defined adequately by sampling 15 times (middle), but not by sampling 5 times across the peak (bottom).

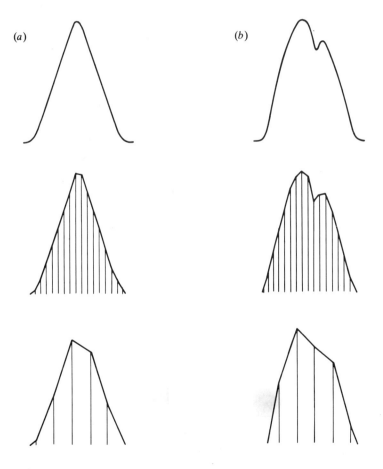

may be examined later for multiplicity, to determine resolution and for checking the operation of the mass spectrometer. A good data system may allow either method to be selected.

The information acquired by the computer, which is usually stored ('written') on a magnetic disc, is a *time/intensity file*. The process of converting a time/intensity file to a *mass/intensity file* (a mass spectrum) is known as *mass calibration* and relies upon the acquisition of data from a *reference (calibration) compound*, the mass spectrum of which is stored in the memory of the computer as a mass/intensity file called a *reference table*. The commonest calibration compounds are perfluorokerosene (PFK), most useful for magnetic sector instruments in the mass range m/z 30–800, and heptacosafluorotributylamine which is suitable for the mass range m/z 40–600 and used most often with quadrupole mass spectrometers. By matching and equating the acquired time/intensity file and the reference table, a mass/time conversion file, called a *calibration table*, is produced. It can be used to convert any subsequently acquired time/intensity files to mass/intensity files as long as the mass spectrometric conditions are stable, that is, as long as the scan is reproducible. In practice, calibration for low-resolution mass spectra will usually suffice for at least one day. Whilst the mechanism of mass calibration is the same for both high- and low-resolution mass spectrometric data, there are special requirements for accurate mass measurement which will be detailed in section 3.4. The process of mass calibration is represented as a flow chart in figure 3.4.

The ideal calibration compound would be reasonably volatile for easy inlet to the ion source and give a mass spectrum of closely spaced, large peaks covering a wide mass range; the m/z values would be quite distinct

Figure 3.3. Calculation of a peak centroid.

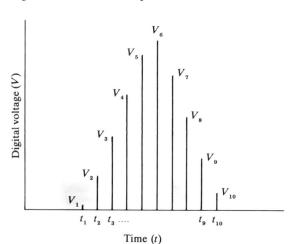

Time (t)

from those of the sample. The compound closest to this ideal for EI ionization is PFK. For calibration to high mass, fomblin oil (a polyfluoroether)
gives reference ions up to m/z 3000. Substituted triazines cover a mass range
up to m/z 1600 and substituted phosphazenes are useful for calibration to a
mass of about 2000 for EI and FD mass spectrometry. The use of PFK for
calibration in CI mass spectrometry is limited. Instead, the mass scale can be
calibrated in the EI mode prior to conversion to CI conditions if a combined
EI/CI source is used. If a double-beam mass spectrometer is used, the source
containing the calibration compound can be operated in the EI mode whilst
the other source is operated under CI conditions for the sample being
examined. Alternatively, mixtures of hydrocarbons or of halogenated aromatic hydrocarbons give useful reference peaks under CI conditions. For
negative-ion mass spectrometry, PFK can be used up to a mass of about
600.

The process of mass calibration allows for any reproducible deviation of
the magnetic scan from the theoretical scan law. Because a data system can
control precisely and reproducibly the scanning of quadrupole mass filters,
some of these instruments do not require conventional calibration compounds. Instead, with any compound in the source, the data system searches
for a known mass peak onto which it can 'lock' and from which it can predict
the rest of the linear mass scale from the theoretical scan law. Alternatively,
the mass scale of quadrupole mass spectrometers can be calibrated as
described above but using the value of the voltages applied to the rods rather
than the time variable.

In chapter 1 it was seen that, except for carbon, elements have masses
that are not exact integer values, that is, they have *mass defects* (table 1.5).
Data acquisition and mass calibration of low-resolution data usually result
in a mass scale accurate to four figures and rounding off to integer mass is
necessary. This is not a simple process since some elements (e.g. halogens)
give ions with masses just less than and others (e.g. hydrogen) just greater
than the appropriate integer value. The situation is most critical at higher
masses because in a large molecule the individual mass defects of the elements

Figure 3.4. Flow chart for mass calibration.

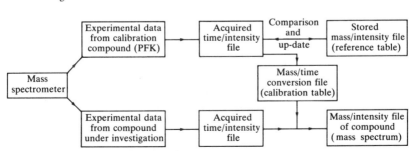

often add to give masses significantly different from the integer mass. For instance, ions detected at m/z 492.6 and 493.7 would appear to be separated by one mass unit, but if the former were the molecular ion of the hydrocarbon $C_{35}H_{72}$, the mass should be rounded off to a nominal mass of 492, whereas if the latter were one of the molecular ions of fully chlorinated biphenyl ($C_{12}{}^{35}Cl_{10}{}^{+\cdot}$), its integer mass is 494. It is improbable that such dissimilar types of sample would be present in the source at the same time and, as long as this situation does not arise, rounding off to nominal masses can be brought about successfully in a variety of ways by assuming that there will be no sudden large changes in the mass defect throughout the mass range. The predominance of carbon and hydrogen in most organic compounds ensures that samples usually have masses slightly greater than their respective integer values and is the reason for preferring mass-deficient, polyfluorinated calibration compounds in which mass peaks are well separated, and easily distinguished, from sample peaks.

A normal mass spectrum will contain metastable ions (chapter 8). These ions are of low abundance (usually less than 1 per cent of the base peak) and many will be disregarded by the data system because they fall below the threshold value of the ADC. In addition, a computer with a maximum peak width parameter as a device for rejecting low-frequency noise will discard many metastable ion peaks because they are broader than peaks from normal ions. Any remaining metastable ions are 'lost' in the procedure for rounding off to nominal mass. There are techniques for acquisition of metastable ion data which rely on different methods of scanning the mass spectrometer; these will be detailed in chapter 8.

3.2.2. Off-line operation

In off-line systems the individual steps for data acquisition are the same as those described above for on-line operation, but they are interrupted at some stage prior to transmission of the data to the computer for evaluation. Two possible levels of interrupt can be distinguished. Firstly, the mass spectral data may be stored temporarily prior to digitization. This situation has already been encountered and described in the acquisition of photographically recorded data (section 2.5). Alternatively, the analogue signal from the detector of a mass spectrometer may be recorded on magnetic tape at different attenuations (usually ✕1, ✕10 and ✕100) together with timing pulses. The tape recording is played back to a computer at any later time for digitization and processing. The tape can be played back at a lower speed than that used for recording, thereby reducing the sampling-rate requirement for data from fast-scanning or high-resolution mass spectrometers. Secondly, the data may be recorded magnetically on tape after digitization, a method which makes for shorter tape lengths, particularly if only the time centroid and size of each peak are recorded.

3.3.   Control of the mass spectrometer

Most data systems are not passive; they not only acquire and process data but are used interactively to control several functions of the mass spectrometer. For instance, microprocessor-controlled units are available which automatically optimize the ion source conditions for best sensitivity, and many on-line data systems will control the scanning of the mass spectrometer. The latter requires a digital-to-analogue converter (DAC) to provide the mass spectrometer with the necessary analogue voltages from the digital scan instructions generated by the computer.

The DAC output can be applied directly to the rods of a quadrupole mass spectrometer. At any one value of this voltage only ions of one specific mass can pass through the filter to the detector. The relationship of mass to rod voltage is computed from a calibration spectrum after which the operator needs only to specify the masses at which the scan is to start and end, the type of scan and the scan rate. Then the computer and DAC generate the voltages required to 'drive' the instrument as desired. For a quadrupole mass filter there are two types of scan. The mass scale may be scanned linearly and continuously by application of a voltage varied continuously and regularly, or the applied voltage may be stepped, providing a discontinuous scan. The latter method is used for monitoring ions of a few selected masses for sensitive, quantitative mass spectrometry (chapter 6) or for acquiring full mass spectra by 'jumping' from one peak to the next and not scanning the gaps between the peaks. Since quadrupole mass filters are scanned electronically and the settling time of the low voltages involved is very short, extremely fast sweeping and switching over the entire mass range is possible.

The control of the scan of magnetic sector mass spectrometers is more complex and requires a mass marker, usually a Hall-effect sensor which is placed between the poles of the electromagnet. The sensor measures the magnetic flux and its amplified output is calibrated against mass by use of a calibration compound. The sensor relates mass-to-charge ratio of ions to the magnetic field strength which deflects them onto the detector. The mass count it provides may be displayed directly, marked on the mass spectrum if recorded on uv-sensitized paper or used, like the calibration of the quadrupole rod voltage against mass, to allow computer control of the scan. Once the operator specifies the mass range, the scan rate and the type of scan (most frequently exponential), the computer uses feedback from the magnetic field sensor to program the DAC to produce the voltages required to change the magnetic field strength. Since most magnets in current use are prone to some hysteresis effects and require far greater settling time than the rod voltages of a quadrupole mass filter, mass 'jumping' by switching the magnetic field discontinuously is relatively slow and hence unpopular. Instead, for monitoring selected ions with a magnetic sector instrument, the accelerating voltage (V) is switched rapidly, with the magnetic field strength constant, to focus ions of one mass after those of another according to equation (2.2). The operation is readily controlled by a data system as long as the accelerating voltage has been mass calibrated with a reference compound such as PFK and there is a second DAC to control the voltage, V. The same provisions allow for voltage scanning which can be useful when a narrow mass range is required to be scanned at a rate faster than that possible with magnetic scanning. However, technological advances mean that newer magnets with virtually insignificant hysteresis and much faster settling times are capable of faster continuous scanning and rapid switching throughout the whole mass range for selected ion monitoring. The advantages and disadvantages of the various techniques for selected ion monitoring and their applications will be discussed in chapter 6.

An advantage of a magnetic sector instrument over a quadrupole mass filter is that metastable ions can be studied. Since these ions are of low abundance, they are best observed in the absence of the relatively abundant normal ions. Normal ions can be eliminated specifically by one of several techniques involving scanning of the potential (E) of the energy-focussing electric sector. Such scanning methods can be brought under the control of a data system; these techniques and their applications are discussed in chapter 8.

In the foregoing discussion, several applications of computer control have been mentioned. In nearly all cases, at the heart of the process is the principle of repetitive scanning, a technique realistic only when a computer with a large data storage capacity is available. Repetitive scanning is most useful for combined chromatography/mass spectrometry but is also employed with simpler methods of examining samples such as with the direct insertion probe. As soon as the sample is placed in the inlet of the mass spectrometer, the data system is instructed to start the scan program for full spectra, selected ion monitoring or metastable ion analysis. When the first scan ends, the program returns the analyser to its starting point and a second scan is initiated, and so on. In this way the scanning is cycled, every 0.1–10 s in many applications, until the analysis is complete. Compared with non-repetitive, operator-initiated scans, the procedure allows for the unexpected appearance of components in the sample, that is, materials present in the source for short periods only. The technique provides a surer analysis of small amounts of sample and requires a less skilled operator. It also permits access to a vast data processing capability which is detailed below.

3.4. Data processing

The better data systems are capable of data processing whenever the signals they are acquiring concurrently fall below threshold. This facility is often called *foreground/background operation* and data of the current analysis or of a previously acquired analysis can be processed. In a correctly set system, acquisition of data occupies only about 1–10 per cent of total computer time. The data system must not lose any data by operating in the processing mode when it should be in the acquisition mode, so there is a strict order of priorities in which timing pulses from the clock, data acquisition and control of scanning all take precedence over data processing.

Different commercial data systems offer different processing programs and some allow the operator to write his own programs. Therefore, a large array of data processing routines is available. The most useful programs, forming the common core of many modern data systems, are discussed in the following subsections. Where appropriate, there is a brief discussion of some more specialized, less common routines which, with further development, may attain wider application in the future.

3.4.1. Accurate mass measurement

To calculate the elemental composition of an ion, its mass must be measured correct to three or four decimal places. This requirement usually necessitates a double-focussing mass spectrometer (but see later discussion). In the absence of a computer, accurate mass measurement is performed manually by a process called *peak matching*, in which a reference compound such as PFK and the compound under investigation are introduced into the ion source at the same time. To measure the mass of a sample ion, M^+ (often but not always the molecular ion), an ion (R^+) of known mass, close to that of M^+, is selected from the reference spectrum. The peak from the ion R^+ is displayed on a cathode ray oscilloscope and, by changing the accelerating voltage at the ion source in small steps, the peak from ion M^+ is displayed on the cathode ray tube alternately with R^+. The voltage is changed till R^+ and M^+ coincide on the screen. At constant magnetic field, the mass (m) of singly charged ions is inversely related to the accelerating voltage (V) according to equation (2.2). Therefore, the accurate mass of M^+ is calculated simply from equation (3.4) where the subscript r refers to the reference compound and s to the sample.

$$m_s = (V_r/V_s)m_r \qquad (3.4)$$

When the two peaks on the screen are matched, the voltage ratio V_r/V_s is obtained from the control panel of the mass spectrometer.

The process of peak matching has been automated with either on-line or off-line data systems. In some cases the operator judges when the peaks are matched, in others the computer ascertains the matching condition.

The peak-matching technique is accurate but time-consuming, especially if several ions of the sample need to be examined. Measurements cannot be performed on compounds present in the ion source for short periods of time. Hence, chromatographic effluents and materials of which only a small quantity is available are excluded, a situation often obtaining in the study of natural products. For these reasons, data systems which offer alternative faster procedures for accurate mass measurement have been developed. The mass spectrometer is operated at moderately high resolution (often 10 000) with magnetic scanning over the relevant mass range. The scan rate has to be relatively slow, usually about 10 s/decade although this figure depends on the performance characteristics of the particular instrument and the available sampling rate of the ADC. A few scans of the spectrum of a calibration compound are acquired and used for mass calibration. Then the sample is admitted into the ion source so that, for at least one scan, both it and the calibration compound are ionized simultaneously. The data system is programmed to locate the reference peaks in the mass spectrum of the mixture using the first calibration table and to recalibrate to take into account any variation in the scanning. The accurate masses of all reasonably abundant ions of the sample are computed by interpolation between adjacent reference ions. Computer subtraction from the mixture of those ions due to the calibration compound leaves the accurate mass spectrum of the sample. Unlike the peak-matching method, in one brief experiment the accurate masses of all major ions of the compound are measured. The technique is suitable for GC/MS, although if the chromatographic peaks are narrow (widths of 5–10 s are common with the better gas chromatographic columns), some compromises must be made (Meili, Walls, McPherron & Burlingame, 1979). The data system may be on-line or off-line. The accuracy of the method is not as good as that of peak matching, but its greater applicability and convenience render it more popular.

The accurate mass information is frequently given as a mass list as illustrated in table 3.1 for methyl octadecanoate (methyl stearate). Typically, the list of accurate masses is accompanied by the ion abundance, a possible elemental composition for each ion and the error between the measured mass and that calculated for the given elemental composition. The data system computes the elemental composition of an ion by checking, against its experimentally determined mass, the calculated accurate masses of all elemental possibilities for the particular nominal mass. Such calculation is prohibitively time-consuming unless the operator limits the range by stating the type and maximum number of atoms to be expected as well as a maximum value for the error beyond which the potential match is considered incorrect (Dromey & Foyster, 1980). Some data systems are programmed to reject unlikely elemental compositions, such as $C_{20}H$ for an ion at m/z 241, by setting a limit on the number of double-bond equivalents.

It should be appreciated that the accuracy of mass measurement is a function of the product of resolution and sensitivity and does not depend

Table 3.1. *Computer print-out of measured masses and elemental compositions of some ions in the spectrum of methyl octadecanoate*

Measured mass	Peak height	Elemental composition							Error in mass (mmu[a])
		$^{12}C/^{13}C$	H	N	O	Others			
299.2919	835	18/1	38	0	2	0	0	0	1.4
298.2883[b]	4919	19/0	38	0	2	0	0	0	1.1
271.2610	290	17/0	35	0	2	0	0	0	−2.7
		16/1	34	0	2	0	0	0	1.7
270.2558	1819	17/0	34	0	2	0	0	0	−0.1
269.2474	1222	17/0	33	0	2	0	0	0	−0.6
268.2835[c]	59	No composition calculated							
268.2716	377	17/1	35	0	1	0	0	0	−0.6
267.2687[d]	1745	18/0	35	0	1	0	0	0	−0.1
256.2358	1070	15/1	31	0	2	0	0	0	0.0
255.2306	4734	16/0	31	0	2	0	0	0	−1.8
		15/1	30	0	2	0	0	0	2.6

[a]mmu, millimass units.
[b]This is the molecular ion, $^{12}C_{19}H_{38}O_2^{+\cdot}$; the measured mass at 298.2883 differs from the mass calculated for this composition by 1.1 millimass units. Note the $(M + 1)^{+\cdot}$ ion at 299.2919 of composition, $^{13}C_1{}^{12}C_{18}H_{38}O_2$.
[c]This is a small peak in the mass spectrum for which no composition could be calculated from the given parameters. It probably represents a small amount of impurity.
[d]This ion, $^{12}C_{18}H_{35}O^+$, arises by loss of $CH_3O^{\cdot}$ from the molecular ion.

solely on the attainment of high resolving powers (Smith, Olsen, Walls & Burlingame, 1971). For statistical reasons, the greater the number of ions making up the approximately Gaussian-shaped mass peak, the more precisely can its centroid be measured. If a peak comprises only a few ions, they may not accurately define the peak shape. In this case the accuracy of mass measurement is limited by *ion statistics*. The divergent beam of ions from the source is passed through adjustable slits to reduce its spread. For a resolution of 1000, the slits are wide and transmission of ions high; to achieve a resolution of 10 000, the slits must be narrowed so that the mass peaks too are much narrower but with the penalty of reducing transmission to about one-tenth of its value at a resolution of 1000. Therefore, this increase in resolving power is achieved at the expense of a drop in sensitivity of about 90 per cent. It is a fallacy to suppose that increased resolving power necessarily leads to increased accuracy of mass measurement. Once the resolution is sufficiently high to separate ions of the sample from ions of the calibration compound at the same integer mass, it is pointless to increase the resolution further since both the accuracy and sensitivity of the analysis will deteriorate. In a double-beam mass spectrometer the reference and sample ions are physically separated, irrespective of the resolution. With such an instrument it has been shown that, at low resolution (1500) and with a small quantity of material, the accuracy of mass measurement was better than that obtained with a single-beam instrument at high resolution (10 000) using a much larger quantity of sample.

For a single-beam instrument, accurate masses can be measured at the lower resolving powers which preserve good sensitivity, although the technique is more cumbersome. Mass calibration with a compound such as PFK is carried out at the required resolution. Since ions of the sample under investigation and PFK would be unresolved if analysed simultaneously, the calibration compound is pumped away. Then the sample is ionized together with a *secondary reference compound* which has in its mass spectrum only very few ions, each of which is well removed from any sample ions (tetraiodoethylene, or a mixture of triiodomethane and phenanthrene are frequently used for this purpose). The data system is programmed to locate these secondary reference ions of accurately known mass and align their arrival times with the primary calibration table. In effect, an accurate mass scale is superimposed on the sample spectrum, allowing accurate mass measurement by interpolation. The same technique can be used for determination of elemental compositions with quadrupole mass filters which are restricted to low resolution because of a drastic loss of sensitivity with increasing resolution. The method is equally applicable to EI, CI or FD mass spectrometry and has the advantage that, at the lower resolving powers, the scan rate can be sufficiently fast to allow ready determination of the accurate masses of ions of eluates from gas chromatographic capillary columns.

3.4.2. *Combined chromatography/mass spectrometric data processing (multiscan data processing)*

The following programs are typical for the processing of GC/MS and LC/MS data and will be illustrated with reference to the former. It should be remembered that they are equally applicable to any repetitively scanned (*multiscan*) analyses. A sample which contains several components of significantly different volatilities can be analysed on the direct insertion probe by temperature programming the probe heater (Franzen, Küper & Riepe, 1974). In practice, the data processing of this fractionally evaporated mixture is the same as that of GC/MS. Similarly, programs developed for GC/MS can often be used to remove the ions due to impurities from the mass spectrum of a directly inserted contaminated compound. Data processing specific to selected ion monitoring and studies of metastable ions will be described in chapters 6 and 8 respectively.

Figure 3.5 shows an analysis by GC/MS of the trimethylsilyl (TMS) ether derivatives of five sterols, the last two of which co-elute. The trace is called the *total ion current chromatogram* and is equivalent to the chromatogram obtained from a gas chromatograph with a simple detector. It comprises the summed ion abundances of each spectrum scanned, normalized to the largest of these values and plotted against time and scan number. The computer usually has a program for quantification of the chromatographic peaks based on the heights or areas of the peaks.

Any spectrum of the chromatogram can be identified by its scan number and recalled immediately for examination. The mass spectrum of the first peak (scan number 127), which is due to 3-trimethylsilyloxycholest-5-ene, is shown in figure 3.6a. The spectrum also contains ions of *column bleed* (decomposition products of the liquid phase of the chromatographic column), for example, m/z 75, 197, 315, 393, 451 and 529. This problem can be overcome by a numerically simple program allowing the operator to select a spectrum, or a number of spectra, which contains only background ions and to subtract this background from the spectrum of interest. When the spectrum of figure 3.6a is processed in this way, the result is a 'clean' spectrum of the steroid derivative (figure 3.6b). Note in particular that the ions at m/z 315, 451 and 529 due solely to column bleed, disappear and that the molecular ion at m/z 458 becomes clear. The same program can be used to remove ions of a calibration compound if one were 'bled' into the source throughout the analysis for the purposes of accurate mass measurement.

A *mass chromatogram* is a plot of the abundance of ions of a specified mass against time or scan number. The technique is often termed *selected*

Figure 3.5. The total ion current chromatogram for the analysis of sterol derivatives by GC/MS. Mass spectra were recorded every 3.5 s; the peaks are identified by the number of the scan in which they maximize.

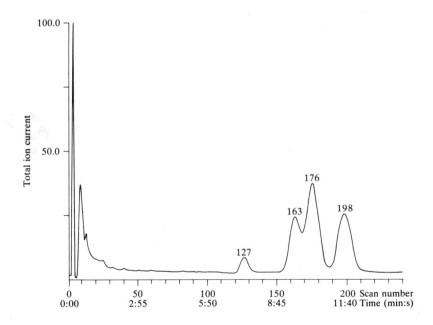

Figure 3.6. Comparison of the mass spectrum of 3-trimethylsilyloxycholest-5-ene (*a*) before and (*b*) after removal of 'background' ions by computer. The spectra are obtained from the analysis by GC/MS shown in figure 3.5.

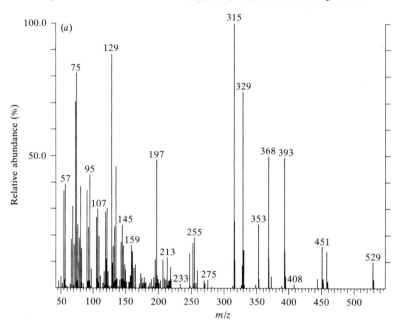

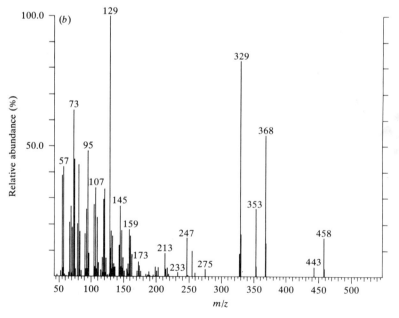

ion retrieval but should not be confused with selected ion monitoring since the whole mass range is scanned. It is used to advantage to test the homogeneity, or otherwise, of chromatographic peaks. When an effluent enters the source, the ions which comprise its mass spectrum will all increase in abundance to a maximum simultaneously, at the retention time. If a chromatographic peak contains more than one compound, provided that each component has in its mass spectrum at least one distinguishing mass peak, mass chromatograms will reveal the true peak profiles of the components despite being unresolved chromatographically. Such a situation is illustrated in figure 3.7 in which a peak, apparently homogenous by inspection of the total ion current chromatogram, is shown by selected ion retrieval to be composite. The figure shows part of an analysis of a urine sample after chemical derivatization. The spectrum of scan number 90 contained some ions which were inexplicable in terms of the assigned structure, the methyl ester of acetylphenylalanine (1). One of the contaminating ions (m/z 193) and an ion of the amino-acid derivative (m/z 162) were chosen for selected ion retrieval with the result shown in the figure. Since the two peaks maximize

Figure 3.7. Part of the total ion current chromatogram of an analysis of a urine sample by GC/MS (bottom). Mass chromatograms of ions at m/z 162 (top) and 193 (middle) are also shown. Note that the mass chromatograms maximize in different scans.

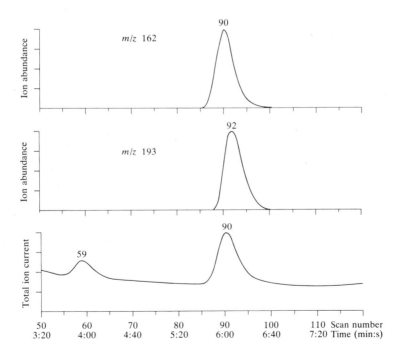

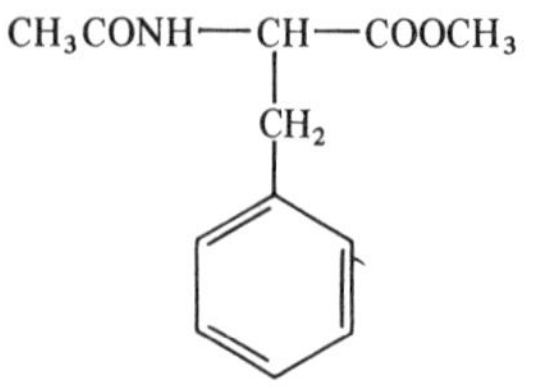

$$CH_3CONH-CH-COOCH_3$$

$$\overset{\mid}{C}H_2$$

(1)

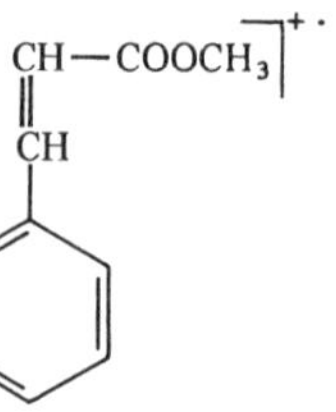

$$\left[CH-COOCH_3\right]^{+\cdot}$$

$$\overset{\mid}{C}H$$

m/z 162

two scans apart, they could not originate from the same compound. The second component was identified as the methyl ester of benzoylglycine (2), the molecular ion of which has mass of 193 (see section 4.2).

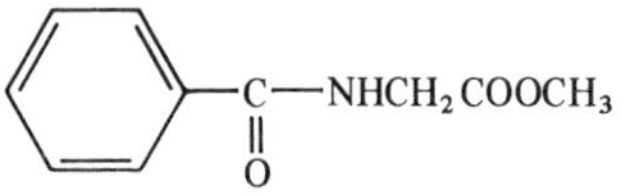

$$C-NHCH_2COOCH_3$$
$$\overset{\|}{O}$$

(2)

It is possible to specify for this procedure not just single masses but mass ranges too. Its applications are discussed more fully in chapter 4.

In the Biller–Biemann enhancement technique, the data system examines the mass chromatogram for every integer mass in the range scanned (Biller & Biemann, 1974). When several rise to a maximum coincidentally, a chromatographic peak is defined irrespective of the total ion current. This is done by ignoring all ions that do not maximize collectively and the result is seen in figure 3.8. The lower trace shows a portion of the total ion current chromatogram from figure 3.5 and, immediately above it, the result of the Biller–Biemann enhancement. The upper trace is called the *mass-resolved gas chromatogram*, consisting of the summed ion current of all maximizing ions. The artificial increase in chromatographic resolution is striking; partially resolved components of scan numbers 163 and 176 are completely resolved. Before enhancement, the mass spectrum of the compound eluting at scan number 163 was contaminated with ions from the succeeding eluate and from column bleed (figure 3.9*a*) but, since these do not maximize in the same scan, they are computationally erased (figure 3.9*b*). The enhanced spectrum is called a *reconstructed mass spectrum* in which the ordinate can be a percentage of relative abundance (as shown in figure 3.9) or a percentage of reconstructed ion current (%RIC). Since the constant background ions of column bleed or calibration compound never maximize during the analysis, the enhancement eliminates these peaks altogether. However, deconvolution of the mass spectra of truly co-eluting components cannot be effected since all ions will maximize in the same scan. This outcome is seen in figure 3.8 where the last peak (number 198) is due to the TMS ethers of both lanosterol and sitosterol which co-elute under the chromatographic conditions employed.

To be resolved by this technique, chromatographic peaks must maximize at least three scans apart. Hence, the faster the scanning of the mass spectrometer, the more likely it is that the method will succeed for closely eluting compounds.

The method is computationally fast and works particularly well for locating unresolved components of chromatographic peaks, even if a component is minor and masked by a more abundant one. The technique is exemplified later (section 4.2). It performs markedly less well when the unresolved components are of similar structure (e.g. isomers) and have similar mass spectra. For these cases, more sophisticated enhancement programs are available which require powerful computing facilities and more processing time (Chapman, 1978, pp. 89–91). In one method, the peak position of every maximizing mass chromatogram is estimated to one-third of the scan time by a least-squares interpolation about several ion current readings either side of the maximum. This procedure resolves components which are separated by 1½ scans or more (Dromey, Stefik, Rindfleisch & Duffield, 1976).

The techniques discussed above reduce multiscan data to a single mass spectrum or a series of mass spectra, each of a pure compound and suitable for further processing, if required, by single scan programs.

Figure 3.8. Enhancement of GC/MS data. The original total ion current chromatogram (figure 3.5) is shown (bottom) along with the mass-resolved gas chromatogram obtained after computer enhancement (top).

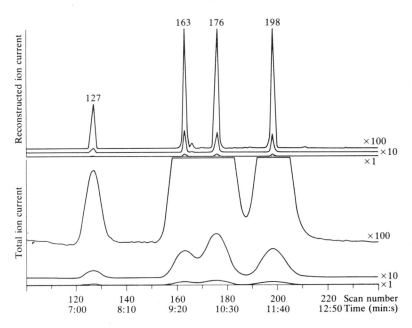

3.4.3. Single-scan data processing

Determination of elemental composition by accurate mass measurement has already been detailed. The number and type of atoms present in some compounds may be estimated by examination of the natural isotope pattern in the molecular ion regions of their low-resolution mass spectra (sections 1.3 and 12.2). The technique requires that the sample gives a molecular ion and contains elements such as chlorine, bromine, boron or silicon, which have abundant isotopes. The computer compares the experimentally determined isotope abundances with those computed for various elemental combinations; the goodness of comparative fit is the criterion by which it is decided whether or not the correct elemental composition has been found. The method is limited, but may also be used to calculate monoisotopic mass spectra from isotopically complex spectra (section 1.3).

The most popular computerized method of determining the structure of an unidentified, but not novel, compound is to search through a library of low-resolution mass spectra for a match between its spectrum and that of a compound in the library. The technique of *library searching* has been reviewed in considerable detail (Chapman, 1978, pp. 101–49). Most commercial data systems have such a library, usually stored on a magnetic disc, which may contain as many as 50 000 electron impact mass spectra.

Figure 3.9. Comparison of mass spectra of 24-methyl-3-trimethylsilyloxy-choles-5-ene (*a*) before and (*b*) after enhancement. The spectra are obtained from the analysis by GC/MS shown in figure 3.8.

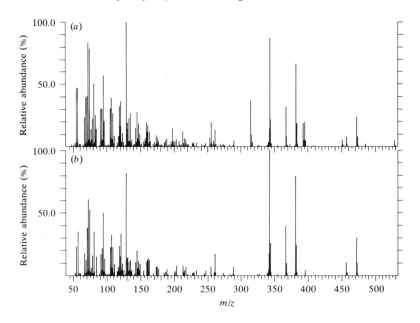

Alternatively, remote large libraries may be searched over telephone lines and sometimes via satellite. Libraries of mass spectra obtained by other methods of ionization are not so advanced, the main reason being that electron impact has been by far the most popular method of ionization. Chemical ionization and field desorption give spectra which are prone to considerable variation with experimental conditions and are less suited to matching. There are so many different approaches to library searching that only a limited description can be given here. A typical library does not contain complete mass spectra since too much storage space would be needed and excessively long search times would result. Instead, the spectra are abbreviated (*encoded*). A simple way of doing this is to store, with the name of each compound, the masses and abundances of only the 5–10 largest peaks in the spectrum although, in this and most other encoding procedures, the important molecular ion is included even if it is of low abundance. Of this type of encoding procedure, the most well known is the 'Eight-Peak Index' which, in book form, is also used by many chemists for manual assignment of structure (see section 12.5). A particular disadvantage of the method is that, with compounds of higher molecular weights, the most abundant ions tend to be of low mass and low diagnostic value because they are common to many different structures. There are several ways of circumventing this problem. Firstly, there can be 'weighting' towards the more diagnostic high-mass ions by defining a new parameter, the product of the original peak height and the mass of the ion. Secondly, starting at low and progressing to high mass, the n largest peaks every m mass units are stored. The best known of these procedures is the encoding of the two ($n = 2$) largest peaks every fourteen ($m = 14$) mass units (Hertz, Hites & Biemann, 1971), but other variants are used ($n = 1$, $m = 7$ or 14; $n = 2$, $m = 25$; $n = 3$, $m = 14$). Lastly, attempts may be made to assign and store automatically the peaks of greatest structural significance (Grönneberg, Gray & Eglinton, 1975; Pesyna, Venkataraghavan, Dayringer & McLafferty, 1976) or information content (Wangen, Woodward & Isenhour, 1971; van Marlen & Dijkstra, 1976). In the more rigorous encoding methods, elements from all three approaches are evident. In another type of encoding, some other characteristics of the mass spectrum are stored along with the largest peaks (Kwok, Venkataraghavan & McLafferty, 1973). These characteristics are chosen to be highly diagnostic of structure, as with: (*a*) losses of small neutral molecules, such has H_2O, HCN and H_2CO, from the molecular ion, (*b*) series of ions occurring every 14 mass units (e.g. m/z 43, 57, 71, . . . for $C_nH_{2n+1}^+$ or $C_nH_{2n-1}O^+$) and (*c*) some low-mass ions characteristic of functional groups, e.g. m/z 30 ($CH_2 = \overset{+}{N}H_2$) indicating a primary amine. With this method some interpretative aspects are brought into play as well as simple 'fingerprinting'.

It is not possible to reproduce mass spectra exactly, but good methods of encoding spectra minimize anomalies. Quadrupole and magnetic sector instruments give somewhat different spectra for the same compound and, even with the same mass spectrometer, the spectrum of a sample can vary with the method of introduction into the ion source, the condition of the source and so on. To some extent a library search is adversely affected by peaks due to impurities in the spectrum of the unknown compound, and enhancement by one of the techniques described in the previous section is imperative. This situation applies particularly to spectra obtained by GC/MS in which contamination by column bleed might otherwise prevent a correct match. Frequently, it is useful to construct a library of one's own by encoding and storing mass spectra of standard compounds to ensure that the content of the library and the conditions under which the spectra are recorded are relevant to the work carried out.

A typical search of a large library of several thousand spectra may take 15–30 s, a time which can be reduced markedly with a data system which allows additional information to be stored with the encoded spectra. Molecular weight and elemental composition are often included and the search time can be lessened by instructing the computer to consider only certain library entries. For example, the search would only concern itself with reference spectra of nitrogen-containing compounds if the unknown compound were an amine. Other properties which may be stored and used to reduce search times are melting or boiling points, nuclear magnetic resonance spectral details or gas chromatographic retention times. A large library may be classified according to structure so that, in effect, it is a series of sublibraries of steroids, ketones, amines and so on. In many studies, the class of the compound under investigation is known and then only one sublibrary need be searched. For specialized laboratories it may be unnecessary and wasteful to have a large library. In these cases, storage of a small amount of relevant spectral data can be most useful and complete mass spectra may be stored since demands on storage space are not so great. For compounds with very few differences in their mass spectra, such as isomeric terpenes, reference to complete spectra may be necessary to distinguish them (Adams, Granat, Hogge & von Rudloff, 1979).

The matching routine involves calculation by the data system of a *similarity index, match factor* or *purity* between the unknown spectrum and library (reference) spectra (McLafferty, 1977). Common scales are 0 for a complete mismatch to 1 or 1000 for a perfect match. To avoid time-consuming calculations of the purity against all entries in a large library, there is usually a *pre-search* or *filter* which rapidly eliminates grossly dissimilar spectra by requiring that a candidate reference spectrum must contain at least some of the largest peaks of the unknown spectrum. Only the entries which show the most matches in this test are selected for the more rigorous calculations of purity of match. The technique by which an unknown and reference spectrum are compared varies amongst different data systems (Chapman, 1978, pp. 120–39). Often, at each m/z value in turn, the ratio (R) of the mass peak heights in the normalized unknown and reference spectra is calculated. The smaller height is always divided by the larger so that $0 \leqslant R \leqslant 1$ (where $R = 0$ if there is a peak in one spectrum but not in the other). The individual R values are summed and the average value is normalized to produce the final purity figure for the particular reference spectrum. The entries with the highest figures are displayed as the result of the library search.

In the search techniques described above, all of the peaks in the unknown spectrum are compared with all of the peaks in the reference spectrum. If the only peaks considered for the search are those in the reference spectra, the technique is called *reverse searching*. The so-called *fit* is analogous to and calculated in the same way as for purity but peaks present only in the unknown spectrum are ignored. Therefore, extraneous peaks due to impurities or unresolved components will not prevent a correct match if the compound is in the library. The technique should be used with caution since a high fit will occur if the unknown and the reference compound are not identical but have a substructure in common. For instance, a reference spectrum of benzoic acid matches well with the spectrum of ethyl benzoate because the molecular ion of the ester loses ethylene to give an ion that fragments further like the molecular ion of the acid. It is occasionally useful in analyses by GC/MS or LC/MS to perform a mass spectral search for a specific compound thought to be present in the sample being investigated. The technique, sometimes called *inverse searching*, involves computer comparison of a reference spectrum of the material in question with each spectrum of the repetitively scanned analysis. If a plot of the calculated similarity indices against time or scan number maximizes at the retention time of the compound, its presence in the sample is indicated.

The enhanced spectrum of the first-eluting component in figure 3.5, when subjected to a library search, gave the result tabulated in figure 3.10. The report shows that the library contains 25 409 reference spectra and that, of these, only 174 pass the pre-search requirement. For each of these candidate spectra, the similarity to the unknown spectrum is evaluated by the data system on the scale of 0 for a complete mismatch and 1000 for a perfect match. The search is conventional in that the entries are ranked in order of highest purity, but the fit values for these matches are also calculated. In the figure, both values are high for the entry ranked first and the assignment of 3-trimethylsilyloxycholest-5-ene is correct. Had the compound not been in the library, a result affording some structural information about the unknown compound would still have been gained since compounds of similar structure would have been retrieved. This gain in structural information can be seen in figure 3.10 where the entry ranked second is an isomer of the correct compound and the three entries ranked third to fifth are all structurally related to cholest-5-en-3-ol. Whatever the fit or purity values, the result of a library search should be assessed by careful visual examination of the matching spectra when presented as shown in figure 3.11 for the highest ranked library entry of figure 3.10. A match against an uncontaminated sample spectrum should be judged correct only if the differences between the two spectra can be attributed to peaks eliminated in the encoding of the library spectrum. This is the case in figure 3.11; notice that virtually all the peaks in the reference spectrum are present also in the spectrum of the sample, and that the relative heights of the matching peaks are approximately the same.

When an unknown compound is novel or not present in the library, the methods discussed above are of limited use. There are two approaches to computerized structural analysis which do not require there to be a standard spectrum of the unknown compound. Both methods are far less popular than library searching and discussion of them will be correspondingly brief. Several groups of workers have attempted to crystallize into computer-compatible rules the approaches a mass spectrometrist would use for structure elucidation (chapter 10). A data system programmed with these rules has interpretative ability and performs as well as, but faster than a trained mass spectrometrist in some instances (Slagle, 1971; Chapman, 1978, pp. 187–219). The technique (*automatic spectral interpretation*) is said to use *artificial intelligence* or the *heuristic method*, but the latter name is confusing since it implies self-learning which is used only in *learning machine* methods, to be described later. Artificial intelligence requires powerful computational facilities, large storage areas and a high-level

Figure 3.10. Computer print-out of result of library search on first-eluting component in figure 3.5.

LIBRARY SEARCH   DATA: ENHANCED SPECTRUM NUMBER 127

SAMPLE: TMS DERIVATIVES OF STEROLS

25409 SPECTRA IN LIBRARY SEARCHED FOR MAXIMUM PURITY
174 MATCHED AT LEAST 3 OF THE 16 LARGEST PEAKS IN THE UNKNOWN SPECTRUM

RANK	NAME
1	3.BETA-TRIMETHYLSILYLOXYCHOLEST-5-ENE
2	3.BETA-TRIMETHYLSILYLOXYCHOLEST-4-ENE
3	3.BETA, 5.ALPHA-BIS-TRIMETHYLSILYLOXYCHOLESTANE
4	3.BETA, 6.BETA-BIS-TRIMETHYLSILYLOXYCHOLESTANE
5	3.BETA-ETHOXYCHOLEST-5-ENE

RANK	FORMULA	MOL. WT	PURITY	FIT
1	C30.H54.O.SI	458	780	966
2	C30.H54.O.SI	458	658	934
3	C33.H64.O2.SI2	548	649	899
4	C33.H64.O2.SI2	548	495	721
5	C29.H50.O	414	470	722

programming language with consequent slow processing. Both metastable ion
and accurate mass measurement data can be used if these have been acquired
(see earlier). Accurate masses have been used to determine the amino acid
sequence of peptides by a program which follows the same steps as would
be taken by the human interpreter (see example I in chapter 11). The best
known automatic spectral interpretation program is DENDRAL in which
the functional groups present in an unknown compound are ascertained
from its mass spectrum by way of a qualitative theory of fragmentation
processes, any available metastable ion data and the empirical or molecular
formula. Potential structures are generated, taking into account any other
information supplied such as the nuclear magnetic resonance spectrum.
For each candidate structure generated, a mass spectrum is predicted, using
known fragmentation processes or a program which generates possible
fragmentation mechanisms by examining standard spectra of similar, known
structures. These theoretically predicted spectra are checked against the
experimentally determined spectrum of the unknown compound and those
which are consistent are ranked in order of best matching and printed out.
A fairly successful program designed to predict the mass of a molecular ion,
whether it is present in the spectrum or not, is also available. As before,

Figure 3.11. Comparison of (*a*) the sample spectrum and (*b*) the abbreviated
mass spectrum of 3-trimethylsilyloxycholest-5-ene stored in the library
(bottom).

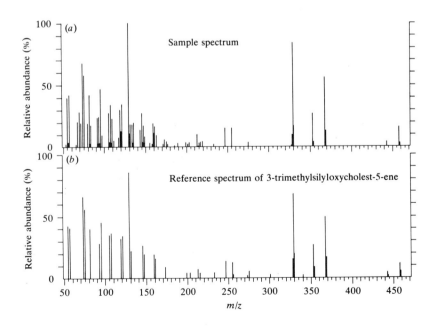

this program operates in much the same way as a mass spectrometrist does, only more rapidly (Dromey, Buchanan, Smith, Lederberg & Djerassi, 1975).

The last method of automated structure elucidation discussed here is that of *pattern recognition* or *machine learning* (Meisel, 1972; Isenhour, Kowalski & Jurs, 1974; Chapman, 1978, pp. 150-86). The mass spectra of known compounds (a *training set*) are represented as points in multidimensional ·space (an *n*-fold vector), the co-ordinates of each point being the mass and abundance of each ion. It is reasonably argued that since a mass spectrum is characteristic of structure, the points of compounds with similar structure will tend to cluster in hyperspace. When the spectrum of an unknown material is similarly treated, it is hoped that it will fall within one of the clusters, thereby categorizing its structural type. There are many ways in which this cluster analysis has been performed (Justice & Isenhour, 1974; Mellon, 1975, pp. 127-32). Structural features which have been identified readily by the method are: the presence of nitrogen or oxygen, the ratio of carbon to hydrogen, the phenyl group (although an experienced human interpreter would usually recognize this feature from the mass spectrum at a glance) and the phosphonate and carbonyl groups. Interesting studies include the analysis of isomeric compounds (e.g. sugar derivatives, alkylbenzenes), the spectra of which are too similar to be distinguished by other methods, the classification of bacteria by subjecting the mass spectra of their pyrolysis products to cluster analysis, and determination of pharmacological activity of a limited range of drugs. In this last case, it is assumed that just as a mass spectrum is related to structure by an obscure function, so structure is related to pharmacological activity by a complex function. Thus, the application of pattern recognition has been validated to some extent, and some successful predictive ability exhibited.

3.5. Data display units

Interaction between the data system and operator is through a *visual display unit* (VDU) or a keyboard such as a teletype. All commands to and communications from the computer can pass through these units. The mechanically based teletype is far slower than the electronically based VDU, but, unlike the VDU, does provide a permanent presentation of results, called the *hard copy*. A mass spectrometer produces and a computer processes data at high rates. For example, in an analysis by GC/MS, the printing out of the results is likely to be the most time-consuming part of the analysis so that a fast VDU for almost immediate display of results and a fast output device for hard copying of those results needing to be printed, is recommended. Ideally, the data system is programmed to allow unsupervised data processing and hard copying so that these steps can be carried out at times when the mass spectrometer is not in use for data acquisition. In such circumstances, the speed of the hard copy unit is of lesser importance.

4 Combined chromatography and mass spectrometry

4.1 Introduction

It is possible to collect (trap) compounds as they elute from gas or liquid chromatographic columns and afterwards to obtain their mass spectra. Such gathering of mass spectrometric data is cumbersome, especially for complex mixtures, and has been superseded by coupling the chromatograph directly to a mass spectrometer. The present chapter is concerned with the unified, continuous processes of gas chromatography/mass spectrometry and liquid chromatography/mass spectrometry, and there will be no discussion of the trapping of chromatographic effluents prior to mass spectrometric analysis. Discussion of the chromatographic processes themselves is outside the scope of this book. GC/MS is an established technique for the analysis of complex mixtures, holding a prime position in analytical chemistry because of its combination of sensitivity, wide range of applicability and versatility. LC/MS instruments show great promise for materials too involatile or thermally unstable to pass unchanged through a gas chromatograph. Both techniques are capable of obtaining complete mass spectra of a few nanogrammes of each component. Below this level, analysis is still possible but special scanning techniques, detailed in chapter 6, are necessary. A further method for the analysis of mixtures, involving separation by mass rather than by chromatography, is discussed in chapter 8.

The amount of material in the ion source changes constantly during the emergence of a chromatographic peak, so that the mass spectrometer must be capable of scanning a spectrum quickly; otherwise, ion abundances would be grossly distorted by the change in concentration of the sample during the scan. This problem only arises with older mass spectrometers that scan more slowly, very narrow chromatographic peaks (as obtained from capillary columns) or mass spectrometers operating at high resolution. Modern mass spectrometers are usually adequate to meet these more demanding tasks. If necessary, compensation for changes in quantity of sample during the scan is possible by reference to the change in total ion current in the same time interval, although this is not usually needed. At high mass-spectrometric resolution, the accurate masses of the ions, and not their relative abundances, are of prime importance.

References to GC/MS and LC/MS are to be found in the preceding chapters and, in particular, computer programmes to process the data of such analyses are described in section 3.4. In this chapter it will be assumed that the combined GC/MS instrument is coupled to a data system which is now regarded as a necessity for thorough analyses.

Gas chromatographs operate at atmospheric pressure but the ion sources of most mass spectrometers operate at low pressure so that, in general for a combined instrument, the gas chromatographic effluent must be transferred to the mass spectrometer with concomitant rejection of most of the carrier gas. Transfer of liquid chromatographic effluents to conventional mass spectrometers also requires a pressure-reducing device, and there is the additional difficulty of removing the relatively involatile solvent. For both GC/MS and LC/MS, the ideal device would remove all of the chromatographic mobile phase and transfer all of the solute to the ion source without degrading the chromatographic resolution. In the following sections, different ways of combining (*interfacing*) chromatographs and mass spectrometers are discussed, with an appraisal of how closely they approach the ideal.

4.2. Gas chromatography/mass spectrometry

This combination of two gas-phase analytical techniques can be applied to a wide range of problems. With a good interface, any compounds which elute from a gas chromatograph may be examined by GC/MS. Chemistry, medicine, biochemistry, pharmacology, environmental pollution control, food science, geochemistry and gas analysis are some of the areas which have benefitted from GC/MS. Information on the full range of applications is best obtained by reference to the biennial reviews in *Specialist Periodical Reports on Mass Spectrometry* (Brooks & Middleditch, 1973, 1975, 1977, 1979). The technique has been described extensively in a book (McFadden, 1973) and recent reviews (Brooks & Edmonds, 1979; ten Noever de Brauw, 1979).

The thermal stability must be reasonable and the volatility low for a gas chromatographic liquid phase; otherwise, a large amount of the phase would bleed into the ion source and detract from the analysis by giving abundant 'background peaks' in the mass spectra. The proportion of liquid phase in a packed column must be low, certainly under 10 per cent and preferably under 5 per cent, because the higher the loading, the greater the column bleed.

4.2.1. The GC/MS interface

The simplest interface is a direct line from the end of the chromatographic column to the mass spectrometer. The coupling is frequently constructed of glass-lined metal tubing to prevent contact between the eluted organic compounds and metal which, at high temperature, tends to act as a catalytical surface for decomposition. A mass spectrometer with efficient pumps operating in the electron impact mode can cope adequately with a flow of carrier gas from the column of about 2 ml/min or less, values commonly encountered with capillary columns. If the flow is much greater than this, special conditions for ionization are necessary. With gas chromatographs coupled to mass spectrometers that have chemical ionization sources, it is possible to pass the carrier gas directly into the ion source where it may act also as the CI reactant gas. Methane and hydrogen are most often used for such purposes, but the method is not particularly popular because it restricts the choice of CI reactant gas to those gases which are suitable also as GC carriers. Methods of direct coupling using electrohydrodynamic or atmospheric pressure ionization sources (section 2.3) have not been widely developed. A simple alternative interface is the *splitter* which is attached to the column exit and allows a small fraction of the total GC effluent into the source. The remainder is passed to a parallel detector such as a flame ionization detector or into the atmosphere. This procedure prevents the pressure in the ion source from rising to unacceptably high values but it is wasteful of sample and hence insensitive. Further, there is no increase in concentration of the solute relative to the carrier gas in the effluent passed to the ion source, viz. there is no *enrichment*.

Enrichment devices are called *molecular separators,* of which there are three main types. The *Llewellyn–Littlejohn (membrane) separator* is illustrated in figure 4.1. The chromatographic effluent is passed over a thin silicone rubber membrane interposed between chromatograph and ion source. At the membrane surface, organic molecules dissolve in and diffuse through the membrane into the ion source. Enrichment occurs because the inorganic carrier gas, almost exclusively helium, scarcely dissolves in the rubber and so passes to atmosphere without permeating the membrane. The membrane separator provides good levels of enrichment but the overall efficiency varies widely, depending on the structure of the solute because different compounds have differential solubility in the rubber and therefore permeate the membrane with various degrees of difficulty. Some compounds, like primary amines, even react with the membrane. The separator can affect adversely chromatographic resolution due to the finite time required for a compound to diffuse through the membrane so that, sometimes, a residual amount of material from one chromatographic peak is caught up by the next peak eluted. This 'memory' effect causes some post-chromatographic mixing of components previously resolved on the column. The membrane must operate at the same temperature as the gas chromatographic column and this places a further limitation on the separator because the membranes degrade rapidly at high temperatures. Also, the membranes must be thin if permeation is not to be too slow and so they are prone to breaking.

Another device is the *Watson–Biemann (effusion) separator*, shown in figure 4.2. It uses the principle of differential molecular effusion through a porous partition to achieve enrichment of the organic phase entrained in the helium carrier gas. Because of its low molecular weight, helium effuses faster than organic solutes, particularly those of high molecular weight. The helium that has effused is pumped away together with some of the solute whilst the

Figure 4.1. Membrane separator showing enrichment of organic material (O) in carrier gas (●).

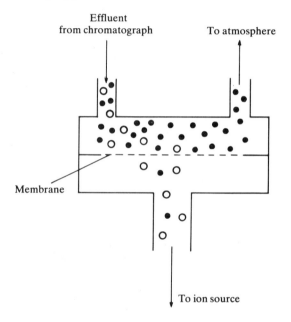

Figure 4.2. Effusion separator showing enrichment of organic material (O) in carrier gas (●).

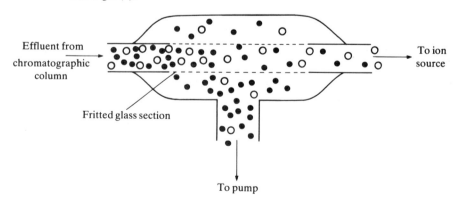

enriched effluent containing relatively little helium is passed into the ion
source. The separator is only efficient within a narrow range of gas flow.
Decomposition or adsorption of polar organic molecules on the sintered
glass forming the porous partition is common unless its surface has been
deactivated chemically. The device does not perform well with lower
molecular weight materials which effuse through the frit at similar rates
as the helium and it is restricted in temperature range but it is robust and
easy to manufacture. A modification of the effusion separator uses a variable
slit instead of a porous partition. For each different carrier gas flow rate,
the very narrow slit along the length of the gas flow is adjusted for optimum
efficiency of the separator.

The *Rhyage (jet) separator* has displaced the effusion separator as the most
widely used enrichment device. The jet separator relies upon differential
diffusion of solute and carrier gas after the effluent has passed through a
small jet into a vacuum. Helium, the carrier gas, diffuses faster than organic
molecules from the expanding stream of gas emanating from the jet. As
shown in figure 4.3, the gas stream is directed towards a small orifice, about
1 mm away from the jet, and when it reaches this point the majority of the
carrier gas, as well as some organic material, has diffused from the stream
and has been pumped away. Usually the enriched effluent is transferred
directly into the ion source but, with some devices, notably the older ones,
a second jet is used to increase further the degree of enrichment. Good
enrichment is obtained with the single-stage separator, but much of the
organic material, particularly if it is of low molecular weight, is also pumped
away so that the overall efficiency of transfer of sample from the gas
chromatographic column into the ion source is not high. The fine orifice of
the jet is prone to blocking. The device does have several advantages. It can

Figure 4.3. Jet separator showing enrichment of organic material (O) in carrier
gas (●).

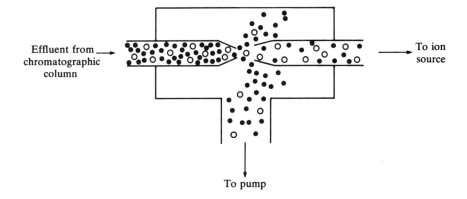

Effluent from →
chromatographic
column

To ion
source

To pump

be constructed entirely of glass, thereby reducing the incidence of sample decomposition, and has a temperature range as great as that of most chromatographic columns. The alignment of the orifices and the distance between them is critical, the efficiency decreasing markedly with any deviation from the optimum relative positions. To make impossible any misalignment, some jet separators are constructed from a single piece of glass capillary tube with a partial cut, through which the helium diffuses. In some respects this inflexibility of the one-piece separator is a disadvantage. The distance between the orifices is variable in some jet separators, so that for each analysis the gap can be optimized for the particular flow rate and the molecular weight range of the sample. When set correctly, these latter systems can attain relatively high efficiencies of transfer.

It is usual for a solution of the mixture under investigation to be injected onto the chromatographic column, but it is not desirable to admit to the ion source the vaporized solvent since this may cause unacceptably high pressure and damage to the ion source. Therefore, separators are fitted with some means of deflecting the solvent away from the source. This may be a by-pass valve, a vent to atmosphere or, in the case of the jet separator, a mobile metallic strip which can be moved into the gap between the orifices during the elution of the solvent to block off the inlet line to the source. The vapour is then pumped away with the helium. The methods of ionization suitable for GC/MS with a molecular separator are electron impact, field ionization, and negative-ion and positive-ion chemical ionization. For analysis by CI mass spectrometry, the removal of the majority of the helium by means of a separator allows a choice from the whole range of CI reactant gases. The reactant gas is let into the ion source through an independent inlet.

4.2.2. Applications

Discussion of quantitative aspects of GC/MS is presented in chapter 6. Here, some of the ways in which a mass spectrometrist can extract information from GC/MS data are illustrated by reference to a typical example concerning the analysis of dipeptides. This example will be discussed in some detail and then a few other interesting applications of GC/MS will be outlined.

Dipeptides in the urine of a patient with a disease which causes hydrolysis of endogenous collagen had been shown by mass spectrometric analysis to contain proline as the second amino-acid residue (Johnstone *et al.*, 1974). The dipeptides were not sufficiently volatile for GC, so they were derivatized

$$CH_3C—NH—CH—C—N \underset{O}{\overset{}{\underset{}{\overset{}{|}}}} C—OCH_3 \qquad (1)$$

by acetylation of free amino and hydroxyl groups and methylation of
carboxyl groups. These derivatives (1), and other compounds present in the
urine extract, were analysed on a conventional packed column, with a GC/MS
system employing a jet separator and, initially, electron impact ionization.
The extraction and purification of the urine samples produced a mixture
containing some amino acids and a few miscellaneous compounds as well
as the dipeptides so that a complex total ion current chromatogram resulted
(figure 4.4*b*). There are several ways of approaching the problem of identifying
the constituents of the mixture. It is possible to recall mass spectra for
examination whenever the total ion current chromatogram indicates that
there is a chromatographic peak, but this does not take account of unresolved
components. A better starting point is to perform a Biller–Biemann enhancement (section 3.4) to locate unresolved peaks (figure 4.4*a*). By this procedure,
the chromatographic resolution is increased artificially and over 40 components are revealed. Note in particular that the large single peak near scan
number 200 actually comprises four compounds and the peak at scan number
90, two compounds. In the latter case, the two components are not totally
resolved in the mass-resolved gas chromatogram (figure 4.4*a*) because they
maximize only two scan numbers apart and this is insufficient for separation

Figure 4.4. Total ion current chromatogram of (*b*) a urine extract and (*a*) the
corresponding mass-resolved gas chromatogram. Note that the upper trace
shows much greater resolution of peaks.

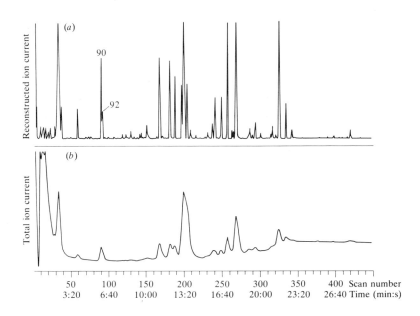

by the relatively simple enhancement procedure. In this instance, and in other similar ones, recourse is made to a more rigorous enhancement technique able to deconvolute components that elute closely together (section 3.4). Figure 4.5 shows the spectrum of the minor component at scan number 92 before and after this enhancement. The spectrum in figure 4.5*b* shows many ions, such as *m/z* 43, 91 and 162, which originate from the major component at scan number 90, the methyl ester of acetylphenylalanine. After enhancement, these ions are computationally removed (see figure 4.5*a*), giving a reasonably 'clean' spectrum of benzoylglycine methyl ester despite the fact that in the original chromatogram (figure 4.4*b*) its presence cannot be discerned by visual inspection.

The approach discussed so far is appropriate for analysis of all components of the mixture but, in some circumstances, it is unnecessary to examine them all. To reduce processing time, it is possible to focus attention on only the dipeptide content by the use of mass chromatograms. It was found that the standard spectra of derivatized proline-containing dipeptides all contained abundant ions at *m/z* 70 and 128 which can be drawn as shown (2). The mass chromatograms of these two *m/z* values, along with the total ion current

Figure 4.5. Mass spectrum of scan number 92 from the analysis shown in figure 4.4 before (*b*) and after (*a*) enhancement. Spectrum (*a*) is due to $C_6H_5CONHCH_2COOCH_3$ whilst spectrum (*b*) is contaminated with peaks due to $CH_3CONHCH(CH_2C_6H_5)COOCH_3$.

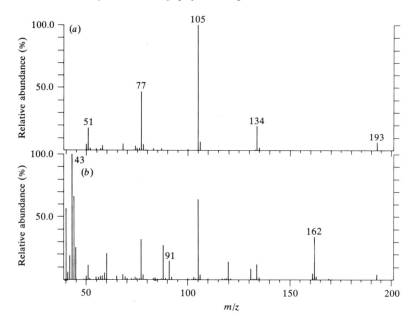

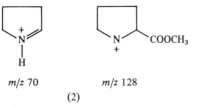

m/z 70 m/z 128

(2)

chromatogram, are shown in figure 4.6. It is seen that, of the large number of chromatographic peaks, the ones due to the compounds of interest become apparent wherever the two mass chromatograms rise and fall in unison, such as at scan numbers 168, 200, 257 and 268. The peak at scan number 34 is not due to a dipeptide, but to acetylproline methyl ester which also has the ions m/z 70 and 128 in its mass spectrum. Alternatively, the computer can be programmed to process the data in a manner which has the combined advantages of both the mass chromatogram approach and the Biller–Biemann enhancement. The mass-specific Biller–Biemann treatment defines a chromatographic peak and records a reconstructed mass spectrum whenever the ions at m/z 70 and 128 rise in abundance to a maximum. Where both maximize simultaneously, the data system marks the position on the trace. This approach is illustrated in figure 4.7 in which small ticks over the upper, enhanced

Figure 4.6. The total ion current chromatogram of the urine extract (bottom) and the mass chromatograms of ions at m/z 70 and 128.

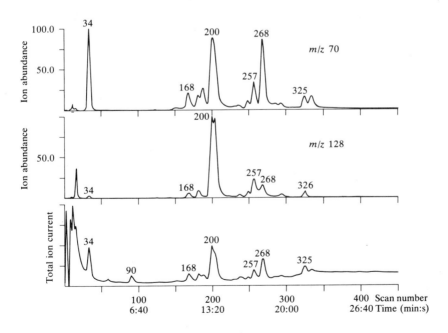

chromatogram indicate the probable occurrence of a dipeptide. Note that the components at scan numbers 168, 200, 257 and 268 are again selected. The enhanced spectrum of the last of these, the methyl ester of acetylprolyl-proline (Ac-Pro-Pro-OMe) is shown in figure 4.8, together with a reference spectrum of the same material. It is clear from the figure that there are no important differences between the two spectra.

Using the techniques described above, many derivatives of amino acids and dipeptides were identified, often by searching through a small, specialized library of reference spectra by computer as illustrated in figure 4.8 for Ac-Pro-Pro-OMe. The structures of other, miscellaneous constituents were elucidated either by examining fragmentation pathways or by searching through a large library of mass spectra. The analysis is not complete because there may be compounds which do not contain proline or are present in such small amounts that they are not picked out by the enhancement procedures. Discussion of mass chromatograms has concentrated so far on the simultaneous detection of several related compounds by use of common characteristic ions in their mass spectra. Mass chromatograms can also be used to search for one specific compound using ions which, ideally, are unique to that compound. Each suitable m/z value is either predicted or selected from a standard mass spectrum of the compound suspected of being present.

Figure 4.7. Total ion current chromatogram (b) and mass-specific enhancement (a) of urine extract. Ticks over peaks on the upper trace indicate where ions at m/z 70 and 128 rise in abundance to a maximum at the same time.

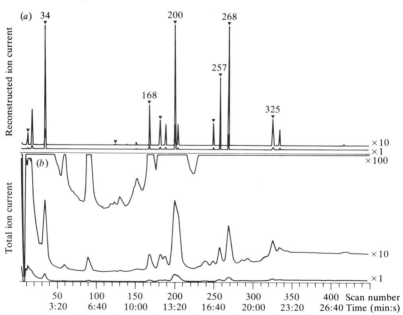

The presence of the methyl ester of diacetylated tyrosine (3) was determined in this way. Ions at m/z 178 (4), which were expected to give a large peak in its mass spectrum, are seen in figure 4.9 to be reasonably specific for this analysis. There are very few constituents which have significant abundance

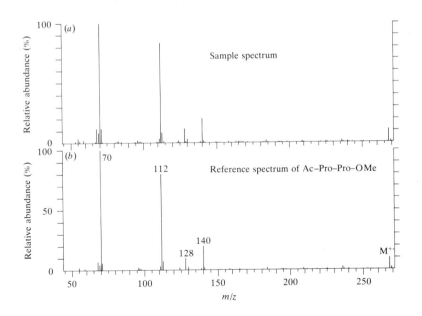

(3)   (4)

of ions at m/z 178 in their mass spectra (see figure 4.9) and only the spectrum at scan number 216 was consistent with the structure of the tyrosine derivative. It should also be noticed from the figure that this compound is a very minor component; its chromatographic peak is not visible in the total ion current chromatogram. The selection of ions specific to compounds of interest in complex mixtures is of prime importance for selected ion monitoring and is discussed in chapter 6. When a standard spectrum of a compound is available, it is possible to perform an inverse search for it

Figure 4.8. Comparison of a spectrum from the analysis of urine extract (*a*) with the mass spectrum of Ac-Pro-Pro-OMe from a small library of spectra of peptide derivatives (*b*).

(section 3.4) which is an examination of the mass chromatograms of all of the peaks in the reference spectrum.

In summary, the utility of mass chromatograms has been stressed as they have been shown to be useful not only for specific detection of a single compound but also for locating simultaneously several members of the same class of compound. For example, the location and identity of several peptides containing hydroxyproline was aided by observing the mass chromatogram of m/z 186 (5), an ion equivalent to the ion of proline-containing peptides at m/z 128 (2). This was important to the analysis since the only protein in

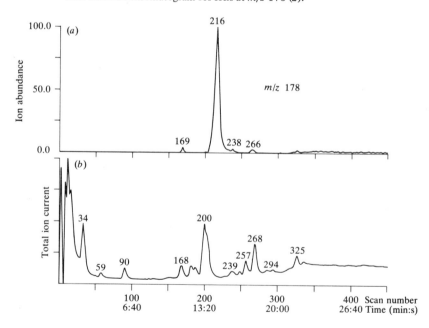

humans known to contain substantial amounts of hydroxyproline is collagen. Thus, the presence of acetylated, carboxymethylated hydroxyproline (6) identifies the protein from which the dipeptides and amino acids originated. In other studies, benzylic compounds could be located in chromatographic

Figure 4.9. Total ion current chromatogram of the urine extract (*b*), together with the mass chromatogram for ions at m/z 178 (*a*).

analyses by examination of the mass chromatograms of m/z 91 ($C_7H_7^+$) and 65 ($C_6H_5^+$), TMS derivatives by looking for m/z 73 ($(CH_3)_3Si^+$) and so on. Enhancements, based on examination of all of the mass chromatograms in the mass range scanned, artificially improve chromatographic resolution and even allow detection and location of unresolved components. The enhancements can be made mass-specific to emphasize some types of compounds. It should be appreciated that a single mass chromatogram which rises to a maximum at the retention time of a particular compound is not sufficient evidence for its identification. Such a signal is used only as a means of locating the mass spectrum or spectra which must be examined to ascertain the presence or absence of the compound. For a compound present in very small quantities, only the major ions in its mass spectrum may be detectable. In such circumstances, if the mass chromatograms of two or more characteristic ions rise and fall in abundance simultaneously at the correct retention time, the presence of the compound is indicated but should not be considered as proven.

When a mass spectrometer is operated so that the masses of all abundant ions are accurately determined, elemental formulae can be calculated. In effect, an accurate mass chromatogram is then a plot of elemental composition against time or scan number. It is a more specific indicator of structure than is the simpler integer mass chromatogram. Measurement of accurate masses makes it possible to distinguish readily background ions (such as those due to column bleed) from sample ions when these occur at the same integer mass but have different accurate masses. The methodology for GC/MS with mass spectrometers of high resolving powers has been reviewed (Kimble, 1978).

Returning to the example of figure 4.4, the structures of dipeptides for which standard spectra are unavailable are difficult to elucidate from EI mass spectra because they contain only low abundances of the informative high-mass ions. In particular, molecular ions are of low or negligible abundance. When the ion source is changed to chemical ionization conditions, the resulting mass spectra almost always contain abundant ions at one atomic mass unit greater than the molecular weight ($(M + H)^+$) owing to proton transfer from the reactant gas; this effect is not confined to peptides but is general to most classes of compound (section 2.3 and 7.2). Several advantages accrue from this emphasis on the higher-mass ions. The structures of unknown dipeptides and unexpected compounds are far easier to elucidate by examination of both CI and EI mass spectra. In the CI mode, the choice of characteristic ions used to search for specific compounds by mass chromatograms is made easy since it can be assumed that $(M + H)^+$ ions will be abundant. The requirement for characteristic ions of high mass arises because low mass ions are relatively non-specific, frequently being common to many different structures. This topic is discussed further in the following two chapters. The reactant gases used for the study described above were ammonia, isobutane and methanol. The last of these is unsuitable as a carrier gas for GC, and therefore helium is used as carrier gas. Differences between and the relative merits of EI and the various CI mass spectra in general are discussed in sections 7.1 and 7.2. Some mass spectrometers are capable of combining the electron impact and chemical ionization approaches in one analysis by alternate EI/CI switching so that, for example, all the odd-numbered scans are EI and the even-numbered scans CI spectra.

The following examples are included to illustrate the wide scope of GC/MS. A recent survey of applications has been published (Brooks & Middleditch, 1979). The Viking mission to Mars carried equipment for analysis by GC/MS of any organic material in the Martian atmosphere and soil (Biemann, 1979). Unfortunately, organic compounds were not detected in the samples despite the high sensitivity. The experiments were carried out with a small magnetic-sector instrument. Space research provided considerable impetus for the development of GC/MS with quadrupole mass filters, which are more robust and much more easily miniaturized than magnetic-sector instruments. For food flavour and odour research, the GC/MS interface is often a simple splitter, with the effluent which is not transferred to the mass spectrometer being passed to a 'sniffing port' at which a human being evaluates the odour of eluates. The structures identified by mass spectrometry are thus related to aroma. The aroma of any food is rarely attributed to a single component and results from a complex mixture. For example, 2-isobutylthiazole, unsaturated aliphatic alcohols and aldehydes, and metabolites of carotenoid compounds have been shown to contribute to the odour of tomatoes. The complex mixtures obtained by heating amino acids, triglycerides and carbohydrates have also been examined by GC/MS in a study of the chemical changes brought about by the cooking of food.

GC/MS is used in forensic laboratories to detect cocaine and naturally occurring cannabinoid and opiate compounds amongst many others. In a remarkable analysis, GC/MS was used to identify the acidic steroids in the gall bladder of a 3200-year-old Egyptian mummy. After taking account of changes induced by bacteria and the environment, the qualitative and quantitative composition of the acids was found to be the same as that in modern man, indicating that the pathways for cholesterol metabolism in man have not changed in that time.

Analyses by GC/MS of gaseous samples are common. Cigarette smoke, human breath and pollutants in air have all been studied. The volatile products resulting from pyrolysis of the polymeric materials of meteorites were shown by GC/MS to include alkanes, alkenes, alkylbenzenes and various thiophens. For on-line pyrolysis/GC/MS, products of pyrolysis are passed immediately to a GC/MS instrument. Such a system is useful for analysis of paint, fibres, greases, adhesives, dye pigments, polymers and bacteria, and has been used for diagnosis of cancer.

4.2.3. Capillary columns

The resolution obtained with packed columns is too low for many studies. There are several types of glass capillary columns which offer far greater resolution. Figure 4.10 shows two chromatograms for the analysis of the same mixture of methyl esters of fatty acids. Note that the lower trace,

Figure 4.10. Gas chromatograms of a mixture of long-chain esters. Note that the lower trace from a capillary column shows considerably more separation than the upper trace from a packed column. The first large peak, due to solvent, in the upper trace is missing from the lower trace because, with the system used, nearly all solvent is removed prior to injection onto the capillary column.

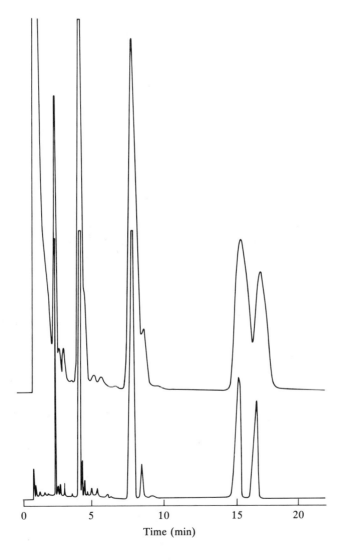

Time (min)

obtained from a capillary column, reveals some small peaks that are not visible in the upper trace from a packed column having lower resolution. Modern instruments for GC/MS are equipped to take either a capillary or a packed column (see figure 2.2*a*) and some allow rapid interchange between columns by having both installed in the chromatograph together with a valving device. The subject of GC with capillary columns is outside the scope of this discussion but both the theory and practice have been well covered in a recent book (Jennings, 1978).

Peak shapes from a capillary column are considerably sharper than those from a packed column (figure 4.10). If the same amount of a compound were eluted through the two different types of column the areas of the resulting peaks would be the same but the height of the peak from the high resolution column would be greater. Therefore, the concentration of the compound in the source would be greater for the capillary column. As long as the mass spectrometer is scanned fast enough to cope with peak widths of the order of 5–15 s, chromatographic columns with the higher resolutions give the greater sensitivity. Also, these columns are usually directly linked to the mass spectrometer so that 100 per cent of sample is transferred from the column to the ion source without enrichment. With a packed column and splitter or molecular separator, the transfer efficiency is usually at the lower end of the range 20–80 per cent, depending on the type of interface. In addition, the amount of column bleed is very low with capillary columns so that small amounts of sample are less likely to be obscured by background. Therefore, there is less need for data enhancement particularly as there will be few unresolved peaks. Disadvantages of capillary columns include expense, degradation of performance over a relatively short period (although this is not as serious as is sometimes stated), the relative difficulty of preparation and manipulation, awkwardness of injection technique in some situations and the requirement for a very fast scanning mass spectrometer. In addition, the load capacity of such columns is low, so that if more than a few tens of nanogrammes are injected, the resulting peaks are distorted and resolution degraded, but this is rarely a serious drawback. All things considered, the balance weighs heavily in favour of the use of capillary columns. The data processing programs which have been described above, and in the previous chapter with respect to GC/MS with packed columns, are applicable to columns of higher resolution. Operation of the mass spectrometer at high resolution for GC/MS with capillary columns is difficult because of the conflict between the slow scan of the spectrometer in this mode and the narrow peak widths, but it is possible (Meili, Walls, McPherron & Burlingame, 1979). The technique of accurate mass measurement at low resolving powers is more amenable to GC/MS with capillary columns.

For analyses involving highly complex mixtures, the components of which have similar structures and chromatographic properties, high-resolution gas

chromatographic columns are necessary. Such mixtures are frequently
obtained when studying the metabolism of drugs or natural compounds and
when monitoring environmental pollution. For this reason, capillary columns
are widely applied in medicine, biochemistry and environmental science.

4.3.   Liquid chromatography/mass spectrometry

The solute emerging from a liquid chromatographic column is not
only in the condensed phase but also highly diluted (components eluting
from a gas chromatographic column are relatively concentrated). Often,
involatile inorganic compounds may be present in the eluate (as buffers,
for example) and, if gradient elution is used, the composition of the solvent
changes during the analysis. Hence, the technical difficulties of combining
(interfacing) liquid chromatography and mass spectrometry exceed those for
the GC/MS combination. This section is concerned mostly with a discussion
of interfaces for LC/MS. There will be no need to examine in detail any one
particular application because the approaches to the processing of data
obtained by GC/MS, discussed in the previous section, are relevant also for
LC/MS and additional computer programs are not required.

The technique of LC/MS is undergoing a rapid expansion, similar to that
undergone by GC/MS some years ago. Its future depends to some degree on
the development of liquid chromatographic capillary columns which are
capable of high chromatographic resolution at very low flow rates, making
them less incompatible with mass spectrometers and allowing direct coupling.
This situation parallels that of GC/MS when capillary columns were first
coupled to mass spectrometers.

Unless there are extenuating circumstances, there is little point in analysing
by LC/MS a mixture which is amenable to GC/MS since the latter method is
better developed and the necessary equipment is more widely available and
less expensive. LC/MS is best reserved for mixtures which are impossible or
difficult to analyse by GC/MS because of thermal instability or involatility
of the constituents. Many otherwise troublesome compounds may be made
suitable for GC/MS by chemical derivatization (chapter 5) but, if there are
no suitable derivatives or if the yield of the derivatization reaction is low,
LC/MS can be applied with advantage to the underivatized mixture.

Recent reviews of LC/MS are concerned mainly with methodology
(Arpino, 1975; Games, 1979; McFadden, 1979; Zerilli, 1979), but a
survey of applications and selected examples has been published (McFadden
et al., 1980).

Typical flow rates for LC are 0.5-5 ml/min which, for common solvents
after vaporization, are equivalent to gas flows in the range 100-3000 ml/min,
compared with gas flows of 0.5-50 ml/min for GC. The commoner ion
sources, employing electron impact, chemical or field ionization, cannot
tolerate such large volumes of vaporized solvent so direct coupling of liquid

chromatographic columns to these types of mass spectrometer is precluded. Mass spectrometers employing electrohydrodynamic or atmospheric pressure ionization are amenable to direct coupling. The former method requires that the sample to be ionized is dissolved in a solvent and the latter ionizes sample molecules by a complex series of ion/molecule reactions at atmospheric pressure. When these methods are used for LC/MS, there are restrictions on the LC solvent and difficulties in the case of involatile buffers and solutes of low vapour pressure. Both methods are inherently very sensitive but await considerable technical improvements.

A splitter can be used to direct about 1 per cent of the effluent to a chemical ionization mass spectrometer in which the LC solvent is used as the CI reactant gas. In its simplest form, the choice of LC solvent is restricted to those liquids which are suitable reactant gases when vaporized and gradient elution causes complications in that the composition of CI reactant gas changes during the analysis. In more advanced systems, these problems are overcome by continuously passing into the source a relatively large amount of a gas such as ammonia, suitable for CI. Some mass spectrometers require a cryogenic trap (cold-fingers around the ion source) to condense some of the solvent, thereby supplementing the pumping system. The method does not work very well with samples of very low vapour pressure and involatile substances are not acceptable. Electron impact spectra cannot be obtained but LC/MS is of most use for thermally unstable compounds for which CI spectra are usually more informative. The interface is commercially available and has been used in the analysis of mixtures of peptides, triglycerides and metabolites. Despite the inherent insensitivity resulting from splitting the effluent, analyses in the low nanogramme range are possible.

As with the splitter for GC/MS, the interface discussed above is hardly ideal since the efficiency of transfer is low (about 1 per cent) and there is no enrichment. A silicone rubber membrane has been used to enrich LC eluates in a manner analogous to the membrane separator for GC/MS. The membrane is permeable to non-polar solutes, but impermeable to the polar solvent, an effect which enables enrichment but, at the same time, imposes restrictions on types of solute and solvent. Like its GC/MS counterpart, it has many drawbacks, such as temperature limitation and a tendency to degrade chromatographic resolution. Another LC/MS interface derived from GC/MS methodology is the *jet separator* in which the eluate is forced through a fine jet and directed towards a small orifice which leads into the ion source. One successful system uses a laser beam focussed on the liquid stream emerging from the jet to effect vaporization and expansion of the solvent. The region between the two orifices is evacuated to remove some of the diffused, vaporized solvent as with the analogous GC/MS interface (figure 4.3). An ultrasonic 'atomizer' may be used in place of the laser. The technique allows EI, CI or both methods of ionization, but it is relatively insensitive and some molecules are pyrolysed before reaching the ion source. The method has been applied to hydrocarbon mixtures and nucleic acid components. Another interface based on the jet separator is discussed below.

Continuously moving, endless wires or belts provide the most successful LC/MS interfaces to date. The *moving belt interface*, suitable for quadrupole or magnetic sector mass spectrometers, has been developed into a useful, commercially available system. Its mode of operation is illustrated in figure 4.11. The LC eluate is applied continuously to the moving belt, which is a few millimetres wide and has a far greater load capacity than has a moving wire. The solvent is subsequently evaporated by an infrared heater and application of a vacuum. The solute remains on the belt and is mechanically transported through a system of vacuum locks to the ion source where it is evaporated by a powerful heater. The sample may be carried right into the ion source before it is vaporized, ensuring best results with thermally unstable compounds. Any residue remaining on the belt is removed by another heater after it leaves the ion source. The method is suitable for operation of the mass spectrometer in the EI or positive- or negative-ion CI mode without severe restrictions on the choice of LC solvent or CI reactant gas. The use of involatile salts as buffers is permitted since they can be scrubbed off the belt as it returns to the inlet chamber, and gradient elution is not problematic. The interface does not cause significant loss of chromatographic resolution and the efficiency of transfer can be as high as 80 per cent, but 40 per cent is more typical. A few nanogrammes of sample provide complete mass spectra. The device goes a long way towards

Figure 4.11. Schematic diagram of the moving belt interface for LC/MS.

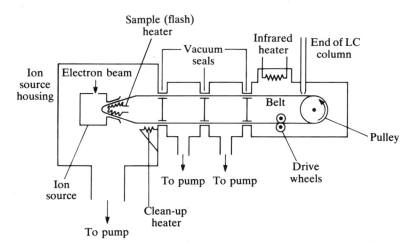

satisfying the criteria given in section 4.1 for an ideal interface. In particular, the degree of enrichment is very high because virtually all of the solvent is removed before introduction of the sample into the ion source. One disadvantage of the system is that the more volatile components of the sample tend to be vaporized and pumped away with the solvent. Flow rates compatible with the moving belt interface depend on the volatility of the solvent: the more volatile the solvent, the higher the acceptable flow rate. The maximum flow rate of aqueous solutions is about 0.5 ml/min whereas that of non-polar solvents is about 2 ml/min. The most notable applications of this interface have been to porphyrins, coumarins, aflatoxins, nucleosides, antibiotics, complex mixtures of alcohols and acids in waxes from the eye, and drug metabolism studies. The nucleoside, adenosine, when analysed by LC/MS using a moving belt interface, afforded the chemical ionization mass spectrum shown in figure 4.12. By proton transfer from the isobutane reactant gas ions, abundant $(M + H)^+$ ions are formed. Fragment ions at m/z 135 and 136 contain the purine base and result from ejection of the sugar moiety with concomitant hydrogen transfers.

A promising technique involves bombarding the solutes on the moving belt with ions, 'hot' atoms or laser beams. The energy deposited in the sample is sufficient for vaporization and ionization of substances with low vapour pressure (see section 12.3).

Extensive modification of the properties of the chromatographic eluate prior to transfer to the mass spectrometer can be a useful adjunct to

Figure 4.12. Partial chemical ionization mass spectrum of adenosine with isobutane. The sample was eluted with methanol (1 ml/min) through a liquid chromatographic column and introduced into the ion source via a moving belt interface. Figure reproduced with permission of VG Analytical Ltd, Altrincham, England.

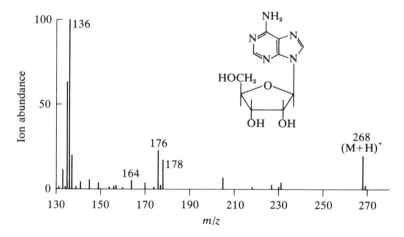

interfaces. For instance, the solutes in an aqueous eluate may be extracted by an on-line process into a non-polar solvent which is immiscible with water (e.g. dichloromethane). The method causes surprisingly little degradation of chromatographic resolution, removes inorganic salts because they are retained in the aqueous phase, permits the use of gradient elution and removes the limitations of moving belt interfaces with aqueous solutions. In another method, the eluate is applied to a perforated metal belt and, after removal of the solvent, the residue is carried into a heated reaction chamber where components of low volatility are reduced in an atmosphere of hydrogen. The resulting hydrocarbons are vaporized readily and swept into the ion source for detection and analysis. The method has been applied to lipids and the sensitivity of the device was good, but its general applicability has yet to be proved and, of course, much structural information concerning the compound before reduction is lost.

Another interface, based on the jet separator and incorporating a new method of ionization, has the advantage that there are no moving parts. The effluent from the column is heated rapidly to a very high temperature, forming a stream of vapour and aerosol. The vapour comprises about 95 per cent of the solvent whilst the larger particles of the aerosol (droplets) carry a charge and contain most of the solute. The solvent vapour is removed by a jet separator whilst the resulting beam of droplets is accelerated towards a hot metallic probe. When the charged droplets strike the probe, they break up and some molecules are ionized by proton transfer between solute (M) and residual solvent, giving quasi-molecular ions, usually $(M + H)^+$; negative ions are also formed. With flow rates in the region of 0.5–1 ml/min, at least 80 per cent of solutes with low vapour pressures is transmitted by the jet separator into the mass spectrometer, together with about 5 per cent of the solvent. Compounds present at the nanogramme level may be analysed. The resulting mass spectra show little fragmentation, rather like field desorption mass spectra (section 7.6). The interface is applied advantageously to nucleosides, nucleotides, peptides and oligosaccharides (Blakley, Carmody & Vestal, 1980).

5 Uses of derivatization

5.1. Introduction

Most forms of mass spectrometry require the sample under investigation to be in the gaseous phase for analysis. In the case of studies by GC/MS, samples spend considerable time in the vapour phase during their elution through the chromatographic column and transfer into the ion source of the mass spectrometer. To be suitable for mass spectrometric analysis, compounds must be both volatile and stable to decomposition and rearrangement in the gaseous phase. Unfortunately, many substances do not meet these requirements and are not directly amenable to the common mass spectrometric techniques. In such cases the analysis may be approached in three distinct ways. Firstly, formation of a derivative which imparts volatility and gas-phase stability on a troublesome compound may render it suitable for conventional mass spectrometry. Secondly, recourse may be made to specialized techniques for the analysis of involatile compounds (sections 7.6 and 12.3). Lastly, and usually less satisfactorily, the volatile pyrolysis products of the involatile material may be analysed to gain some information about the original structure. The present chapter is concerned with the first of these approaches, derivatization. It is assumed in the following discussion that a mass spectrometer suitable for examination of involatile substances is not available and that intact molecules (as opposed to pyrolysis products) are to be analysed. The term 'involatile' refers to a substance which does not have significant vapour pressure at the temperature and pressure of an ion source or, when referring to analysis by GC/MS, is not sufficiently volatile to pass through a gas chromatographic column.

Derivatization as an aid to the analysis of organic compounds is a very old concept, possibly the best known application being the confirmation of a structural assignment by measurement of the melting point of a derivative of an unknown compound. For this determination, derivatization is a necessity if the unknown substance is a liquid. If it is a solid, derivatization may not be necessary but it can be advantageous. An analogous situation exists in the field of mass spectrometry. To examine involatile materials, derivatization is a necessity, but for some compounds which are directly amenable to mass spectrometry, prior derivatization may offer several advantages. It is a versatile process because the type of derivative may be chosen such that its properties are more suitable for the ensuing analysis than those of the original substance. Derivatization not only affects the volatility and stability of a molecule. The derivative, having a structure different from that of the original structure, will have a mass spectrometric fragmentation pattern quite different from that of the parent. This change in fragmentation pattern can be advantageous or disadvantageous and is discussed more fully in later sections.

At an ion source pressure of about 10^{-6} torr, many directly inserted compounds which would be classified as non-volatile at atmospheric pressure, have vapour pressures high enough to give good mass spectra. The requirements for derivatives for GC/MS are more stringent than those for directly inserted samples because the compounds are in the gaseous phase for a longer time and at far higher pressure than are compounds inserted into the ion source by direct probe. The sample also comes into prolonged contact with surfaces (the stationary phase and the walls of the column) on which it may adsorb or decompose. Derivatives suitable for GC/MS should not only possess long-term thermal stability and volatility but also good chromatographic properties. Figure 5.1 shows the gas chromatograms obtained for 6-methylpurine and its trimethylsilylated derivative. The underivatized sample adsorbs strongly on the column, resulting in an unsatisfactory peak shape and poor chromatographic resolution. In comparison, the derivatized compound gives rise to a sharp peak which preserves chromatographic resolution. Derivatives commonly used in the field of GC are usually useful also for GC/MS. Undoubtedly, trimethylsilylation (see below) is the most ubiquitous derivatization procedure for GC/MS, the only serious disadvantage being that the resulting derivatives, such as that shown in figure 5.1, are susceptible to hydrolysis by traces of water. The problem is often overestimated and is overcome by carefully avoiding contact of the sample with sources of water. A short review of derivatization for GC/MS has been published (Brooks, Edmonds, Gaskell & Smith, 1978). Derivatives for all forms of chromatography, including GC, and analytical derivatization reactions have been described in detail in handbooks by Blau & King (1978) and Knapp (1979), respectively. The latter contains considerable information pertinent to mass spectrometry and GC/MS in particular.

The derivatization reaction itself must satisfy several criteria, particularly for analyses by GC/MS in which the quantity of sample is very often small. With amounts in the picogramme to microgramme range, manipulation of the sample should be minimized to prevent accidental or unavoidable losses.

For instance, transfer of the sample between several different vessels is best avoided. If possible, the reaction should be designed to require little or no work-up as long as reagents or by-products do not interfere with subsequent analysis. Trimethylsilylation is an example of this since the reaction mixture may be injected directly onto the gas chromatographic column. Ideally, derivatization should proceed rapidly and quantitatively to one product. If components of a complex mixture afford more than one product, the analysis is made much more difficult. The reagents for derivatization must affect molecules in a predictable and reproducible way. Reagents may be either general or selective in their interaction with functional groups. Alcohols, phenols, acids, amides and amines are all readily trimethylsilylated (Pierce, 1968), whereas diazomethane reacts selectively with phenols and acids to

Figure 5.1. Gas chromatograms for 6-methylpurine (top) and its trimethylsilyl derivative (bottom).

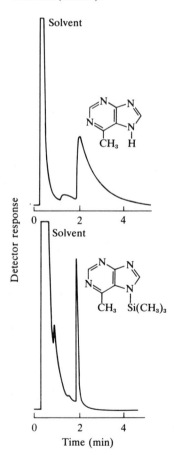

give methyl ethers and esters respectively. In spite of the hazards associated with diazomethane, it is an excellent derivatizing reagent: it is not only volatile itself but the only by-product of its quantitative, almost instantaneous reactions is nitrogen. Some reagents which react with several different functional groups may be made selective by proper choice of reaction conditions. Acyl halides and anhydrides react with alcohols, phenols and amines to give, respectively, aliphatic esters, aromatic esters and amides but, under controlled conditions, amino-alcohols can be derivatized to amido-alcohols. Reactions which may not normally be thought of as derivatizations are often useful. For example, reduction of acids and amides with lithium aluminium hydride converts them into the more volatile alcohols and amines.

In the following sections, it is not possible to present a comprehensive survey of derivatives used in mass spectrometry because the topic is very large. Rather, the aspects discussed and exemplified are those which determine the circumstances in which derivatization is advantageous and the criteria by which suitable derivatives are chosen.

5.2.   Imparting volatility and thermal stability

Involatility of a compound is caused by its having high molecular weight or strong intermolecular forces or both. For substances with high molecular weight, the common derivatives only exacerbate the situation inasmuch as they add to the molecular weight but, if at the same time they decrease intermolecular forces, the compounds are nevertheless made more volatile. Hydrogen bonding and the electrostatic associations of zwitterionic compounds are common causes of involatility. The effects of hydrogen bonding can be prevented by removing active hydrogen atoms. For example, the strongly hydrogen-bonding carboxylic acids and amides may be alkylated (scheme 5.1), involatile poly-alcohols like sugars may be acetylated or trimethylsilylated (scheme 5.2) and the zwitterionic amino acids may be acetylated and methylated (scheme 5.3). Generally, derivatization should

$$RCOOH \xrightarrow{\text{diazomethane}} RCOOCH_3$$
$$RCONH_2 \xrightarrow{\text{base/iodomethane}} RCON(CH_3)_2 \quad \Big\} \ (5.1)$$

$$ROH \xrightarrow{\text{acetic anhydride}} ROCOCH_3$$
$$ROH \xrightarrow[\text{(BSA)}]{\textit{bis}\text{-trimethylsilylacetamide}} ROSi(CH_3)_3 \quad \Big\} \ (5.2)$$

$$\overset{+}{N}H_3 - CH - \overset{-}{C}OO \xrightarrow[\text{(2) diazomethane}]{\text{(1) acetic anhydride}} CH_3CONH - CH - COOCH_3 \quad (5.3)$$
$$\underset{R}{|} \underset{R}{|}$$

convert the polar groups of a molecule into functionalities of lesser polarity which are unlikely to form strong intermolecular associations. By increasing their volatility, compounds of quite high molecular weight can be vaporized in the ion source. A decapeptide is unlikely to be sufficiently volatile to give a mass spectrum without considerable thermal decomposition. However, acetylation of the terminal amino group, methylation of the terminal car-boxyl group and prevention of hydrogen bonding by methylation of all the amide (peptide) bonds affords a derivative that is readily volatile even with a molecular weight of about 1500 (scheme 5.4). With compounds of high

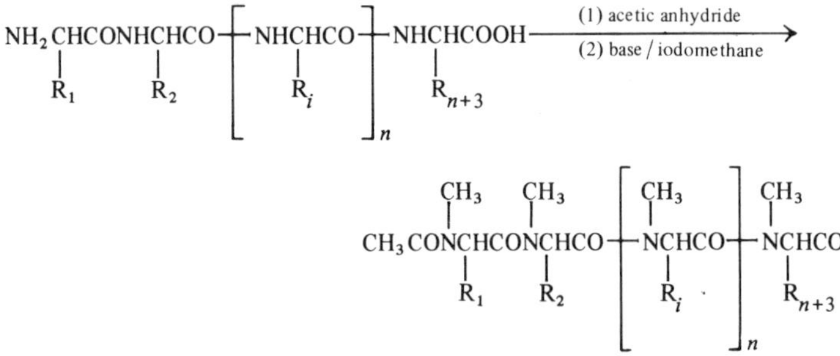

$$NH_2CHCONHCHCO \left[NHCHCO \right]_n NHCHCOOH \xrightarrow[\text{(2) base / iodomethane}]{\text{(1) acetic anhydride}}$$

with substituents R_1, R_2, R_i, R_{n+3}

$$CH_3CONCHCONCHCO \left[NCHCO \right]_n NCHCOOCH_3 \quad (5.4)$$

with CH_3 on each N and substituents R_1, R_2, R_i, R_{n+3}

molecular weight, it is important, particularly for GC/MS, that derivatization does not cause an inordinate increase in molecular weight. Derivatives which form cyclic products from bifunctional molecules are helpful in this respect. The increases in molecular weight caused by the derivatization of 1,2-diols are 144 on trimethylsilylation (5.5a), 40 on formation of the cyclic ketal (5.5b) and 24 for the cyclic boronate (5.5c), and thus the last would be favoured.

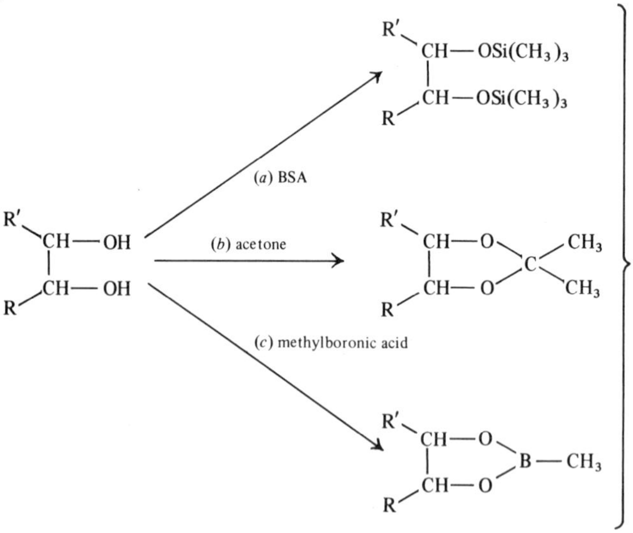

$$\begin{array}{c} \text{(a) BSA} \\ \text{(b) acetone} \\ \text{(c) methylboronic acid} \end{array} \quad (5.5)$$

Because heat may cause rearrangement or decomposition and adds to the internal vibrational energy of a molecule, the lower the temperature required to volatilize the sample, the less likelihood there is of changes in structure or of excessive fragmentation after ionization. Therefore, it is preferable to convert compounds, especially those having thermal instability, into volatile derivatives which require less heating for vaporization. Instability, like involatility, is often associated with active hydrogen atoms which can form reactive sites in the structure, so derivatization to confer volatility frequently also confers thermal stability on a compound. Sterols with several hydroxyl groups are prone to decomposition through dehydration on gas chromotographic columns, so conversion to a polyether by trimethylsilylation or methylation, or to a polyester by acetylation is advisable.

5.3.   Modelling the molecule for the analysis

Judicious selection of derivatives is of inestimable value for maximizing the efficiency of many analyses. Several examples are discussed individually below to illustrate this principle.

5.3.1.   Structure elucidation

The molecular ion is one of the most informative ions for elucidating the structure of an unknown compound from its mass spectrum but, for many substances, the molecular ion fragments as soon as or soon after it is formed so that it is not seen in the spectrum. In such cases, once the class of compound is known, a derivative may be selected which preferentially gives abundant molecular ions. Unknown polyhydroxylic compounds should not be acetylated in general because the resulting polyacetates afford complex electron impact mass spectra which rarely contain molecular ions of significant abundance because of a facile fragmentation involving elimination of acetic acid. Etherification through methylation or trimethylsilylation yields derivatives with molecular ions of greater stability and hence greater abundance. Generally, chemical ionization is more likely to yield molecular weight information than is electron impact mass spectrometry (section 7.2) so, when chemical ionization is employed for this purpose, the choice of derivative is not so critical. Other methods of enhancing the abundance of molecular ions are described in sections 10.2 and 10.3.

When derivatives are prepared specifically to induce a particular mode of ion decomposition, it is known as *directed fragmentation*. The positions of substitution in a cyclic ketone can be determined from the masses of the fragment ions derived by cleavage of ketal derivatives (scheme 5.6). The location of double bonds in underivatized alkenes and unsaturated lipids is difficult or impossible because of the ease with which these bonds appear to migrate in the molecular ion before fragmentation. Several derivatives (scheme 5.7) have been used to accomplish this analysis by 'fixing' the

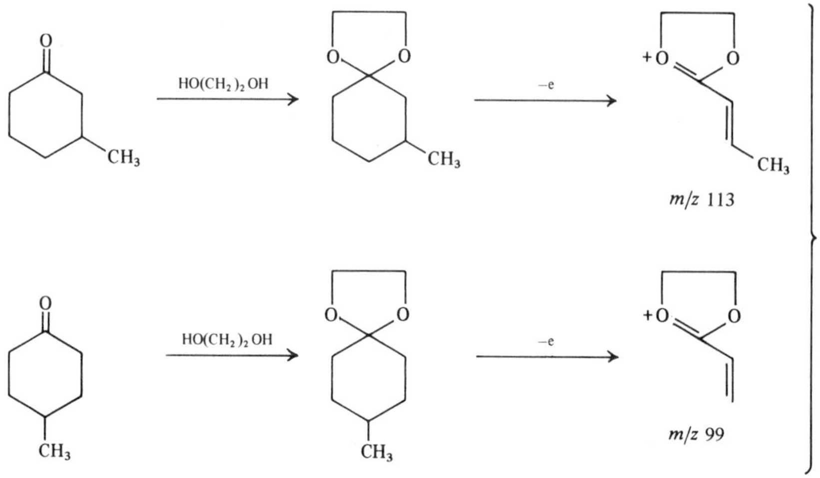

$$\left.\begin{array}{c} \end{array}\right\} (5.6)$$

$$\left.\begin{array}{c} \end{array}\right\} (5.7)$$

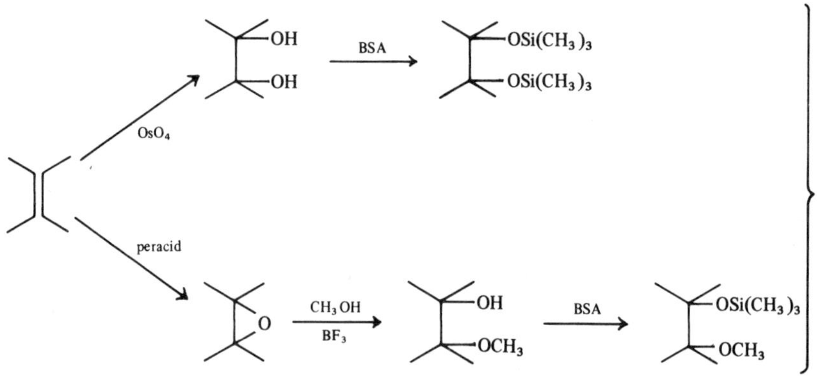

double bond so that the fragment ions in the mass spectra define the position of the double bond in the original molecule. Unsaturated fatty acids can be converted to pyrrolidide amides which fragment in a manner indicative of the position of unsaturation. Methods for the determination of the position of the double bond have been reviewed (Minnikin, 1978). A further method for locating double bonds relies on specialized chemical ionization methodology and is more suitably described in section 7.2.

Methylation of peptides (scheme 5.4) not only increases volatility by preventing hydrogen bonding but also causes directed fragmentation. The breaking of the bonds shown (1) is more favoured for tertiary amides than for the original secondary amides. The resultant fragment ions define the amino-acid sequence of the peptide.

$$CH_3CO\!-\!\!\!-NCH\!-\!\!\!-CO\!-\!\!\!-NCH\!-\!\!\!-CO\!-\!\!\!-NCH\!-\!\!\!-CO\!-\!\!\!\sim\!\!\sim$$

with CH_3, R_1 on the first NCH; CH_3, R_2 on the second; CH_3, R_3 on the third.

(1)

5.3.2.   Selected ion monitoring

The technique of selected ion monitoring, which is almost always performed by GC/MS, is described in detail in the next chapter. The mass spectrometer is tuned to focus on ions of one mass or a small number of ions of different masses, characteristic of the compound or compounds under investigation. Each m/z value monitored must be as specific as possible to one compound. In particular, ions which are present as 'background', say from column bleed, would not be a wise choice. The analysis is often carried out on natural products, in the range of picogrammes to nanogrammes, which may form only a small percentage of the total sample. Therefore, there is a clear requirement for the derivatization to be quantitative so as to preserve the sensitivity of the analysis. Proper choice of derivatives greatly enhances the value of studies by selected ion monitoring by imparting volatility, thermal stability and good gas chromatographic properties on the sample and by increasing the sensitivity and specificity of detection.

Because the presence and amount of a substance is measured by monitoring selected ions, the greater the proportion of the total ion current that resides in those ions, the more sensitive is the analysis, viz. the smaller the amount of sample that can be detected. A derivative is required which affords particularly stable and therefore abundant ions, preferably of high mass (see below). The loss of $C_4H_9\,^\cdot$ from tertiarybutyldimethylsilyl ethers (scheme 5.8) and C_5H_{11} from the derivative of prostaglandin $F_{2\alpha}$ shown (scheme 5.9) give suitably abundant fragment ions. With many compounds, the trimethylsilyl ether grouping tends to give abundant molecular ions which are frequently useful for selected monitoring.

Many materials in the lower molecular weight range (up to about 300) pose two special problems. Firstly, they are likely to elute rapidly through gas chromatographic columns, emerging shortly after the solvent together with the many impurities which may be present, such as fatty substances in natural products, plasticizers, impurities in the solvent and so on. Secondly, structurally important ions in the mass spectra of compounds of low molecular weight are of relatively low mass and may well be common to those of the impurities. Both of these two related sources of interference are overcome by selecting a derivative which adds considerable mass to the sample, but not to the impurities, and increases the retention time on the

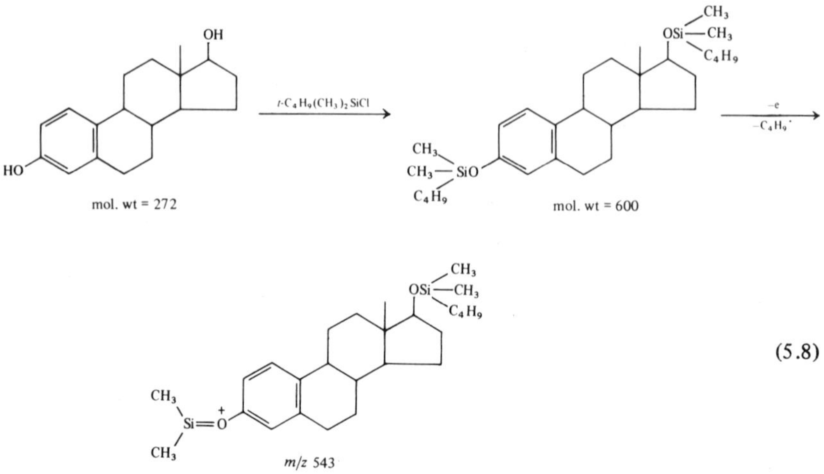

(5.8)

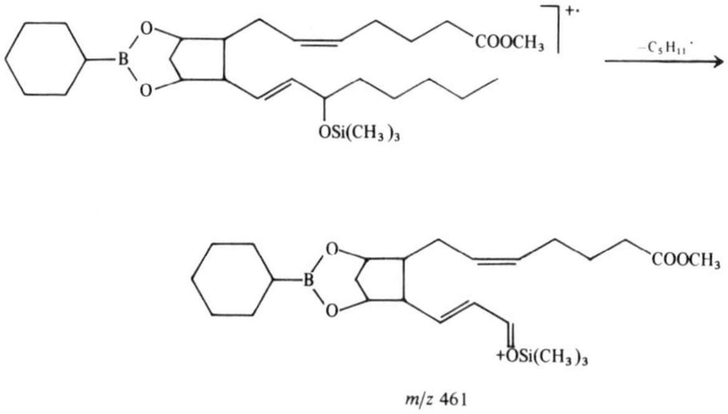

(5.9)

column. Tertiarybutyldimethylsilyl and perfluoroacyl derivatives of amines, alcohols and phenols (see, for example, schemes 5.8 and 5.10) have been used to increase their molecular weights so that they are removed from the region containing interfering compounds and thereby the specificity of the analysis is improved.

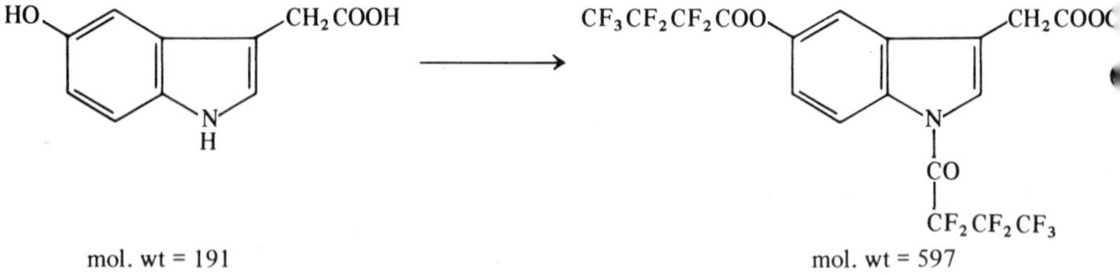

Another means of increasing the specificity of selected ion monitoring involves the operation of the mass spectrometer at high resolution. Accurate, rather than integer, masses of ions are monitored. The method is specific in that only one elemental composition is detected at each mass. For example, at the integer mass 282, three possible elemental compositions are $C_{20}H_{42}$, $C_{18}H_{34}O_2$ and $C_{22}H_{18}$, having accurate masses 282.3284, 282.2557 and 282.1408, respectively. These three mass peaks are separable with a mass spectrometric resolution of 4000 or over, so by monitoring only m/z 282.2557 at high resolution, analysis of the heteroatomic species can be effected even in the presence of the hydrocarbon ions. In conjunction with this technique, a derivative containing elements which are not common to the impurities or background are useful because they shift the accurate mass of ions containing those elements to distinctive values. Frequently, a natural product is contaminated with compounds of high hydrocarbon content (lipids). Derivatization of the natural product with a polyfluorinated reagent, such as $(C_nF_{2n+1}CO)_2O$, to give compounds like that shown in scheme (5.10), enables ions of the sample to be distinguished from those of the background at the same integer mass.

5.3.3. Isotope studies

Many selected ion monitoring methods require as a standard an isotopically labelled compound which is otherwise identical to the substance under examination (chapter 6). The label may be introduced by derivatization through the use of isotopically labelled reagents like $(CD_3CO_2)O$ for acetylation, CD_3I or $^{13}CH_3I$ for methylation and fully deuteriated silylating reagents for the introduction of $(CD_3)_3Si$-groups.

Isotopically labelled derivatizing agents are often used to reduce ambiguity of results. Methylation of peptides (scheme 5.4) sometimes results in undesirable reaction at active C–H groups. For example, if this occurs with a peptide with a terminal glycyl residue the product is

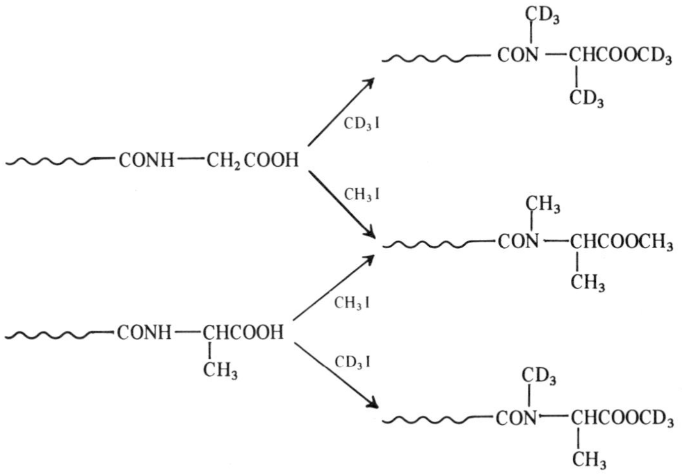

(5.11)

indistinguishable from the equivalent peptide terminating with alanine. The problem is solved by derivatization with CD_3I in place of CH_3I because the molecular ion of the chemically altered glycyl peptide then occurs at three atomic mass units higher than that of the alanyl peptide (scheme 5.11). For a polyfunctional natural product which has been derivatized by acetylation prior to analysis, the possibility exists that one or more of the acetyl groups was present naturally before derivatization. The use of $(CD_3CO)_2O$ would allow ready determination of the natural or other origin of the acetyl groups.

Another useful technique is to add to a sample a selective derivatizing reagent containing elements with isotopes having characteristic abundance ratios. In section 1.3, elements with readily recognizable isotope abundance ratios were discussed, the most familiar ones being chlorine, bromine, sulphur and boron. Examination of the resulting mass spectrum for the abundance of isotopes in the molecular ion region reveals whether a compound has been derivatized by such a reagent. This is useful in the analysis of mixtures containing compounds of different classes because, for each component, the type of derivative formed will lead to identification of the functional group present. Distinctive isotope abundance ratios may be introduced artificially by use of a mixed reagent such as that containing equal proportions of $(CH_3CO)_2O$ and $(CD_3CO)_2O$ for selective acetylation of amines. In the ensuing analysis, observation of mass spectra containing a 'doublet' of equally abundant ions, three mass units apart, immediately confirms the presence of derivatized amines. In a related method, the precursor for a metabolic study is isotopically labelled, usually with ^2H, ^{13}C or ^{15}N, at a site which is not affected by and does not significantly affect the metabolic pathways of interest. After incubation, products derived from the precursor may be unambiguously assigned because their mass spectra will contain the same isotopes in the same ratio as for the parent compound (see also section 12.2).

Isotope labelling may be used also to investigate mechanisms of mass spectrometric fragmentations but this does not involve derivatization in the context of the present chapter. The topic is discussed in section 12.2.

5.3.4. *Metastable ion studies*

Some compounds tend to give relatively abundant metastable ions when they fragment. Such ions are often associated with functional groups which may be introduced by derivatization for analysis of samples by examination of their metastable ions. The method may be considered as a specialized application of directed fragmentation and is suitable for aiding structure elucidation of unknown compounds (chapter 8) and for selected metastable ion monitoring (chapter 6). An example of the latter technique is the specific and sensitive analysis of tertiarybutyldimethylsilyl ethers of sterols by tuning the mass spectrometer to monitor the metastable ion which

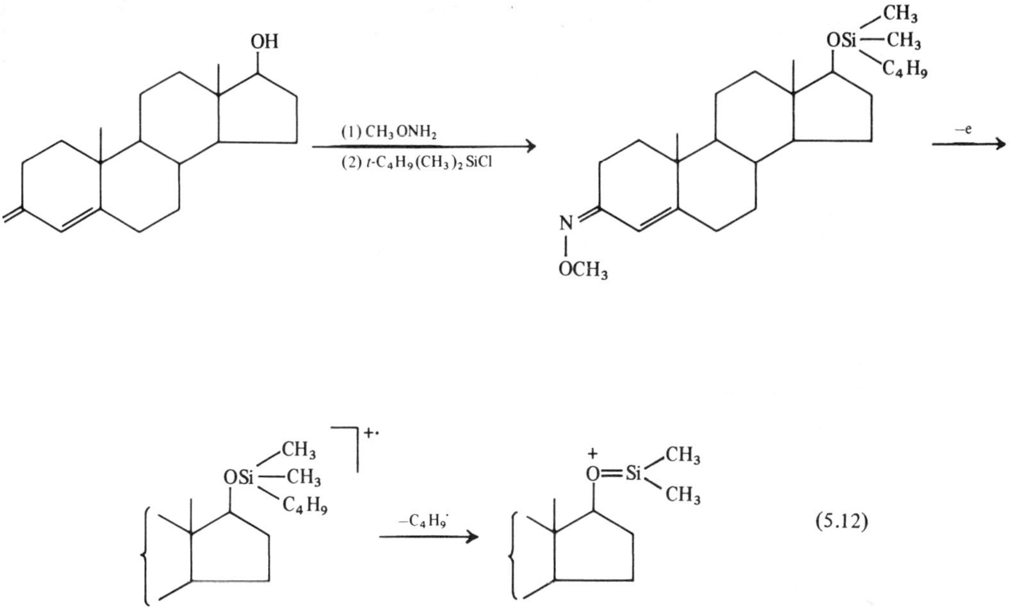

(5.12)

is found when molecular ions fragment by loss of C_4H_9 · (scheme 5.12). In this study, the keto group was derivatized by formation of the methyl oxime to prevent enolization and subsequent silylation. For studies by GC/MS, formation of oximes from ketones is popular but often unnecessary because the keto group does not usually impart involatility, thermal instability or poor chromatographic properties on the substance.

5.3.5. Negative-ion mass spectrometry

Typical derivatizations for the analysis of amines by positive-ion mass spectrometry are acetylation and benzoylation (scheme 5.13).

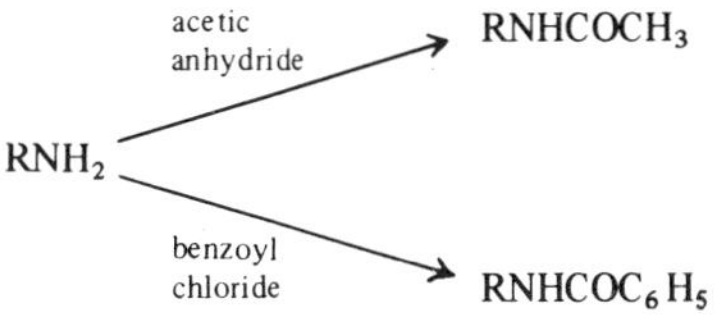

(5.13)

The same derivatives would not be ideal for negative-ion studies since they are relatively inefficient at stabilizing a negative charge. To take best advantage of negative-ion techniques, electronegative derivatives should be employed because they increase ionization efficiency and hence sensitivity. For amines, the derivatives shown in scheme (5.14) are more satisfactory than those shown earlier (scheme 5.13).

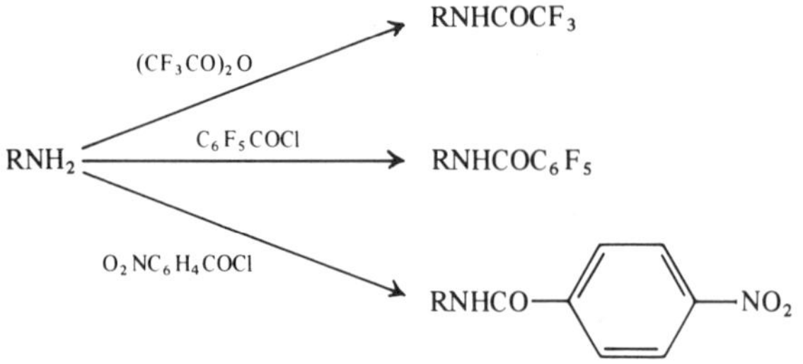

$$RNH_2 \begin{cases} \xrightarrow{(CF_3CO)_2O} RNHCOCF_3 \\ \xrightarrow{C_6F_5COCl} RNHCOC_6F_5 \\ \xrightarrow{O_2NC_6H_4COCl} RNHCO\text{—}\langle\text{aryl}\rangle\text{—}NO_2 \end{cases} \qquad (5.14)$$

5.3.6. Enantiomeric samples

On conventional chromatographic columns, enantiomers co-elute. Derivatization with a chiral reagent affords diastereoisomers which, because of their different physical properties, may be differentiated by GC/MS or LC/MS.

5.4. Inorganic compounds

Whilst it is true that much mass spectrometry is concerned with organic compounds, inorganic substances are also analysed frequently. Volatile inorganic compounds may be analysed directly, examples being rhombic sulphur (S_8 rings), the hydrides of boron and elements of groups IV and V, and liquids such as arsenic trichloride and mercury. Involatile compounds such as metallic salts may be analysed by special techniques (section 12.3) or derivatized to volatile organometallic compounds. Some diketone or ketoester chelates of metals are so volatile that they can be distilled, passed through gas chromatographic columns or introduced into an ion source at room temperature. The propionylacetonates, for example that of copper (2), and trifluoroacetonates are useful for obtaining spectra of metal ions. Generally, as the proportion of organic groups in the organometallic compound increases, so does volatility. Thus, diketones with bulkyl alkyl groups are useful derivatizing agents. 8-Hydroxyquinoline

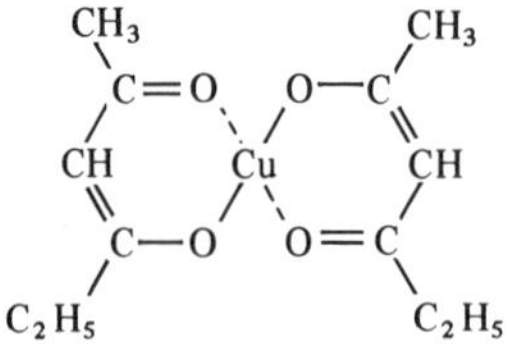

(2) 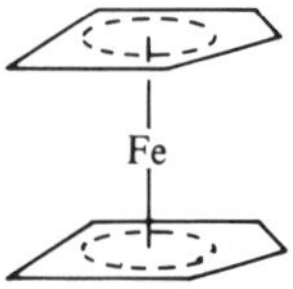   (3)

(oxine) is well known in solution chemistry for precipitating most metal ions. These metal oxinates give good mass spectra. Transition metals may be studied by use of such derivatives as the carbonyl (e.g. $W(CO)_6$ and $Mn_2(CO)_{10}$), and π-cyclopentadienyl complexes (e.g. ferrocene, 3). The mass spectrum of such a compound is discussed in chapter 11 (example H).

Organometallic derivatives pose some problems not normally encountered in organic chemistry. By several mechanisms, polymerization may occur, so that a metal complex, $MetL_2$ where L is a ligand and Met is the metal, may show a 'mixed' mass spectrum of $MetL_2$, Met_2L_4 and Met_3L_6. When determining the components of any mixture, it is hoped that, if several substances volatilize at the same time, the resulting mass spectrum comprises the superimposed spectra of individual components. This is not necessarily so with organometallic compounds because of reactions occurring in the ion source yielding compounds not present in the original mixture. Exchange reactions are quite common as shown in scheme (5.15). Misleading results may also be due to associations such that two complexes, $MetL_2$ and $Met'L$, give rise

$$
\begin{array}{ccc}
Na^+ & & K^+ \\
CF_3CO\bar{C}HCOCH_3 & & CF_3CO\bar{C}HCOCH_3 \\
+ & \longrightarrow & + \qquad (5.15) \\
C_3F_7CO\bar{C}HCOC_4H_9 & & C_3F_7CO\bar{C}HCOC_4H_9 \\
K^+ & & Na^+
\end{array}
$$

to $Met-Met'L_3$. An awareness of such reactions should prevent erroneous conclusions.

6 Quantitative mass spectrometry

6.1 Introduction and principles

Estimation of quantities of substances by mass spectrometry is not straightforward because mass spectrometric measurements are not exactly reproducible (section 1.1). The response to a sample at the detector depends on several parameters that are difficult or impossible to control, including the condition, temperature and pressure of the ion source and the condition of the detector. It is important to recognize that equimolar amounts of different compounds do not give an equal response because the proportion of the total number of molecules which are ionized (the *ionization efficiency*) depends on molecular structure and vapour pressure. Despite these complications, quantification by mass spectrometry is possible because, when a compound is ionized, the absolute abundance of any of its ions is related to the amount of that substance, albeit by a complex function which varies during day-to-day operation of the mass spectrometer and which cannot be applied to any other compound. To achieve quantification, the mass spectrometer must be calibrated with known amounts of the compound under investigation either just before the assay is carried out on the true sample (on the assumption that instrumental conditions do not change significantly during calibration and analysis) or in a manner which makes the measurement independent of instrumental variability (section 6.2).

For quantification, the abundance of ions may be measured in three distinct ways. During the residence of the compound in the ion source, repetitive scanning with a mass spectrometer (section 3.3) can provide several complete mass spectra. When the absolute abundance of ions of a selected m/z value is plotted against time or scan number, a *mass chromatogram* is obtained (section 3.4, figure 3.7). The height or area of a peak in the mass chromatogram is a measure of the quantity of substance giving rise to that peak. The method normally necessitates a computer for data processing. On the other hand, even the most unsophisticated mass spectrometer may be tuned to monitor continuously ions of only one mass by applying an appropriate stationary magnetic field or analyser voltage. This second method of measuring a mass peak, reported as early as 1959, is more sensitive than the first because, for the entire time that a compound resides in the ion source (typically 10–60 s), the detector records only the selected ions. For a typical complete mass spectral *scan*, in which mass spectra may be recorded every 2 s, any one m/z value would be sampled 5–30 times, being focussed on the detector each time for a few microseconds only, depending on the resolution (equation 3.1). Therefore, by monitoring ions of one specific mass instead of the whole spectrum, a thousand-fold increase in sensitivity is attained. Single-ion monitoring is capable of measuring amounts of substances in the pico-gramme or, in favourable cases, femtogramme range whereas a nanogramme or more of sample is required to obtain a complete mass spectrum.

Between the monitoring of ions of a single mass and the recording of complete mass spectra, there are several techniques of intermediate sensitivity involving measurement of a limited number of mass peaks. This third group includes repetitive, continuous scanning of a relatively small mass range and rapid, discontinuous switching to bring into focus, in turn, ions of a few different masses characteristic of the compound or compounds under investigation. Generally, the sensitivity of the analysis increases as the number of mass peaks monitored decreases. Unfortunately, a great deal of analytical information is lost at the highest sensitivities because only one m/z value or a few m/z values are observed. The consequences of this are discussed in section 6.3.

The discussion above is applicable to quadrupole mass filters and most magnetic-sector mass spectrometers. The less common magnetic-sector double-focussing instruments of Mattauch–Herzog geometry detect simultaneously all ions of a sample (section 2.4). Therefore, they record complete mass spectra with a sensitivity as high as that of mass spectrometers of Nier–Johnson geometry which monitor only one mass peak. They are used to good effect in the quantification of traces of metals by spark source ionization (Chapman, 1978, pp. 246–51). The inconvenience of, and difficulty of quantification from, the photographic plates usually used as detectors for such instruments makes them relatively unpopular.

Because of the high sensitivity available, quantification of a compound is often brought about by monitoring just one mass peak of its mass spectrum, together with a mass peak from a chemically similar reference compound called an *internal standard* (see next section). The methodology for monitoring a few selected mass peaks was reported first in 1966, when it was termed *mass fragmentography*. This is not an ideal name because mass peaks from molecular as well as from fragment ions may be monitored. As the technique developed and proliferated, different researchers gave it different names, with corresponding acronyms. Along with mass fragmentography, some common names to be found in the chemical literature are the *integrated ion current technique, multiple ion detection, single ion monitoring* (when the mass spectrometer is tuned to one m/z value only), *multiple ion monitoring, multiple peak scanning* and *selected ion monitoring* or *detection*. The preferred term is *selected ion monitoring* because it describes the method most aptly and because it is applicable irrespective of the number of mass peaks monitored. The popular abbreviation SIM is not recommended because it may be confused with acronyms of other techniques.

Selected ion monitoring is especially useful in the fields of biochemistry, medicine and environmental science and in those cases where the compound or compounds to be quantified are present in complex mixtures. The vast majority of applications utilizes GC/MS to fractionate samples prior to analysis. Substances introduced directly into the ion source, as on a direct insertion probe, may also be quantified (Millard, 1978*a*, pp. 91-115), a method useful for compounds of low volatility and thermal stability. These more intractable samples may be analysed quantitatively also by LC/MS. Alternatively, it may be possible to form derivatives that are suitable for GC/MS. By careful selection of the type of derivative, quantitative measurements are often improved considerably (section 5.3).

Many of the mass spectrometric techniques described in this book are amenable to quantitative analysis, but only the most useful of these are discussed in this chapter. Further information on the subject of quantitative mass spectrometry may be found in reviews (Lehmann & Schulten, 1978; Bjoerkhem, 1979; Chen, 1979; de Leenheer & Cruyl, 1980) and in a detailed book (Millard, 1978*a*). Also, the published proceedings of annual symposia on quantitative mass spectrometry in life sciences present many interesting reviews and papers (de Leenheer & Roncucci, 1977; de Leenheer, Roncucci & van Peteghem, 1978). Sources of errors and criteria for the selection of internal standards for quantitative measurements have been treated sensibly by Millard (1978*a,b*).

6.2. Calibration and internal standards

One method for relating the quantity of a substance to the signal it causes at the detector is to assay successively different, known amounts of the substance. A plot of signal intensities against quantities then serves as a *calibration graph* for subsequent determination of unknown amounts. The compound is said to be an *external standard* because it is not added to the sample to be measured. The method, and variants of it, is often used with non-chromatographic methods of introducing the sample into the mass spectrometer. Techniques based on external standards are prone to error since changes in the condition of the instrument, such as a slight variation in pressure of the ion source, affect the magnitude of the signals obtained. For true quantification, the efficiency of any preparatory steps prior to the mass spectrometric analysis must be measured. Samples for analysis are frequently worked up by one or more extractions, a purification process and possibly a derivatization reaction, all prior to mass spectral analysis. This makes quantification complex and the chances of error high, since deviations from the measured yields at any of the steps affect the final result.

A better procedure is to add to different, known amounts of the substance to be analysed a constant, known amount of a reference compound called an *internal standard*. The different mass spectrometric responses to the various amounts of substance are measured against the response to the internal standard. The ratio of these two responses is plotted against the amount of substance to obtain a calibration graph (see figure 6.5 below). For actual analysis, the same amount of internal standard is admixed with the crude, unknown sample and both are worked up together; a ratio and not an absolute value is measured by mass spectrometry. As long as internal standard and substance to be quantified behave identically during work-up, unavoidable or accidental losses of sample, due to incomplete extraction or derivatization, or spillage, for example, do not affect the ratio of the two compounds so that the result is still valid. Similarly, instrumental changes are made irrelevant because they affect only absolute values, not ratios. Because all these effects would introduce error or even invalidate the results of methods based on external standards, those based on internal standards are preferred. A typical procedure for quantitative measurement of a sample is presented schematically in figure 6.1 and the method is exemplified in section 6.4.

Figure 6.1. Schematic representation of typical analytical procedure for quantitative mass spectrometry.

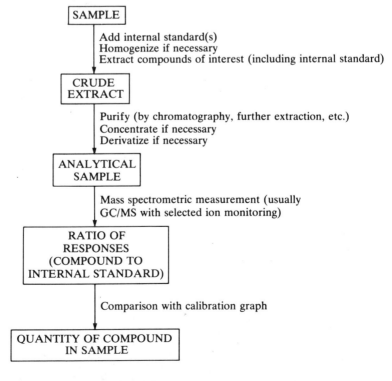

The two commonest types of internal standard are homologues and iso-topically labelled analogues of the compound of interest. Both have distinct advantages and few disadvantages and are approximately equally popular.

An internal standard should compensate for any losses during work-up by behaving identically towards extractive or purification measures applied to the compound to be quantified. Therefore, potential internal standards for a compound are of the same class of compound and structurally similar. Using a standard mixture of a candidate internal standard and a compound under investigation, it is a simple matter to determine if they act identically during work-up (the ratio of the two should be the same, within experimental error, before and after work-up). The tricyclic compound (1) was established as a suitable internal standard for quantification by selected ion monitoring of the antidepressant drug imipramine (2). There is an additional advantage of

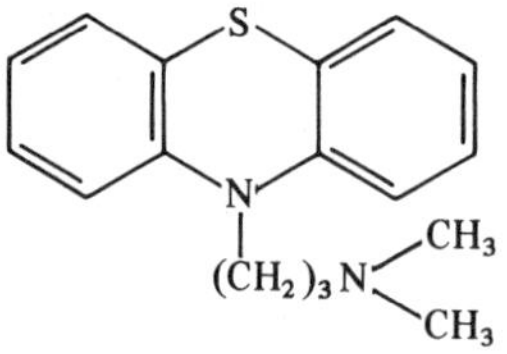

(1) 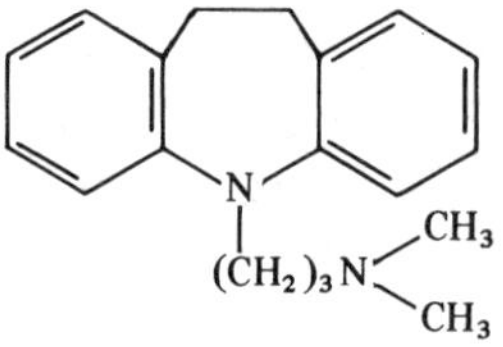   (2)

using this type of internal standard. If it is possible to discover a suitable homologue which contains in its spectrum a mass peak in common with that of the substance to be quantified, only this one mass peak need be monitored to detect both compounds. This procedure affords the advantage of the simplest and most sensitive method of quantification. The trifluoroacetyl derivative (3) is such an internal standard for the derivatized drug (4), for both have the same side-chain which affords a common ion (scheme 6.1).

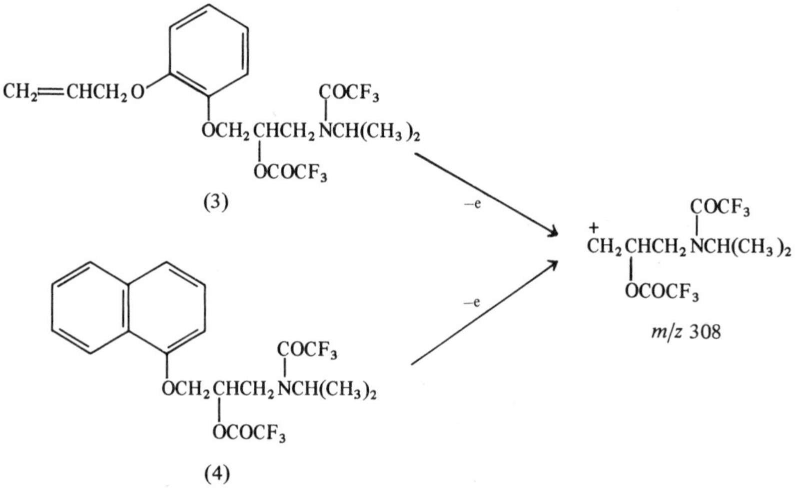

$CH_2{=}CHCH_2O$ ...

$$\overset{\displaystyle COCF_3}{OCH_2CHCH_2NCH(CH_3)_2}$$
$$\underset{\displaystyle OCOCF_3}{}$$

(3)

$$\overset{\displaystyle COCF_3}{\overset{+}{CH_2}CHCH_2NCH(CH_3)_2} \quad (6.1)$$
$$\underset{\displaystyle OCOCF_3}{}$$

m/z 308

$$\overset{\displaystyle COCF_3}{OCH_2CHCH_2NCH(CH_3)_2}$$
$$\underset{\displaystyle OCOCF_3}{}$$

(4)

This method of quantification necessitates the use of GC/MS (or LC/MS) so that the substance and its internal standard are separated prior to mass-spectrometric monitoring of the common ions, otherwise the responses from each compound would be superimposed. For quantification of substances without fractionation at the mass-spectrometric stage, each mass peak monitored should be unique to only one of the unseparated components.

An isotopically labelled internal standard acts like a homologue which does not provide a mass peak in common with the substance to be quantified. For example, the tetradeuteriated analogue (5) is a suitable internal standard for assay of cyclophosphamide (6). Generally, isotopically labelled analogues

(5) mol. wt = 313   (6) mol. wt = 309

are useful as internal standards because they are practically identical in chemical properties to the respective unlabelled compounds whilst being readily distinguishable by mass spectrometry because of their mass difference. The ion of the internal standard which is used as a reference must retain the isotopes that distinguish it from the unlabelled compound. Internal standards labelled with ^{13}C, ^{15}N, ^{18}O and ^{37}Cl are even more similar chemically to the corresponding unlabelled compounds than those labelled with $^{2}H(D)$, but they are not as popular because they are expensive and generally less easy to synthesize.

It was shown in section 1.3 that, for all carbon compounds, there exists a natural isotopic pattern in their mass spectra such that ions 1 and 2 mass units greater than the molecular ion are of significant abundance. The $(M + 3)^{+\cdot}$ ions may be ignored unless the compound contains Cl, Br or other elements with extensive natural isotopes. Even when analysed by GC/MS, an unlabelled compound is hardly separated from its labelled analogue, so the latter should ideally contain sufficient isotopes to shift the molecular weight by at least 3 mass units, so that the mass peaks monitored are not common to both compounds. Likewise, it is disadvantageous but not intolerable for the labelled analogue to contain a small amount of the unlabelled analogue as impurity. Of course, it is possible for both types of interference to occur at the same time as, for example, with a dideuteriated internal standard which contains some unlabelled impurity. As the number of isotopic labels incorporated into a structure increases, the greater is the likelihood that the labelled and unlabelled analogues will behave differently. Therefore, the optimum number of ^{2}H atoms is usually three to five.

When a compound is analysed by GC/MS, some of it may be adsorbed irreversibly on active sites in the gas chromatographic column and the

GC/MS interface. This becomes a problem with small amounts of sample since much or all of it may be adsorbed. The addition of a relatively large amount of an isotopically labelled analogue of the compound is widely assumed to overcome this problem because statistically it is much more likely that the labelled compound will be adsorbed instead and this amount will be made insignificant by the large quantity present. This so-called 'carrier' effect has not been demonstrated satisfactorily for compounds labelled with deuterium and so the use of large amounts of such analogues as both carriers and internal standards is not recommended, particularly because the most precise quantitative measurements are made when the quantity of internal standard and compound of interest are similar. Compounds labelled with ^{13}C, ^{15}N and ^{18}O are more likely to be 'carriers' because the adsorption sites are unlikely to distinguish between ^{12}C and ^{13}C compounds, for instance. Hence, the use of relatively large quantities of these internal standards may be justified.

For a fuller description of calibration, the reader is referred to a more specialized text (Millard, 1978a, pp. 56-90).

6.3. Selected ion monitoring

Quadrupole mass filters offer the advantage of simplicity over magnetic-sector mass spectrometers for selected ion monitoring. The voltage applied to the rods of the quadrupole analyser may be switched very rapidly from one value to any other to detect successively several ions of different mass. The majority of magnetic-sector instruments in current use are not capable of switching rapidly between different mass peaks by changing the magnetic field because it takes a relatively long period for the field to settle to a constant value after switching (hysteresis). Because the mass of an ion is inversely proportional to the accelerating voltage (V; equation 2.2) the magnetic field (B) is kept constant and the accelerating voltage switched rapidly. This voltage scan imposes a restriction which is not found with quadrupole mass filters, namely, that the highest m/z value monitored must be less than about 1.67 times the lowest (for example, within a mass range of m/z 300 to 500) because sensitivity decreases with decreasing acclerating voltage. Magnetic-sector instruments, unlike quadrupole mass filters, permit quantification both with high mass spectrometric resolution and with methods based on the monitoring of metastable ions. The rapid voltage switching is usually under the control of a computer, although a simpler device, the *accelerating voltage alternator* (Sweeley, Elliott, Fries & Ryhage, 1966) can be used instead. Most systems allow switching among up to 25 different mass peaks.

There are two different ways of monitoring a mass peak (figure 6.2). During monitoring, the voltage may be static such that only the top of the peak is focussed on the detector (figure 6.2a), or it may be swept through

a pre-set range so that the whole of the peak is scanned (figure 6.2b). Whilst
the former method provides the greater sensitivity, it is more prone to error
because of instrumental instability. Slight changes in the voltages of the
instrument during an analysis manifest themselves as a drift in the position
of monitoring or of the mass scale (figure 6.2, dashed lines). In the static
mode (figure 6.2a), the result is a large reduction in sensitivity and hence a
large error because only a small proportion of the total number of ions
impinge on the detector, whereas in the scan mode (figure 6.2b), the number
of ions reaching the detector is little changed. The characteristics of individual
instruments determine the mode of selected ion monitoring.

6.3.1.   Selecting ions for monitoring

Before a substance can be analysed quantitatively, its mass spectrum
must be examined for at least one mass peak suitable for selective monitoring.
The spectrum is best not examined in isolation but with a knowledge of the
masses of ions that are likely to be present as 'background'. For studies by
GC/MS, it is useful to have a mass spectrum of 'bleed' from the gas chromatographic column to be used so as to avoid selecting ions common to this
spectrum. Impurities present in the sample which ionize at the same time as
the substance of interest or its internal standard, such as co-eluting components in GC/MS, will not affect the assay as long as they do not have ions
at the m/z values selected for monitoring. An exception to this occurs if the
quantity of the impurity varies from sample to sample and is large enough
to change significantly the pressure in the ion source, because this will then
affect unpredictably the ionization efficiencies of the compounds to be
quantified. Generally, the lower the mass of an ion, the more likely it is
to be present as background. To avoid this interference, ions of high mass
(over m/z 300) should, if possible, be chosen for monitoring. The use of

Figure 6.2. Monitoring a mass peak and the effect of drift (a) in the static
mode and (b) in the scanning mode. The original peak position is drawn as
a continuous line. Instrumental instability causes a drift in the relative
position of monitoring and the mass peak (dashed line).

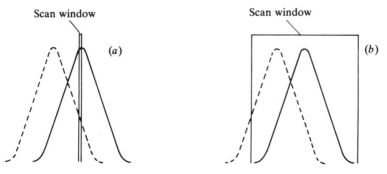

derivatization to impart high molecular weights is described in section 5.3. It is also advantageous to select ions of even mass because ions of odd mass occur more frequently in mass spectrometry, so that the latter are the major contributors to background. The selected ions should be abundant since this will increase the sensitivity of the assay. The size of the mass peak should be gauged in terms of percentage of total ion current rather than percentage relative abundance (section 1.1). A peak which is 80 per cent of the total ion current would dominate a mass spectrum since all other peaks together would constitute only a further 20 per cent. This peak would be a highly favoured candidate for monitoring. A peak which is 80 per cent in relative abundance may be a poor candidate since there could be any number of larger, more suitable peaks between 80 and 100 per cent relative abundance.

Compounds afford more high-mass ions when analysed by positive-ion or negative-ion chemical ionization, field desorption or field ionization than by electron-impact ionization (chapter 7). Frequently, field and chemical ionization mass spectra are very simple, containing only a few mass peaks (figure 7.1). Compared with electron-impact mass spectra, the total ion current is divided amongst fewer ions, allowing a more sensitive quantitative analysis. However, it should be remembered that, if a substance is ionized less efficiently by chemical ionization, for example, than by electron impact, the overall sensitivity may be less for the CI method despite the fact that a peak with a larger proportion of the total ion current is monitored. It is prudent to test different ionization modes for the best response to a standard quantity of the compound of interest before making a final choice. A further, related advantage of chemical and field ionization is that any impurities in a sample will also afford mass spectra with a small number of peaks, thereby reducing the number of potentially interfering ions (i.e. increasing specificity). This effect may be exploited best in the CI mode by careful choice of reactant gas. Ammonia is an excellent reactant gas for amines and carbohydrates, but it does not ionize hydrocarbons or gas chromatographic column bleed (section 7.2). For assay by GC/MS of an alkaloid, for instance, with contaminating hydrocarbon lipids, the use of ammonia CI mass spectrometry would cause no background interference.

6.3.2. *Sensitivity and specificity*

The discussion so far has intimated that sensitivity is of prime importance; this is often so, but not always. When more than a few nanogrammes of the substance to be measured is available, the most sensitive methods are not required. Repetitive scanning of the entire mass range is superior to selected ion monitoring in this case because it provides more analytical information (complete mass spectra) so there is more confidence in the *identification* of compounds. Below the level of a few nanogrammes, quantification requires a selected ion monitoring method (usually by GC/MS), in which the compound of interest is identified on the basis of retention time and the presence of one or more m/z values known to be in the mass spectrum of that compound. When ions of several different masses are monitored, they should have similar relative abundances to those in the standard mass spectrum. The most sensitive quantitative technique, monitoring of ions of a single, selected mass, is most prone to interference since an impurity which has a mass peak at the monitored m/z value and elutes at the same retention time as the suspected substance will cause an erroneous identification. Ideally, before this technique is used, the presence of the compound to be quantified should be proved, not merely anticipated. This is not possible if there is so little of it that there are no other techniques capable of detecting it. In such circumstances, the *specificity* of the analysis should be as high as possible. A gas chromatographic capillary column could be used because retention time is more finely diagnostic of identity in this case than it is with a packed column. For the mass spectrometric analysis, recourse may be made to high resolution since it is then possible to monitor ions of selected elemental composition rather than selected integral mass. The effect is illustrated in figure 6.3 which shows the analysis of a component contributing to the flavour of a complex food extract. To detect the component, ions at m/z 130 (accurate mass 130.083) were monitored at low resolution (figure 6.3*a*) and at high resolution (figure 6.3*b*). The plots of ion current against time are called *selected current profiles*, or sometimes *mass fragmentograms*. It is clear from the figure that the specificity is better at higher resolution because ions with a nominal mass of 130, but different accurate mass (elemental composition) from the ions of interest, are not detected. The area or height of the peak profiles for the flavour component and its internal standard may be measured reliably only at high resolution. Despite the lower absolute sensitivity of high resolution compared with low-resolution operation (section 3.4), analysis of smaller amounts of substance may be possible in the high-resolution mode in such circumstances. The smallest quantity of a compound which may be observed by a particular technique is called the *detection limit*. Generally, it is advisable to use mass-spectrometric resolution just great enough to separate mass peaks of interest from those of interfering substances. Since methods of high specificity are less susceptible to interference by impurities, it is sometimes possible to reduce or eliminate altogether purification procedures prior to analysis. Examples illustrating many of these points are presented in the following sections.

6.4. Applications based on gas chromatography/mass spectrometry

The concentration in plasma of desipramine (7), a metabolite of the drug imipramine (2), may be measured by use of GC/MS in the chemical ionization mode. For calibration, different amounts of compound (7) in the range 25–250 ng/ml of plasma and a standard amount (320 ng/ml) of a tetradeuteriated analogue of (7) were added to drug-free plasma samples. After basification, the aqueous samples were extracted with hexane and the combined organic fractions were evaporated to dryness. Trifluoroacetylation gave a mixture of compounds (8) and (9). These derivatives possess better chromatographic properties and higher mass than the parent secondary amines. They afford electron-impact mass spectra with ions of low abundance above m/z 220, but the base peaks in their chemical-ionization spectra are $(M + H)^+$ ions, formed by proton transfer in the ion source from ions of the methane reactant gas. These $(M + H)^+$ ions occur at m/z 363 and 367 for compound (8) and (9), respectively. Since the quantity of metabolite to be extracted from the plasma of patients taking imipramine was relatively large (greater than 20 ng/ml), the relatively insensitive method of rapid, repetitive

Figure 6.3. Comparison of selected ion current profiles from analysis of food extract by GC/MS. (*a*) Selected ion monitoring of all ions with mass 130 and (*b*) selected ion monitoring at the accurate mass of ions of a flavour component and its internal standard. Figure reproduced by kind permission of VG Analytical Ltd, Altrincham, England.

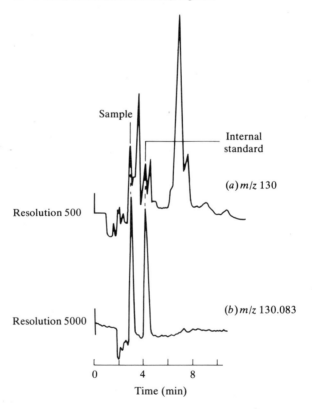

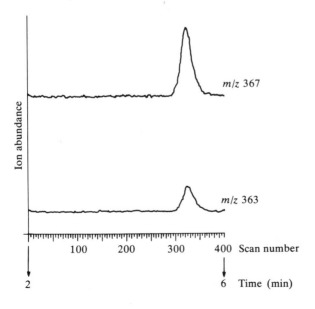

(7)   (8) mol. wt = 362   (9) mol. wt = 366

scanning was applied for quantification. A computer, coupled to a quadrupole mass filter, recorded mass spectra as the standard mixtures eluted from the chromatograph and printed out mass chromatograms of the $(M + H)^+$ ions (figure 6.4). It is noteworthy from the figure that the deuteriated analogue was not separated from the unlabelled compound by gas chromatography. The ratio of the areas of the peaks was plotted against the known concentration of compound (8) in the plasma to give the calibration graph shown (figure 6.5), a straight line in the range under observation. The same amount of internal standard was added to true samples of plasma; work-up and analysis were performed as before to determine the ratios of responses. By reference to the calibration graph, the concentration of the metabolite in the plasma was determined. The measurement provided

Figure 6.4. Mass chromatograms of $(M + H)^+$ ions of compounds (8) and (9). Taken with permission from Claeys, Muscettola & Markey (1976).

clinically important information for adjusting the dose levels of this anti-depressant drug (Claeys, Muscettola & Markey, 1976).

In a very different study (Sen, Miles, Seaman & Lawrence, 1976), very small amounts of highly toxic nitrosamines in cured meats were detected and quantified by electron impact mass spectrometry. These carcinogenic compounds are thought to arise from the interaction of naturally occurring amines and added nitrite salts. In the previous example, each mass peak was specific to one compound, but in this case an ion common to all nitro-samines (NO^+; m/z 30) enabled selected ion monitoring of that single mass peak for all compounds of interest. At low mass-spectrometric resolution (500), the technique would be highly sensitive, but non-specific since the m/z value is so low that it is common to many compounds. Some elemental compositions with nominal mass 30 are NO, ^{13}CHO, CH_2O, H_2N_2, CH_4N, $^{12}C^{13}CH_5$ and C_2H_6 with accurate masses of 29.9980, 30.0061, 30.0106, 30.0218, 30.0344, 30.0425 and 30.0470, respectively. A mass spectro-metric resolution of over 3700 (30/(30.0061−29.9980)) is necessary to separate NO^+ from $^{13}CHO^+$. In fact, a resolution of 5000 was used to monitor ions at m/z 29.9980, allowing specific detection of nitrosamines. The detection limit for the methylated derivative of 3-hydroxy-1-nitroso-pyrrolidine was about 1 ng.

Measurement of low levels of the toxic compound 2, 3, 7, 8-tetrachloro-dibenzodioxin (TCDD) in human tissue requires a specific, sensitive assay. The molecular ion region in the mass spectrum of TCDD (10) is shown in figure 6.6. It is seen that the four chlorine atoms give rise to a complex, characteristic isotope pattern. The eluate from a capillary column was

Figure 6.5. Calibration graph for quantification of desipramine. Redrawn with permission from Claeys, Muscettola & Markey (1976).

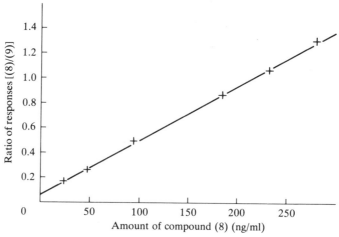

(10)

monitored selectively at m/z 320, 322 and 324 at a resolution of 10 000. Figure 6.7 shows the three selected ion current profiles, offset for clarity, obtained from injecting only 10 pg of TCDD. The specificity of the assay is high because correct identification could be verified by several parameters: (*a*) retention time from high-resolution GC, (*b*) the presence of ions which have reasonably high and even mass, (*c*) the presence of ions of correct elemental composition by high-resolution mass spectrometry and (*d*) three selected ion current profiles of the same relative heights as the corresponding mass peaks in the standard mass spectrum.

Selected ion monitoring with a quadrupole mass filter operating in the negative-ion chemical ionization mode (section 7.3) allowed detection of derivatives of dopamine (11), amphetamine (12) and Δ^9-tetrahydro-cannabinol (13) at the level of 10–25 fg (Hunt & Crow, 1978). This study is notable for its good use of derivatives in increasing both specificity and sensitivity of analysis. The polyfluorinated groups not only impart high electronegativity, and hence high ionization efficiency in the negative-ion

Figure 6.6. Molecular ion region of 2, 3, 7, 8-tetrachlorodibenzodioxin.

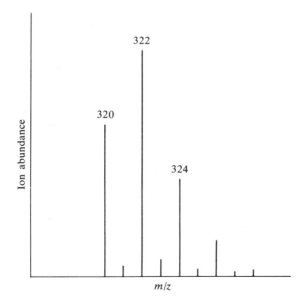

mode, but also add considerable mass. The derivatives stabilize the molecular ions M⁻˙, which then account for a large proportion of the total ion current and are suitable for selected ion monitoring. Natural products which need to be analysed are usually isolated as components of complex mixtures. Therefore, their analysis is often more difficult than that of a pure compound or of a mixture of pure compounds in which the level of other substances is

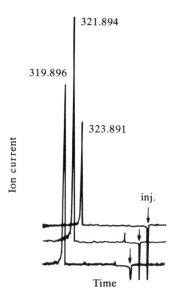

(11) mol. wt = 475

(12) mol. wt = 337

(13) mol. wt = 508

Figure 6.7. High-resolution selected ion current profiles for an analysis of 10 pg of TCDD. The three traces are offset (see injection (inj.) points) for clarity; all three profiles maximize simultaneously. Reproduced by kind permission of VG Analytical Ltd, Altrincham, England.

321.894

319.896

323.891

inj.

Ion current

Time

artificially low. By a similar method, Garland & Min (1979) quantified the drug clonazepam (14) in human plasma by reference to the labelled internal standard (15). The $(M - H)^-$ ions at m/z 314 (for (14)) and 321 (for (15))

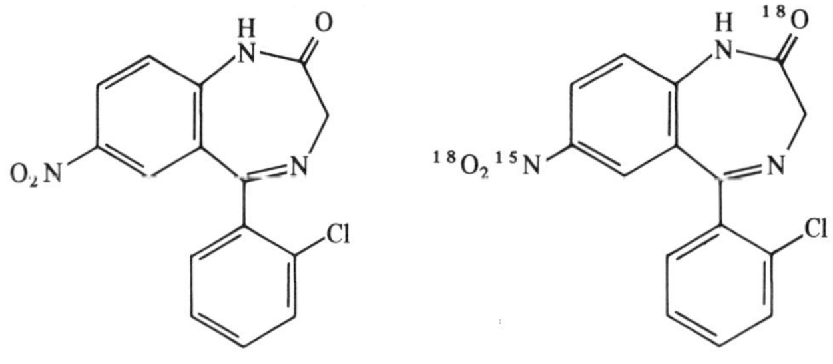

(14)   (15)

were monitored. The detection limit was estimated to be about 100 pg/ml for compound (14) in the original plasma, corresponding to an injection of an aliquot of 4 pg for each assay. The method was about 20 times more sensitive than a similar positive-ion chemical ionization technique developed by the same workers.

6.5. Applications based on direct inlet

Samples for quantification may be inserted directly into the ion source via hot or cold inlets, or on a direct probe (Millard, 1978a, pp. 91–115). The last of these methods is the commonest and forms the subject of most of this section. The same principles and scanning methods are used for quantification by direct probe as for chromatographic methods of inlet, although the former is often given a different name, the *integrated ion current technique*. It was first used to identify and quantify *para*-tyramine (16) in rat brain (Majer & Boulton, 1970). The direct probe inlet has little potential

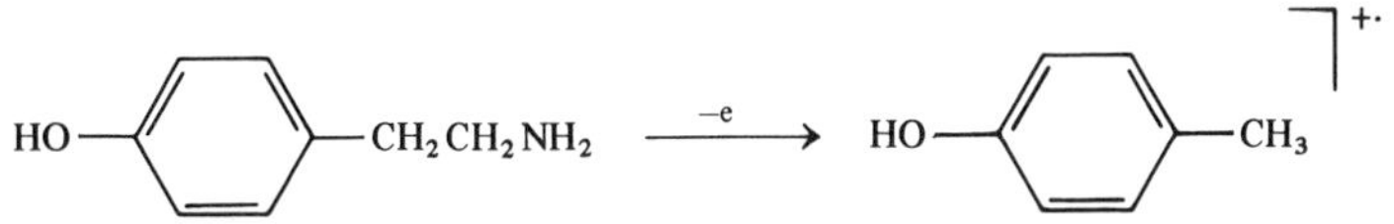

(16)   (17) m/z 108

for fractionating samples unless the components differ widely in volatility. During quantitative analysis, the components of interest in mixtures remain largely unseparated from each other and from impurities, causing many background ions. In the case of *para*-tyramine, the signal at m/z 108 was due to the selected fragment ion (17) ($C_7H_8O^{+\cdot}$ at m/z 108.0575) and to hydrocarbon impurity ($C_8H_{12}^{+\cdot}$ at m/z 108.0939). One way of overcoming this problem of low specificity is to use high mass-spectrometric resolution. A resolution of 3000 is sufficient to separate $C_7H_8O^{+\cdot}$ and $C_8H_{12}^{+\cdot}$ and allow selective detection. Background interference is worst at low mass so that derivatization provides another means of overcoming the problem. The 5-dimethylaminonaphthalene-1-sulphonyl (dansyl) derivative (18) of *para*-tyramine is measurable in the picogramme range by monitoring molecular

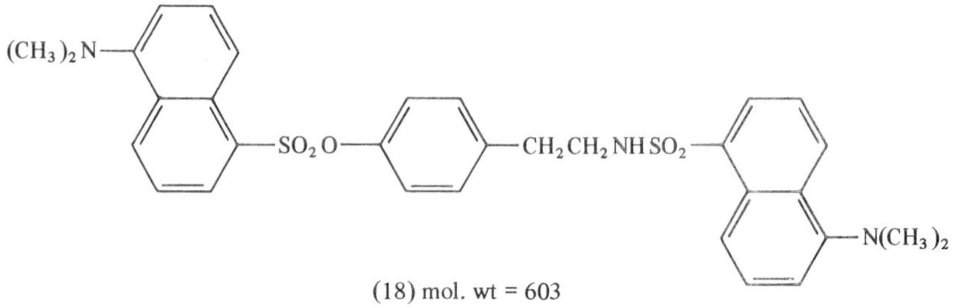

(18) mol. wt = 603

ions at m/z 603.1861. If the quantity of impurities is large enough to affect the pressure in the ion source, the resulting variation in ionization efficiency introduces error in the estimation no matter how specific it is. The cure to this problem is to increase the specificity of the work-up by removing as much of the contamination as possible.

With instruments which allow the direct probe to be heated independently of the ion source, some degree of fractionation is possible by *temperature programming*. Figure 6.8 shows the assay of some steroids from human ovarian tissue by monitoring molecular ions at a resolution of 10 000. Each line in the figure represents the abundance of the selected ion as the compounds evaporate from the probe. The tops of the lines form an *evaporation profile*, the area under which is related to the amount of substance; the relationship is determined by use of external standards. It can be seen that temperature programming of the direct probe, portrayed at the bottom of the figure, causes some separation of the steroids. At the high resolution used, the method was so specific to the steroids that no work-up of the samples was required, a small quantity of the dried tissue being placed directly on the probe (Snedden & Parker, 1976).

Field desorption mass spectrometry is particularly suitable for thermally unstable and involatile compounds. Samples deposited on an electrode are introduced directly into the ion source and subjected to an intense electric field to bring about ionization. Quantification of organic compounds, inorganic cations, especially those of the alkali metals, and organic cations, such as ammonium salts, in the FD mode is technically difficult, but possible when internal standards are used (Schulten, 1979, pp. 111–2). The sensitivity is better for cations than for neutral organic compounds. Field desorption mass spectra are characteristically simple, often consisting only of molecular ions or ions formed by attachment of a cation to the neutral molecules, for example, $(M + Na)^+$. Since ions of high mass account for a large percentage, or all, of the total ion current, the method is both specific and sensitive,

Figure 6.8. Evaporation profiles of four steroids, together with the temperature program for the direct insertion probe. In each case, molecular ions were monitored at high resolution. Adapted with permission from Snedden & Parker (1976).

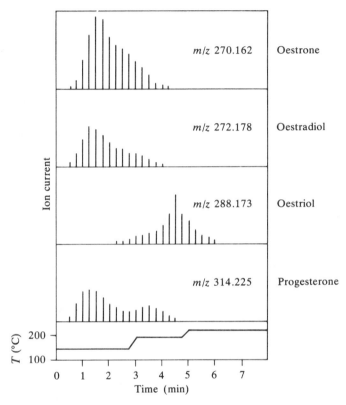

with analysis in the picogramme range for many salts (Schulten, 1979, pp. 234-44). Assays may be carried out at low or high mass spectrometric resolution.

Atmospheric pressure ionization mass spectrometry is a very sensitive technique which allows detection of compounds in the femtogramme range by selected ion monitoring. Compounds in a solvent may be injected directly into the ion source. In the negative-ion mode and using a ^{13}C-labelled internal standard, quantification of the drug diphenylhydantoin (19) was possible by

(19) mol. wt = 252

repetitive scanning of a limited mass range (m/z 245–60) and observing (M–H)⁻ ions (Horning, Stillwell *et al.* 1974).

6.6. Applications based on liquid chromatography/mass spectrometry

The use of LC/MS combines the advantages of direct probe and GC/MS methods because unstable and involatile substances may be examined with efficient fractionation. For instance, lysergic acid diethyl-amide (LSD) in urine may be determined by LC/MS, using the chromato-graphic mobile phase as reactant gas for chemical ionization, by selected ion monitoring of $(M + H)^+$ ions at m/z 324. The detection limit was in the high picogramme range (Kenyon, Melera & Erni, 1980). Being a relatively new technique, LC/MS has had limited use for quantitative measurements, but the methodology is exactly as used for GC/MS, so it is expected that such applications will develop rapidly.

6.7. Selected metastable ion monitoring

Metastable ions were introduced in section 1.4 and are discussed in some detail in chapter 8. Compounds giving metastable ions of significant abundance may be detected and quantified by tuning a magnetic-sector mass spectrometer to monitor the metastable ions selectively, rather than the normal peaks. Since metastable ions are almost always of low abundance compared with normal ions, the technique is relatively insensitive but, like selected ion monitoring at high resolving powers, selected metastable ion monitoring makes up in specificity what it lacks in sensitivity. The molecular

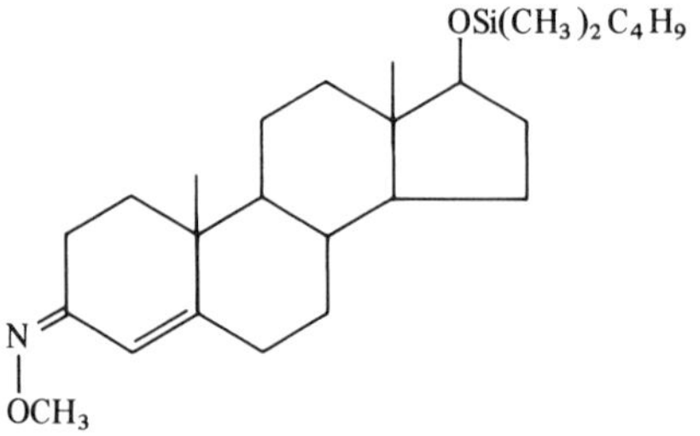

(20) 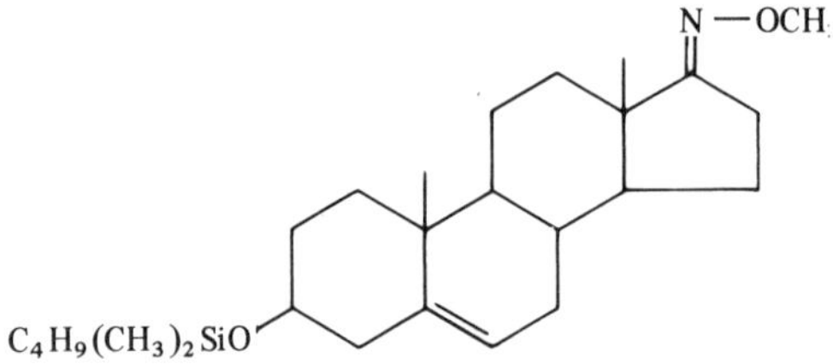   (21)

ions of the methyl oxime, tertiarybutyldimethylsilyl ether derivatives of testosterone (20) and epitestosterone (21) fragment by loss of C_4H_9 ˙ (m/z 431 → m/z 374 in both cases). Metastable ions resulting from this decompo-sition may be monitored by special scanning techniques (chapter 8) as the compounds elute from the chromatograph of a GC/MS system. Figure 6.9 compares the ion current profiles obtained by selected metastable ion monitoring (figure 6.9*a*) and selected ion monitoring of molecular ions at high (figure 6.9*b*) and low resolution (figure 6.9*c*). It is clear that, at low resolution, selected ion monitoring does not allow accurate measurement

of the size of the peaks due to compound (20) and (21) because of an unknown, interfering component. The greater specificity of selected metastable
ion monitoring or monitoring the molecular ions (m/z 431.3220) at a resolution of 8500 enables quantification because the impurity is not detected.
The compound (21) may be used as an internal standard for the determination
of the testosterone derivative (20) at the low nanogramme level in hamster
tissue. The detection limit for the steroid (20) by the method based on
metastable ions was 30 pg (Gaskell, Finney & Harper, 1979). The method
should prove most useful with derivatives, such as the tertiarybutyldimethylsilyl ethers, which favour the production of metastable ions of high mass.

Figure 6.9. GC/MS with (*a*) selected metastable ion monitoring of the reaction
m/z 431→374, (*b*) selected ion monitoring of m/z 431.3220 at a resolution
of 8500 and (*c*) selected ion monitoring of m/z 431 at low resolution.
Testosterone derivative (20), epitestosterone derivative (21) and an unknown
interfering substance (?) are marked. Adapted with permission from Gaskell,
Finney & Harper (1979).

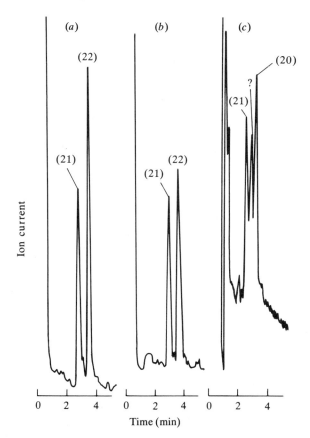

7 Additional methods of ionization

7.1. Introduction

The method used to ionize a substance affects markedly the appearance of its mass spectrum. This factor is illustrated in figure 7.1 which shows mass spectra of talbutal (1) obtained by electron impact, chemical ionization,

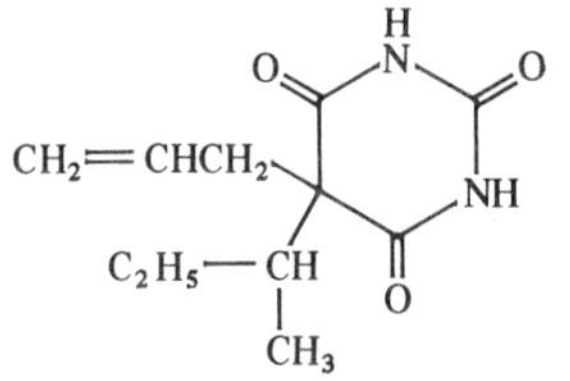

(1) mol. wt = 224

field ionization and field desorption. Throughout this chapter, mass spectra are of positively charged ions unless otherwise stated. The EI spectrum of the pyrimidine (1) contains many mass peaks useful for characterization either by manual interpretation (chapter 10) or library searching, but the diagnostically important molecular ions (m/z 224) are of very low abundance. In the CI spectrum, the base peak comprises protonated molecular ions $(M + H)^+$ formed by reaction of the compound with the isobutane reactant gas in the ion source; there are few fragment ions. Field ionization affords a greater number of fragment ions and abundant molecular ions at m/z 224, whereas field desorption causes no fragmentation at all. This situation is typical for many compounds and implies that no one ionization technique is always superior to the others. If the compound (1) had been unknown, it would not have been possible to be sure of the molecular weight from the EI spectrum. This information would be readily available from the CI, FI or FD mass spectrum, but the smaller number of fragment ions in these modes would have made structural elucidation impossible. If a small quantity of talbutal (1) were required to be quantified by selected ion monitoring, then CI or FD mass spectrometry would appear to be the most suitable methods since both afford spectra containing a mass peak accounting for a large proportion of the total ion current. The method of choice depends on the analysis required and on the type of compound. A most important and general point to be gleaned from figure 7.1 is that, for structural elucidation, examination of the EI mass spectrum in conjunction with any one of the other three types of spectra makes for a much easier interpretation than

Figure 7.1. Mass spectra of talbutal (1) obtained by (*a*) electron impact, (*b*) chemical ionization with isobutane, (*c*) field ionization (FI) and (*d*) field desorption (FD). Taken with permission from Fales *et al.* (1975).

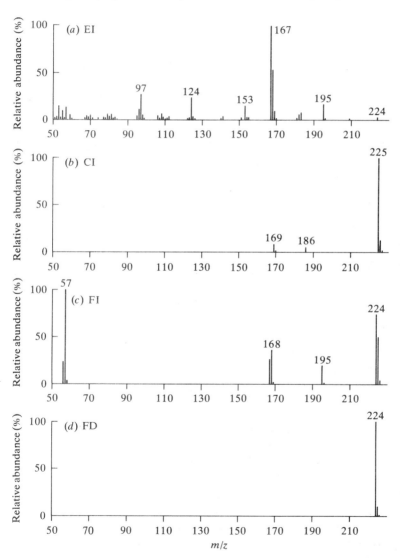

when any one is considered in isolation. The methods of ionization discussed in this chapter are complementary to and not alternatives for electron impact. It is advisable to observe the EI mass spectrum of an unknown substance, even when new and sophisticated techniques are available, because the behaviour of compounds under EI conditions is far better understood than that under the conditions of any other ionization method. It is possible to draw on a vast store of experience and data on EI mass spectrometry (e.g. mass spectral libraries). Such a store is not yet available for other methods of ionization. For instance, assignment of molecular weight by CI, FI and FD mass spectrometry is not always as clear as it is in figure 7.1. Chemical ionization may give $(M - H)^+$, $(M + H)^+$, or $(M + X)^+$ adduct ions, collectively called *quasi-molecular ions*, depending on the structure of M and reactant gas (section 7.2). Field ionization affords $M^{+\cdot}$ or $(M + H)^+$ ions, and field desorption may give $M^{+\cdot}$, $(M + H)^+$, $(M + Na)^+$, $(M + K)^+$ or other adduct ions as well as cluster ions like $(M_n + H)^+$, $(n = 2, 3, \ldots)$. It may not always be obvious which of these ions is being observed under a given set of conditions, making the assignment of molecular weight uncertain.

There are many methods for ionizing compounds (Milne & Lacey, 1974) of which electron impact is the commonest. Additional methods, such as CI, FI and FD, are frequently termed 'soft' ionization techniques because the molecular, or quasi-molecular, ions are formed with less excess of internal energy than in EI ionization. As seen above, soft ionization leads to less fragmentation and greater abundances of molecular or quasi-molecular ions. The full range of ionization techniques has been reviewed by Wilson (1971, 1973, 1975, 1977). In this chapter, the most popular methods of ionization, other than electron impact, are described in separate sections. Some other techniques, developed largely for examining substances of very low vapour pressure, are described in section 12.3.

7.2. Chemical ionization

Ion sources for chemical ionization operate at pressures of 0.2–2 torr compared with about 10^{-6} torr for electron impact sources. At the lower pressure of 10^{-6} torr, the mean free path of a molecule or ion is about 200 mm and ion/molecule collisions are very rare. At the higher pressure of about 1 torr, the mean free path is only about 2×10^{-4} mm and ions collide with neutral molecules many times before leaving the source. The higher pressure is obtained by 'leaking' into the ion source a reactant gas (R) which is ionized, usually by electron impact, to give *primary ions*, $R^{+\cdot}$. Through collision of $R^{+\cdot}$ ions with neutral R molecules, translational, vibrational and rotational energy is equilibrated between the ions and molecules so that, in chemical ionization, the primary ions are generally considered to possess thermal energies, i.e. corresponding to equilibrated ground-state species at the temperature of the ion source. During the collisions, a number of ion/molecule reactions may occur. For structural mass spectrometry, the most useful reaction is the formation of stable *secondary ions*, $(R + H)^+$ or $(R - H)^+$. Generally, ions with an even number of electrons (cations), such as the secondary ions, are more stable than ions with an odd number of electrons (radical–cations) like the $R^{+\cdot}$ primary ions. Methane affords principally CH_5^+, isobutane $C_4H_9^+$, ammonia NH_4^+ and hydrogen H_3^+ ions. It is these thermally equilibrated, stable secondary ions which are the reactant species and which effect ionization of other molecules (M) present by ion/molecule reactions (scheme 7.1). To ensure that a

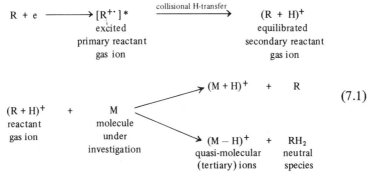

$$(7.1)$$

statistically insignificant number of sample molecules (M) is ionized by electron impact, the ratio of reactant gas to sample should be at least 10^3 at a pressure of about 1 torr. In these circumstances, the stable secondary reactant ions can collide with molecules (M) of the sample to produce *tertiary ions* which are the ions of interest to the mass spectrometrist. The secondary ions are strong Lewis acids and during the formation of the tertiary ions, a proton may be donated by the reactant gas ion or a hydride ion abstracted from it (scheme 7.1). Chemical ionization of acetone in methane is represented by equation (7.2) in which a secondary CH_5^+ ion donates a proton to acetone.

$$CH_5^+ + \underset{\underset{O}{\|}}{CH_3CCH_3} \rightarrow CH_4 + \underset{\underset{^+O-H}{\|}}{CH_3CCH_3}$$

$$(7.2)$$

Quasi-molecular ion

The proton is considered to be attached to the oxygen atom in the quasi-molecular ion. Depending on the acidity of the reactant secondary ion and the basicity of the sample molecule, hydrogen transfer may not occur and instead the secondary ions react with molecules (M) to give adduct ions as shown in equations (7.3) and (7.4).

$$M + NH_4^+ \rightarrow (M + NH_4)^+$$

$$(7.3)$$

$$M + C_2H_5^+ \rightarrow (M + C_2H_5)^+$$
$$M + C_3H_7^+ \rightarrow (M + \dot{C}_3H_7)^+ \tag{7.4}$$

For instance, chemical ionization of an amine in ammonia can afford both $(M + H)^+$ and $(M + NH_4)^+$ tertiary ions, with masses of 1 and 18 mass units greater than the molecular weight, respectively. When using methane or isobutane as reactant gas, in addition to the abundant $(M + H)^+$ or $(M - H)^+$ ions, there are often adduct ions of higher mass and lower abundance. Ion/molecule reactions of methane at 1 torr produce some ethyl and propyl secondary ions which may undergo addition reactions (equation 7.4). Many of these adduct ions are predictable and hence useful for confirming the assignment of molecular weight, but unexpected adduct or cluster ions such as $(2M + H)^+$ may be confusing. The abundance of adduct ions depends on the structure of M, the choice of reactant gas, and the mass-spectrometric conditions, in particular the pressure in the ion source.

In general, quasi-molecular ions obtained by chemical ionization have greater relative abundances than molecular ions obtained by electron impact. Quasi-molecular ions are more stable because they are formed with little excess of internal energy and because they are even-electron species, unlike the radical–cations $(M^{+\cdot})$ formed by electron impact. Chemical ionization may not afford fragment ions, in which case the mass spectrum on its own is of little use for structural elucidation. If the quasi-molecular ions do have sufficient energy, they dissociate by loss of neutral molecules to give even-electron fragment ions. Usually, chemical ionization provides fewer but diagnostically more significant fragment ions than does electron impact. One reason for this is that, after chemical ionization, carbon–carbon bonds tend to cleave only if the products of the dissociation are particularly stable. Frequently, the carbon skeleton remains intact and cleavage is restricted to the bonds of functional groups such as C–O, C–S and C–N bonds. This is certainly not the case with electron impact mass spectrometry (chapter 10). Chemical ionization mass spectra, unlike EI spectra, may reflect fine differences between structural isomers. The loss of water from ketones is found commonly in mass spectrometry, but the CI mass spectra of steroidal ketones exhibit marked selectivity in this reaction. The loss of water from the $(M + H)^+$ ions of 5α-3-ketosteroids is not prominent but is very noticeable in the corresponding 5β-3-ketosteroids.

Chemical ionization allows the actual degree of fragmentation to be controlled by changing the reactant gas. Chemical ionization of acetophenone gives $(M + H)^+$ quasi-molecular ions using either methane, propane or isobutane as reactant gas, but the degree of fragmentation is different in each case. With isobutane, only $(M + H)^+$ ions were observed; with propane $(M + H)^+$ and $C_6H_5CO^+$ ions were found; and with methane $(M + H)^+$,

$C_6H_5CO^+$, $C_6H_5^+$ and CH_3CO^+ ions were formed. The mass spectrum can be changed by using different reactant gases because the amount of excess of energy imparted to an $(M + H)^+$ ion depends on the relative proton affinities of the conjugate base of the reactant ion (CH_4, NH_3 and so on) and the compound (M). As the proton affinity of the conjugate base is decreased (i.e. the acidity of the reactant ion is increased), the amount of fragmentation of $(M + H)^+$ ions increases because more energy is transferred to them during their formation. Since acidity increases in the order $NH_4^+ < C_4H_9^+ < C_2H_5^+ < CH_5^+ < H_3^+$, the degree of fragmentation caused by these common reactant ions increases in the same order as illustrated for acetophenone. If the reactant gas has a proton affinity much higher than that of a compound (M), proton transfer from the reactant ions to M will be energetically too unfavourable. Thus, ammonia protonates selectively compounds of high proton affinity (basicity) such as amines and amides. Hydrocarbons and ethers are not ionized by CI with ammonia, whilst aldehydes, ketones, acids and carbohydrates form only the adduct ions, $(M + NH_4)^+$.

Some infrequently used reactant gases have considerable benefits in special circumstances. The reactant ions of methyl vinyl ether react with alkenes in a manner dependent on the location of double bonds (scheme 7.5).

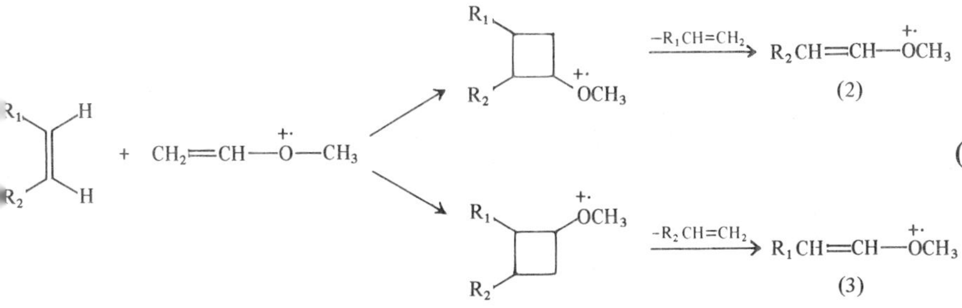

$$(7.5)$$

The masses of ions (2) and (3) indicate the position of the double bond in the original structure. Chemical ionization of dipeptides with methanol affords mass spectra which are more easily interpreted than electron impact spectra or CI mass spectra obtained with ammonia or isobutane as reactant gas. To identify a dipeptide from its mass spectrum, there should be readily recognizable, though not necessarily particularly large, mass peaks for the quasi-molecular ions and prominent peaks defining the amino-acid sequence (section 10.6). Figure 7.2, comparing EI and methanol CI mass spectra of the *N*-acetylated, *O*-methylated derivative of phenylalanylglycine (4), shows that the latter technique meets these requirements. As with all dipeptides

$$CH_3CONHCHCONHCH_2COOCH_3$$
$$|$$
$$CH_2C_6H_5$$

(4) mol. wt = 278

examined to date, reasonably abundant $(M + H)^+$ ions (m/z 279) are accompanied by fragment ions resulting from loss of water (m/z 261) and methanol (m/z 247). These latter are used to confirm the assignment of molecular weight. Mass peaks indicative of amino-acid sequence (marked with an asterisk) are clearer in the CI mass spectrum.

Nitric oxide as a reactant gas provides abundant $(M + NO)^+$ adduct ions with ketones, acids and esters, but $(M - H)^+$ ions with ethers and aldehydes, thereby distinguishing between these two groups of compounds. The reactant gas tetramethylsilane appears to be very useful for analysis of substances which do not give quasi-molecular ions with isobutane CI. Essentially, $(CH_3)_3Si^+$ ions are the only secondary ions and these add to a wide variety of compounds to give $(M + (CH_3)_3Si)^+$ ions, the observation of which leads to the assignment of the molecular weight of M.

Many other reactant gases have been used, including binary mixtures such as argon and water. The aim of some of these binary mixtures is to produce mass spectra with abundant quasi-molecular ions (typical of CI) and many fragment ions (characteristic of EI or charge exchange; see below), making the elucidation of structures easier than when using CI or EI mass spectra alone. For these, more specialized texts should be consulted (Wilson, 1971, 1973; Milne & Lacey, 1974, pp. 77–85; Wilson, 1975, 1977).

7.2.1. Charge exchange

If reactant gas ions are formed from monoatomic species, there are no vibrational degrees of freedom and the ions carry specific amounts of energy. The lowest amount of energy required to ionize an atom or molecule

Figure 7.2. (*a*) Electron impact and (*b*) methanol chemical ionization mass spectra of *Ac–Phe–Gly–OMe* (4). The mass peaks marked with an asterisk are due to ions which characterize the sequence of the amino acids.

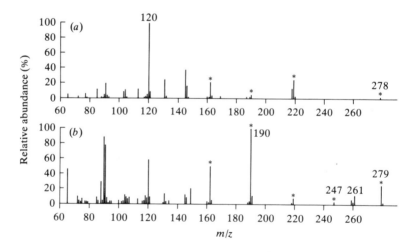

is the ionization energy. Argon, with an ionization energy of 15.755 eV, has just that excess of energy after ionization and there is no possibility of its internal conversion to vibrational energy. In a charge-exchange reaction between reactant argon ions and molecules, M (equation 7.6), the whole of the excess of energy (15.755 eV) is transferred to the molecule, M, which is therefore ionized and given a large excess of energy.

$$A^{+\cdot} + M \quad \to \quad A + M^{+\cdot} \tag{7.6}$$

The molecular ion, $M^{+\cdot}$, does have vibrational degrees of freedom and the excess of energy present causes rapid fragmentation. The behaviour is then very similar to the fragmentation of odd-electron molecular ions produced by electron impact, and is unlike the chemical ionization discussed above in which thermally equilibrated reactant gas ions transfer smaller amounts of energy to form even-electron quasi-molecular ions exhibiting greatly reduced fragmentation. The interest in charge-exchange reactions with monatomic reactant gas ions lies in the knowledge of the precise amount of excess of energy given to the molecular ion. If the ionization energy of the molecule is 10.0 eV and argon is used as the reactant gas, then 15.755 − 10.0 = 5.755 eV excess of energy is imparted to the molecular ion. Such an amount of energy appearing in most of the commoner types of molecular ions leads to rapid fragmentation.

7.2.2.   Conclusion

Chemical ionization mass spectrometry is a versatile technique, compatible with GC/MS and LC/MS, and particularly suited to samples such as hydrocarbons, alcohols, amines, esters, sugars, amino acids, peptides and nucleosides, which are susceptible to excessive fragmentation by the electron impact method. When electron impact gives molecular ions with too great an internal energy to be stable, peaks corresponding to $M^{+\cdot}$ ions are not seen. In such cases, chemical ionization with a carefully selected reactant gas frequently identifies the molecular weight through formation of quasi-molecular ions. With methane and isobutane, $(M + H)^+$ ions are usually abundant unless the compound has very low proton affinity, as occurs with an alkane, in which case $(M - H)^+$ ions are prevalent (see figure 10.10 below). If molecular ions are absent from the EI spectrum because the neutral molecule decomposes thermally at the temperature required to volatilize it, chemical ionization also is unlikely to resolve the problem since it too requires volatilization of the sample. Then the only advantage of chemical ionization over electron impact is that compounds may be volatilized at a somewhat lower temperature because passage of reactant gas over the sample aids evaporation. Substances that are scarcely volatile or are thermally unstable at the temperatures required for vaporization are best examined by *in-beam chemical ionization* (section 12.3), fast atom bombardment or field desorption.

In electron-impact studies, molecular ions often rearrange prior to fragmentation so that unexpected or misleading neutral particles may be ejected (section 10.9). This effect is much less marked with chemical ionization spectra, so the neutral losses tend to be more diagnostic of structure. Many of these points and aspects of structural elucidation with CI mass spectra are presented in chapter 10 and exemplified in chapter 11 (example D).

Preferably, reactant gases do not damage the ion source and filament, are readily and cheaply available, are of high volatility to allow rapid pumping out of the ion source when required and of low mass. Isobutane CI mass spectra are not useful below m/z 60 because of the presence of highly abundant ions from isobutane itself, whereas methane and ammonia CI spectra are largely free of reactant gas ions above m/z 20. An ideal reactant gas would afford only one type of reactant ion rather than a mixture of several different ions.

The large number of ion/molecule collisions occurring in an ion source during chemical ionization causes thermal equilibration to take place. As long as equilibration of vibrational and rotational energy is complete, it is permissible to use the same equations of reaction kinetics that govern the chemistry of ions in solution. For the formation of ions in the gaseous phase, rate constants are seen to obey the Arrhenius equation and activation energies and frequency factors are calculable. Proton affinities are also measurable by observing reaction (7.7) at equilibrium in the ion source at relatively high pressure.

$$(A + H)^+ + B \rightleftharpoons A + (B + H)^+ \tag{7.7}$$

The equilibrium constant (K) for the reaction is calculated through the abundances of $(A + H)^+$ and $(B + H)^+$ at equilibrium and the partial pressures of the two substances (A and B). The value of K is related to the total free energy change $(\Delta G^\ominus)$ by equation (7.8) where R is the gas constant

$$-\Delta G^\ominus = RT \ln K \tag{7.8}$$

and T, the temperature. Since transfer of a proton between relatively large molecules, A and B, involves little entropy change, the total free energy change is virtually equal to the heat of reaction, $\Delta H^\ominus$. For reaction (7.7), $\Delta H^\ominus$ equals the difference in proton affinities of A and B (equation 7.9). If the proton affinity (PA) of A is known, that of B may be calculated.

$$\Delta H^\ominus = PA(B) - PA(A) \approx -RT \ln K \tag{7.9}$$

7.3. Negative-ion chemical ionization

At the low pressures obtaining in ion sources operating under electron impact conditions, formation of negative ions is inefficient relative to that of positive ions. At pressures of about 1 torr, there is a large increase in the number of negative ions produced. There are two mechanisms whereby this occurs, *electron capture* (equation 7.10) and *reactant-ion chemical ionization.*

$$M + e + B \rightarrow M^{-\cdot} + B \qquad (7.10)$$

To form a negative ion by electron capture, a neutral molecule (M) must have a vacant, low-energy orbital in which to accommodate the extra electron. Compounds with double bonds, sulphur, phosphorus or halogens except fluorine will capture electrons, whereas alkanes will not. The latter compounds, methane in particular, may be used as *buffer (moderating) gases* (B) to remove the excess of energy from the negative ion as it is formed (equation 7.10). Since electron capture is a three-body process, the pressure in the ion source should be as high as possible, at least 1 torr. Although the process is not strictly chemical ionization because there is no *reactant* gas, the resultant mass spectra appear like CI spectra with abundant molecular ions and little fragmentation. Strongly electronegative compounds may be analysed in the femtogramme range by high-pressure electron capture.

True negative-ion chemical ionization occurs by reaction of a compound (MH) with thermally equilibrated, negatively charged reactant ions ($R^{-\cdot}$ or R^-). During reactant ion chemical ionization, several types of ion/molecule reaction may occur. One of the commonest, proton abstraction is depicted in equation (7.11).

$$MH + R^- \rightarrow M^- + RH \qquad (7.11)$$

In this process, much of the exothermicity of the reaction is carried away as vibrational energy in the new R–H bond of the neutral species, leaving M^- as a relatively stable ion. In positive-ion chemical ionization, much of the excess of energy in the formation of $(M + H)^+$ ions resides initially in the newly formed bond of the ion. Therefore, negative-ion CI mass spectra show less fragmentation than many positive-ion CI spectra.

Several different types of ion/molecule reaction are observed during negative-ion chemical ionization, depending on the structure of the compound, the reactant gas and the pressure. Some of the reactions are familiar, being common to carbanions in solution, as with nucleophilic displacement (scheme 7.12) and base-induced elimination (scheme 7.13).

$$R^- + \quad \overset{}{\underset{}{>}}C{-}Y \longrightarrow R{-}C\overset{}{\underset{}{<}} + Y^- \quad (7.12) \qquad (7.12)$$

$$R^- + \quad \overset{H}{\underset{}{>}}C{-}C\overset{Y}{\underset{}{<}} \longrightarrow R^- {\cdots} H^+ {\cdots} \overset{Y}{\underset{}{C{-}C}} \longrightarrow R^- {\cdots}H{\cdots}Y + \overset{}{\underset{}{>=<}} \quad (7.13)$$

As the basicity (proton affinity) of a reactant ion increases, the more likely is proton abstraction (equation 7.11). Basicity increases in the order $Cl^- < F^- < O^{-\cdot} < OH^-$. The chloride ion, principal product of ionization of dichloromethane at 1 torr, reacts with hydrogen-bonding compounds such as acids, amides, amines and phenols to give $(M + Cl)^-$ adduct ions rather than $(M - H)^-$ ions. The reactivity of reactant ions increases with the amount of energy available in the ions, that is, in the order of their heats of formation: $F^- < Cl^- < OH^- \ll O^{-\cdot}$. The highly reactive $O^{-\cdot}$ ion, conveniently produced by ionization of nitrous oxide, causes many types of reaction, some of which are shown and exemplified in scheme (7.14). A more useful reactant ion for the elucidation of organic structure is OH^-, generated by ionization of a

H·-abstraction $\quad O^{-\cdot} + CH_3CN \longrightarrow \dot{C}H_2CN + OH^-$

H$^+$-abstraction $\quad O^{-\cdot} + C_6H_5CH_3 \longrightarrow C_6H_5CH_2^- + OH^\cdot$

H$_2$-abstraction $\quad O^{-\cdot} + C_6H_6 \longrightarrow C_6H_4^{-\cdot} + H_2O$

H·-displacement $\quad O^{-\cdot} + \text{(pyridine)} \longrightarrow \text{(pyridine-}O^-\text{)} + H^\cdot$

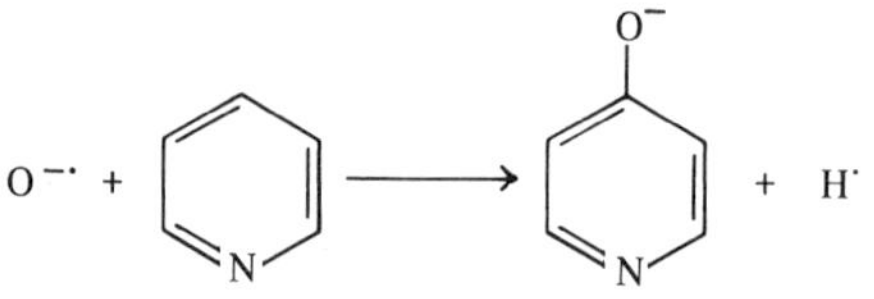

Alkyl-displacement $\quad O^{-\cdot} + C_2H_5COC_2H_5 \longrightarrow C_2H_5COO^- + C_2H_5^\cdot$

(7.14)

mixture of nitrous oxide and methane. Its high proton affinity promotes proton abstraction from virtually all compounds apart from alkanes. Being much less reactive than $O^{-\cdot}$, it does not display a wide variety of reaction types and generally it produces little fragmentation. Stearic acid (octade-canoic acid), 3-ethylpentan-3-ol and 2,4-pentanedione give only $(M - H)^-$ ions whilst ethyl acetate affords $\bar{C}H_2COOC_2H_5$ (60% TIC) and CH_3COO^- (40% TIC).

7.3.1. Conclusion

Negative-ion chemical ionization has many basic similarities to its positive-ion counterpart. Both are soft ionization techniques because molecules are ionized by interaction with charged particles, the excess of energy of which has been dissipated by many collisions. Except for electron capture, the ionization step is a reaction of which there are several types. Under a given set of conditions, more than one reaction may be possible. Both chemical ionization modes allow the degree of fragmentation to be controlled to a degree by use of selected reactant gases. Like positive-ion CI, the negative-ion method is suited to combined chromatography/mass spectrometry. It is frequently useful for structural elucidation to examine both negative-ion and positive-ion chemical ionization mass spectra for they provide supplementary information. Consider the chemical ionization mass spectra of L-phenylalanine shown in figure 7.3. Reactant gas ions, OH^-, are known to abstract protons from sample molecules and thus, the base peak at m/z 164 can be assigned to $(M - H)^-$ ions, giving a molecular weight of 165. In contrast to the negative-ion spectrum, positive-ion chemical ionization with methane affords a base peak at m/z 166. Since $(M + H)^+$ ions are expected in this mode, the assignment of molecular weight is confirmed. Fragmentation in the two spectra is quite different and, as indicated in the foregoing discussion, is less extensive in the negative-ion spectrum. The only significant fragment ions in the latter arise from ejection of H_2 and of C_7H_8 from the quasi-molecular ions. Partly because positive-ion and negative-ion CI spectra supplement each other, instruments capable of measuring both more or less simultaneously have been developed (see next section).

Figure 7.3. Chemical ionization mass spectra of L-phenylalanine. (*a*) Negative-ion chemical ionization with OH^- and (*b*) positive-ion chemical ionization with CH_5^+.

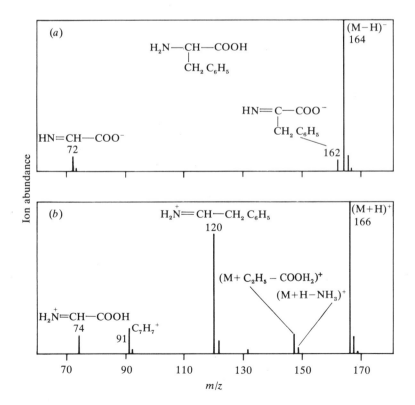

Negative-ion chemical ionization is a useful and sensitive technique for compounds having high electron affinity, such as halides, aromatic nitro-compounds and quinones. It is used to advantage particularly in the analysis of low levels of chlorinated pesticides and herbicides. Compounds with lesser electron affinity are ionized less efficiently or not at all. The opposite situation obtains for positive-ion chemical ionization, so that, for example, hexafluorobenzene is hardly ionized in the positive ion CI mode. The difference in applicability makes inexpedient direct comparisons of the sensitivity of the two techniques with a single compound. For analysis of small amounts of a substance, derivatization with an electronegative group (section 5.3) followed by negative-ion chemical ionization is likely to be two to three times more sensitive than derivatization with a relatively electropositive group followed by positive-ion chemical ionization.

Negative-ion chemical ionization (Jennings, 1977, 1979) and reactions and fragmentations of negative ions of organic, organometallic and co-ordination compounds have been reviewed (Bowie and Williams, 1975; Bowie, 1975, 1977, 1979).

7.4. Pulsed positive-ion/negative-ion chemical ionization

Usually referred to as PPINICI, this technique takes advantage of the fact that the analyser of a quadrupole mass filter passes ions of a given mass regardless of the sign of their charge. The method is not feasible with a magnetic-sector mass spectrometer which is set up to record either positive or negative ions, not both. The ion extraction potential of a conventional quadrupole mass filter is switched rapidly in polarity 10 000 times per s. This has the effect of ejecting from the ion source pulses ('packets') of positive and negative ions. After traversing the analyser, the positive ions are attracted to one detector and negative ions to another. The amplified signals from the two detectors are positive-ion and negative-ion CI mass spectra recorded essentially simultaneously (Hunt, Stafford, Crow & Russell, 1976). Of course, the reactant gas must be suitable for both positive-ion and negative-ion chemical ionization. The use of argon affords charge-exchange mass spectra (rather like EI spectra) in the positive-ion mode along with electron-capture mass spectra in the negative-ion mode. When ionized at a pressure of 1 torr, methane gives reactant ions CH_5^+ (conventional positive-ion CI mass spectra) and electrons (electron-capture spectra). A mixture of methane and nitrous oxide provides CH_5^+ and OH^- as reactant ions for simultaneous chemical ionization.

PFK may be used as a calibration compound for accurate mass measurement by PPINICI (Hunt, Stafford, Shabanowitz & Crow, 1977). The methane positive-ion CI mass spectrum of PFK is very weak relative to its negative-ion CI spectrum. To all intents and purposes, the calibration compound appears only in the negative-ion spectrum and this may be used to calibrate simultaneously acquired positive-ion mass spectra. Accurate mass measurement is possible even at the low resolution of a quadrupole mass filter because, in effect, the calibration compound and sample are separated physically (see section 3.4).

7.5.   Field ionization

Molecules approaching or adsorbed on a surface of high curvature maintained at a high positive potential are subject to potential gradients of about $10^7 - 10^8$ V/cm. Under the influence of these fields, quantum tunnelling of a valence electron from the molecule to the anode takes place in about 10^{-12} s to give a radical-cation. Being positively charged, the ion is accelerated from the anode and is analysed and detected in the same way as for electron-impact ionization.

The probability of a molecule being ionized increases exponentially with the field strength. The field strength (F) may be increased by increasing the voltage (V) applied to the anode or by decreasing the radius (r) of the microscopic tips or fine points on the surface of the anode (equation 7.15).

$$F = V/kr \qquad\qquad (7.15)$$

The value of k, a geometry factor, varies. For a sphere $k = 1$ and for other shapes $k > 1$. A wire or blade is microscopically uneven, but usually produces fields too low for efficient ionization. Anodes are 'activated' by growing on them microneedles ('whiskers'), usually of carbon and about 10^{-5} m in length, which provide on the surface many fine points with radii far smaller than on the original surface. The result, a *multi-point array*, increases efficiency of ionization by one to three orders of magnitude.

For field ionization, the sample is allowed to impinge into the region of high field strength, and this arrangement is suitable for GC/MS operation. Field ionization mass spectra are generally characterized by low degrees of fragmentation compared with electron impact (figure 7.1, and figure 7.4 below). The formation of the molecular ion by tunnelling is thought not to produce ions in highly excited states and therefore with not enough excess of energy for extensive fragmentation. For example, electron-impact ionization of heptane at 70 eV, the normal operating voltage, produces molecular ions with average excess of internal energy of about 6 eV whilst the same ions formed by field ionization have only 0.5 eV excess of internal energy. Sometimes $(M + H)^+$ ions are formed, rather than $M^{+\cdot}$ ions, because proton transfer may occur by ion/molecule reactions in the ionizing regions close to the anode which are relatively dense with molecules (M). The $(M + H)^+$ ions are distinguishable from $M^{+\cdot}$ ions because their relative abundance increases with pressure and decreases with the temperature of the anode. In field ionization, fragmentation can often be explained as cleavage of a molecular ion which is strongly polarized in the electric field, although some ions may not be true fragment ions (derived from $M^{+\cdot}$ or $(M + H)^+$), but the molecular ions of thermal decomposition products of neutral molecules. Fragmentation may also occur through ion/molecule reactions in the condensed phase around the electrode.

Because of steric constraints (entropic factors), rearrangement processes have lower reaction frequencies than do simple bond-cleavage reactions (section 9.4). It has been suggested therefore that, because ions spend a shorter time in an ion source for field ionization than in one for electron impact, there is little time for rearrangement during field ionization and few such processes should be observed. However, rearrangements have been observed in the strong electric fields, partly at least, *before* ionization. For example, $CH_3COCH_2CH_2CH_3$ yields abundant ions at m/z 58 resulting from loss of C_2H_4 prior to ionization.

Field ionization mass spectra are frequently different for isomers and are therefore useful in structural organic chemistry. This is often not the case for electron impact spectra.

7.5.1. Negative-ion field ionization

By applying a high negative voltage to the electrode in ion sources for field ionization (FI), electrons may tunnel from the cathode to a compound of high electron affinity. Negative-ion FI spectra are quite simple. Molecular ions, M^-, of halogenated quinones form virtually all of the total ion current. These ions do not fragment, but ion/molecule reactions may result in $(2M)^{-\cdot}$ ions in low abundance.

7.5.2. Field ionization kinetics

When ionization occurs on the surface of the electrode, maintained at a potential V_0, ions are subject to the full electric field and are drawn into the analyser with a translational energy of zV_0 to be detected as a normal narrow peak in the mass spectrum (z is the charge on the ion). Ionization or fragmentation close to, but not at, the surface gives ions with less than the full kinetic energy. Such ions appear as 'tails' on the low mass side of normal peaks and are called *'fast' metastable ions*. In the case of a molecular ion of mass m_1 decomposing to an ion of mass m_2 some distance from the electrode, its translational energy zV_i is given by equation (7.16) in which V_x is the potential at the position of decomposition.

$$zV_i = (m_2/m_1) z(V_0 - V_x) + zV_x \tag{7.16}$$

Values of zV_i fall between zV_0 for $V_x = V_0$ and $(m_2/m_1) zV_0$ for $V_x = 0$. The translational energy zV_i may be measured accurately with a double-focussing, magnetic-sector mass spectrometer (chapter 8) so the value of V_x is calculable. If the distribution of potential around the electrode and the ion trajectory are calculated, it is possible to determine the time between ionization and decomposition at potential V_x. This value is called the *ion lifetime* and is of the order of 10^{-12}–10^{-9} seconds. At very short ion lifetimes, the method is prone to error because of the uncertainty of the potential and exact position of ionization on the microneedles grown on the electrode. To reduce this error, field ionization may be carried out with unactivated electrodes. In practice, the kinetics of fragmentation of molecular ions at specific ion lifetimes between 10^{-11} and 10^{-9} s can be studied. Within these short lifetimes, the molecular ions are assumed not to have undergone any structural changes from the neutral molecule. The importance of field ionization kinetics lies in the ability to determine rates of decomposition of molecular ions of known structure (Derrick, 1977).

If an ion fragments after leaving the ion source and before detection, it appears as a normal ('*slow*') metastable ion as in electron impact studies (section 1.4 and chapter 8).

Elucidation of fragmentation pathways by field ionization kinetics is aided considerably by isotopic labelling. Fragmentation of 3-phenylpropanal in the FI mode includes loss of C_2H_2O to yield a mass peak at m/z 92. In the case of the labelled analogue (5), the deuterium atom is transferred to the aromatic ring during fragmentation within $10^{-10.2}$ s (scheme 7.17).

At longer lifetimes, the deuterium atom may interchange with the hydrogen atoms in the *ortho*-position of the ring or the benzylic position (Wolkoff, van der Greef & Nibbering, 1978). In electron impact studies, which give an unresolved view of all the processes occurring in the first 10^{-6} s after ionization, the interchange processes leading to hydrogen randomization do not allow observation of the specific transfer shown in scheme (7.17).

7.5.3.   Conclusion

For structure determination, one of the most important facets of field ionization is its ability to give abundant molecular ions as illustrated in figure 7.4 for D-glucose. The EI mass spectrum contains little diagnostic information for the purposes of structural elucidation because of excessive fragmentation, but the FI spectrum has a base peak of $(M + H)^+$ ions (m/z 181) and shows highly diagnostic, consecutive losses of water (m/z 163, 145 and 127). Field desorption, a modified field ionization technique to be described in the next section, occasions even less fragmentation. For molecular weight assignment, FI has the advantage over CI and FD that only $M^{+\cdot}$ or $(M + H)^+$ ions are formed; adduct ions, $(M + X)^+$, in which the nature of X may not be known with certainty, do not usually occur. For the physical chemist, field ionization is a convenient source of kinetic data of fragmentation processes occurring at specified times in a wide time scale. Normal peaks reflect reactions occurring within 10^{-12} s of ionization, fast metastable ions cover the range 10^{-11}-10^{-9} s and slow metastable ions, 10^{-8}-10^{-6} s.

Field ionization has been described in a book (Beckey, 1977) and reviewed (Milne & Lacey, 1974; Derrick, 1977; Schulten, 1979).

Figure 7.4. (*a*) Electron impact, (*b*) field ionization and (*c*) field desorption mass spectra of D-glucose (mol. wt = 180).

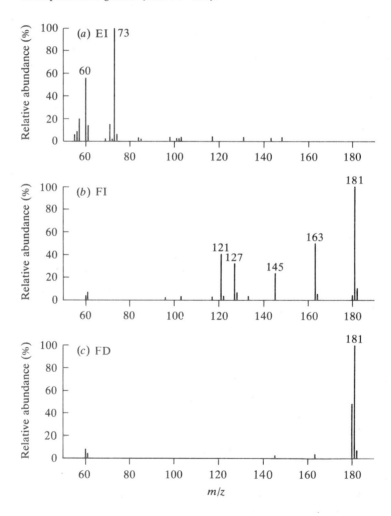

7.6. Field desorption

The technique of field desorption has been reviewed with emphasis on applications in the life sciences (Milne & Lacey, 1974; Beckey & Schulten, 1975; Derrick, 1977; Schulten, 1977, 1979). It is also described in a book (Beckey, 1977). Unlike the other techniques described in this chapter, it is not suitable for combined chromatography/mass spectrometry. For field ionization, a compound in the gas phase is ionized near or at the surface of the field electrode. This can result in at least partial decomposition of thermally labile compounds. In the field desorption method, the ionization step occurs by the same mechanism as for field ionization but the substance is first deposited onto the activated electrode (*emitter*) as a thin layer by dipping the emitter in a dilute solution of the compound or by applying the solution to the electrode with a syringe. Once the solvent has evaporated, the emitter is inserted into the ion source and samples are ionized in the condensed phase without heating them prior to ionization, thereby reducing greatly any possible thermal decomposition. For samples of very low vapour pressure, mild heating of the electrode may be required to 'desorb' them. As with field ionization, $M^{+\cdot}$ or $(M + H)^+$ ions may be formed; the pyrimidine derivative (1) gives the former (figure 7.1) whilst D-glucose yields the latter (figure 7.4). Field desorption may also give rise to adduct ions such as $(M + Li)^+$, $(M + Na)^+$ and $(M + K)^+$ when traces of metal salts are present in the sample. This *cationization* may be induced purposely by adding to the sample a salt such as LiCl or, if the compound is sufficiently acidic, by making a salt of the sample. The uncertainty in molecular weight assignment is thereby reduced since adduct ions of known constitution are anticipated.

When heating is required to cause desorption, it is brought about directly by passing a current through the emitter, or indirectly by radiation from a separate heater or with the aid of a laser. The last of these methods has allowed observation of the molecular ion of the complex organometallic porphyrin, vitamin B_{12} (m/z 1354) along with several fragment ions. With direct heating of the emitter, the *best anode temperature* for a particular compound is the temperature at which molecular or quasi-molecular ions are most abundant. Above this temperature, thermal degradation may occur giving rise to 'fragment ions' which are, in fact, molecular ions of pyrolysis products. For many compounds, there is a very narrow range of temperature over which a reasonable FD mass spectrum is obtained. This can give rise to difficulties of reproducibility.

Talbutal (1), D-glucose and compounds (6), (7) and (8) are examples of compounds which have EI mass spectra with molecular ions of low or insignificant abundance. By field desorption, all five compounds afford molecular and/or quasi-molecular ions with little, if any, fragmentation. Whilst molecular weight assignment is very important, structural identification necessitates some fragmentation. This information may be obtained in

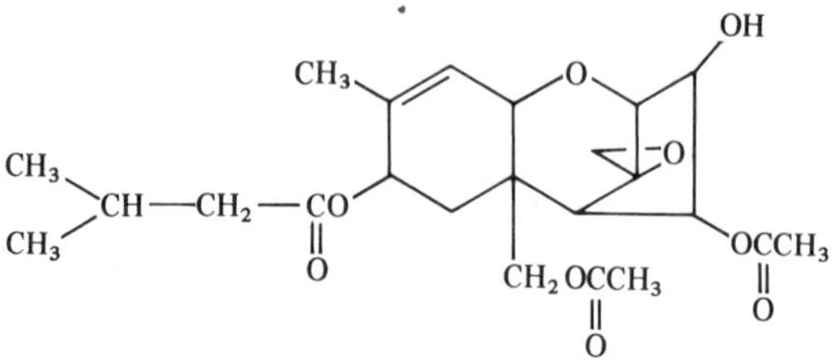

(6) Mycotoxin (mol. wt = 466)

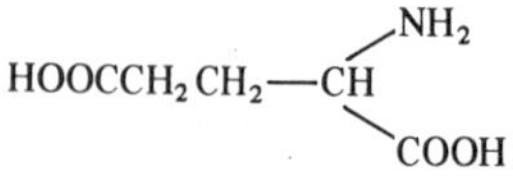

(7) Glutamic acid
(mol. wt = 147)

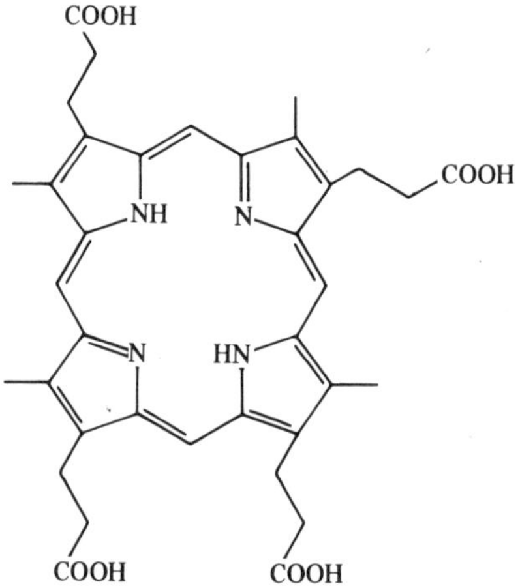

(8) Coproporphyrin-III
(mol. wt = 654)

any of three ways: (*a*) by raising the temperature of the emitter to observe thermal fragmentation, (*b*) by inducing fragmentation of desorbed molecular ions by collision with a neutral gas or (*c*) by measuring the EI mass spectrum. In method (*b*), *collisional activation*, a region of a magnetic sector mass spectrometer is held at relatively high pressure by 'leaking' into it a gas such as argon. As molecular ions pass through this region of relatively high gas pressure, ion/molecule reactions give rise to fragmentation from which the products are observed as metastable ions. The technique imparts versatility on the already powerful FD technique and is especially suitable for natural products. Collisional activation is described more fully in the next chapter.

7.6.1. Conclusion

Currently, field desorption is the method of choice for determining the molecular weight of thermally labile molecules such as carbohydrates, nucleosides, nucleotides, porphyrins and triglycerides. Unlike EI, CI and FI mass spectrometry, samples need not be volatilized by heat prior to ionization. Large molecules of low volatility may often be analysed. For instance, ions up to *m/z* 11 000 have been observed in the FD mass spectrum of polystyrene. Since compounds are not volatilized prior to ionization, organic salts and inorganic compounds as well as non-ionic compounds may be examined.

7.7   Fast atom bombardment

A beam of ions such as $Xe^{+\cdot}$ can be produced by ionizing xenon atoms and accelerating the resulting ions through an electric field. The resulting (fast) ions are directed through a xenon gas chamber when charge-exchange occurs (equation 7.18) to give fast atoms (most of the original direction and kinetic energy of the fast ions is maintained in the resulting fast atoms).

$$Xe^{+\cdot} \text{ (fast)} + Xe \text{ (thermal)} \to Xe \text{ (fast)} + Xe^{+\cdot} \text{ (thermal)} \qquad (7.18)$$

Excess of fast xenon ions can be deflected leaving a beam of fast atoms. Other gases can be used to produce a beam of fast atoms as for example with helium and argon. It has been shown that, when such beams of fast atoms impinge onto (bombard) a metal plate coated with a sample of the substance being investigated, the large amount of kinetic energy in the atoms is dissipated in various ways, some of which lead to volatilisation and ionization of the sample. By maintaining a suitable electric gradient from the plate, either positive or negative ions can be directed into the analyser of the mass spectrometer. Usually, best results are obtained by coating the plate with a relatively involatile polar liquid such as glycerol and mixing the substance under investigation into the glycerol. Early results with this *fast atom bombardment* (FAB) source are exciting considerable attention and suggest that this method will rival FI or FD in its ability to provide mass spectra of difficultly volatile substances such as sugars, peptides and nucleotides, and other thermally unstable compounds. A considerable advantage of this source is its simplicity and robustness. Further, although either positive or negative quasi-molecular ions, $(M + X)^+$ or $(M - X)^-$ where X can be H, Na, K etc., can be formed in large abundance, there are usually sufficient fragment ions to give valuable structural information. Further consideration of FAB appears in section 12.3.

8 Metastable ions

8.1 Metastable ions in conventional mass spectrometers

The term 'metastable' has been applied to those ions in a mass spectrometer which, at the detector, have less than the full kinetic energy originally imparted at the ion source. To understand the origin of metastable ions, consider a conventional double-focussing, magnetic-sector mass spectrometer as depicted in figure 8.1, and a mass spectrometric reaction in which an

Figure 8.1. Path of main ion beam through a conventional mass spectrometer, showing the first, second and third field-free regions.

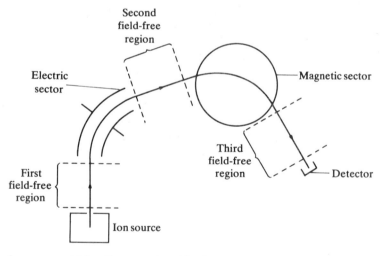

ion M^+ of mass m_1 yields a fragment ion A^+ of mass m_2 and a neutral part N (equation 8.1):

$$M^+ \rightarrow A^+ + N \qquad (8.1)$$
$$(m_1/z) \quad (m_2/z) \quad (m_1-m_2)$$

As was shown earlier, ions M^+ and A^+ will be present in the ion source and, after acceleration from there, each will possess translational energy equal to zV where V is the accelerating voltage and z the charge on the ion. By suitable adjustment of the magnet current, ions M^+ and A^+ are collected at the detector according to the formula, $m/z = B^2r^2/2V$ (section 2.4); these may be called 'normal' ions.

When a substance is ionized, ions are produced with a range of internal energies. Some M^+ ions will possess sufficiently large internal energy to fragment to A^+ whilst still in the ion source, giving rise to normal A^+ ions. The smaller the excess of internal energy of an ion, the slower is its rate of decomposition. Therefore, ions with a small excess of internal energy may reach the detector before they have time to fragment and are detected as normal M^+ ions. Ions of intermediate excesses of internal energy may fragment *after* leaving the ion source and *before* reaching the detector. The ionic products of these reactions occurring during flight through the instrument are 'metastable' ions. Typically, for electron impact, ions spend 10^{-6} s in the ion source and take about 10^{-5} s after leaving the source to reach the detector. Normal ions can have large excesses of internal energy and hence high rate constants for decomposition ($> 10^6$ s^{-1}) so that they fragment in the ion source or, at the other extreme, small excesses (rate constants $< 10^5$ s^{-1}) so that they are stable within the lifetime of the analysis. Reactions with rate constants in the range 10^5-10^6 s^{-1} give rise to metastable ions in conventional mass spectrometry.

When reaction (8.1) occurs outside the ion source affording a metastable A^+ ion and a neutral species, N, the translational energy in M^+, namely zV, must be shared between A^+ and N in accordance with the law of conservation of momentum. Thus, metastable A^+ ions will have only a part of the original energy zV and will not be collected at the detector at the same position as 'normal' A^+ ions with the full translational energy, zV. The metastable A^+ ions have therefore the same mass as normal A^+ ions but are collected differently because they do not have the same translational energy, i.e. if the new translational energy is equivalent to V' (where $V' < V$), then $m/z = B^2r^2/2V'$. Since m/z is the same for normal and metastable ions and $V' < V$, a weaker magnetic field is required to focus the metastable ions. The actual place in the mass spectrometer where the loss of translational energy occurs, i.e. where the reaction $M^+ \rightarrow A^+ + N$ occurs, determines how the metastable ions may be detected. It is necessary to consider distinct areas which may be called the first, second and third 'field-free' regions (Figure 8.1).

8.1.1   First field-free region

This lies between the ion source and the electric sector of the double-focussing instrument; it does not exist in the single-focussing instrument. Any A^+ ions formed in this region have the wrong energy to pass through the electric sector because the latter is an energy-focussing device designed to transmit only those ions of the correct translational energy. Therefore, metastable A^+ ions with only a fraction of the full translational energy will not be focussed by the sector and will not be observed in the mass spectrum. If one wishes to investigate metastable ions in the first field-free region, it is necessary to alter the accelerating voltage at the ion source or the voltage of the electric sector so as to allow the passage of the metastable ions. If an ion M^+ of mass m_1, after acceleration through a potential (V), decomposes to give a metastable ion A^+ of mass m_2 and a neutral particle N of mass (m_1-m_2), the initial translational energy (zV) of M^+ must be shared between A^+ and N as in equation (8.2):

$$zV = [(m_2/m_1)zV] + [((m_1 - m_2)/m_1)zV] \qquad (8.2)$$

If the accelerating voltage is increased to V' without changing the voltage of the electric sector such that $V' = (m_1/m_2)V$, the translational energy of the metastable ion of mass m_2 becomes $(m_1/m_2)z \times (m_2/m_1)V = zV$, i.e. it then has the correct translational energy to pass through the electric sector and be detected. Changing the voltage at the analyser in this way prevents passage of the normal ions. Because the normal ions are defocussed, the term *defocussing technique* is used for this method of investigating reactions giving rise to metastable ions (Barber & Elliott, 1964; Futrell, Ryan & Siek, 1965; Jennings, 1965). Perhaps it ought to be termed a focussing technique since the metastable ions being investigated are focussed and the normal ions defocussed. This method for examining fragmentation occurring early along the flight path of ions is useful for those cases where metastable ions are not observed in the routine mass spectrum (see Second field-free region, section 8.1.2), or where the metastable ion peaks lie partly hidden under large normal ion peaks.

In using this method, suppose a normal ion A^+ is selected and that A^+ may be formed from any or all of the precursor ions M_1^+, M_2^+, M_3^+, M_4^+ (figure

Figure 8.2. Fragment ion A^+ is formed by decomposition of any or all of the ions, M_1^+, M_2^+, M_3^+ and M_4^+. To find which fragmentation path(s) is followed, a search is made for metastable ions, A_1^+, A_2^+, A_3^+ and A_4^+. (a) Possible precursor ions M_1^+, M_2^+, M_3^+ and M_4^+ of normal fragment ion A^+ and (b) possible fragmentations in first field-free region giving metastable ions A_1^+, A_2^+, A_3^+ and A_4^+.

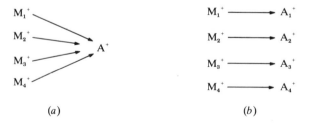

(a)   (b)

8.2a). It is necessary, therefore, to look for possible metastable ions A_1^+, A_2^+, A_3^+, A_4^+ (figure 8.2b). After adjusting the instrument to focus on the detector only ions A^+, the accelerating voltage is gradually increased whilst maintaining a constant voltage on the electric sector to focus the possible metastable ions A_1^+, A_2^+, A_3^+ and A_4^+. If any of these are detected, then the corresponding fragmentation must occur. The process has been semi-automated and the metastable ion peaks can be recorded directly onto chart paper. As an example, some of the metastable ions in the mass spectrum of *N*-phenylsuccinimide found by defocussing are shown in figure 8.3; metastable ions observed in the routine spectrum are also indicated. It can be seen that more metastable ions are found by the defocussing method than appear in the routine spectrum. The defocussing technique gives more definitive information than the metastable peaks appearing in a routine spectrum (see below). The positions of the latter peaks depend on the masses of the precursor (m_1) and product ions (m_2) which may not be exactly definable in a routine spectrum containing many ions, i.e. there can be ambiguity as to which ions are precursors to a given product. In the defocussing method, product ions are examined individually and the precursor ions giving rise to them are found precisely.

An alternative method of focussing the metastable ions is to keep the ion-accelerating voltage constant and to vary the voltage of the electric sector. During routine operation of the mass spectrometer, ions are focussed at the split placed between the electric sector and the magnetic field (figure 8.4) but metastable ions are not focussed. By altering the voltage of the electric sector, the normal ions are defocussed but the metastable ions can then be focussed. This method of examining metastable ion formation has been termed ion kinetic energy spectroscopy (IKES) (Beynon, Caprioli, Baitinger & Amy, 1970; Beynon & Cooks, 1975). A second detector which can be raised or lowered into place behind the slit (figure 8.4), and an arrangement

Figure 8.3. Some pathways elucidated by observation of metastable ions by the defocussing technique (d) and in the routine (r) mass spectrum of *N*-phenylsuccinimide.

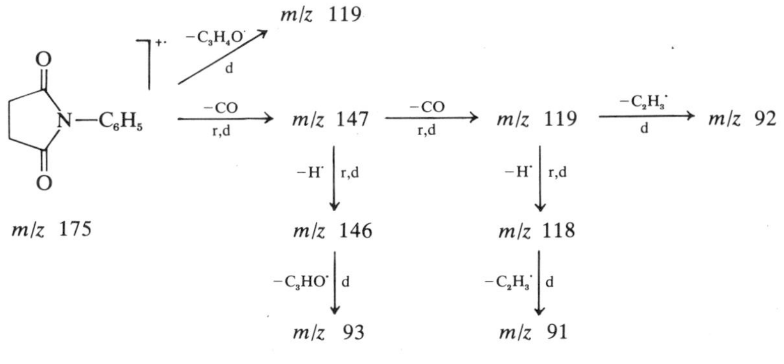

for continuously varying the voltage of the electric sector plates without affecting the ion-accelerating voltage are the extra requirements for using this method.

In general, an ion of mass m_1 with x positive charges may decompose to an ion of mass m_2 with y positive charges in the first field-free region either by a 'unimolecular' reaction or as a result of collision with a neutral particle. Any ion passing through the electric sector is constrained into a curved path of radius R given by $R = 2V/E$ (see equation 2.1) where E is the electric-sector voltage, and V the ion-accelerating voltage. An ion of mass m_1 with x positive charges on decomposition in the field-free region produces the ion of mass m_2 (y positive charges) with kinetic energy equivalent to $(m_2/m_1) \, xV$ rather than V (see above). The radius of curvature (R_1) of the ion of mass m_2 in the electric sector is given by equation (8.3).

$$R_1 = (m_2 x/m_1) \, (2V/E) \qquad (8.3)$$

Since the ion of mass m_2 has y positive charges, the effect of the potential E is magnified y times above that of a singly charged ion so that the radius of curvature (R_2) must be corrected for this as shown in equation (8.4).

$$R_2 = ((m_2/m_1)(x/y))(2V/E) \qquad (8.4)$$

Therefore, to constrain the ion of mass m_2 into a radius of curvature (R) for focussing, the radius R_2 must be multiplied by the factor $(m_1/m_2)(y/x)$. Changing the electric sector voltage from E to $(m_2/m_1)(x/y)E$ will change the radius of curvature of the ion of mass m_2 from R_2 to R and focus it at the slit. Examination of the expression for the required voltage, $(m_2/m_1)(x/y)E$ leads to a number of possible results. If $x = 2y$ and $m_2 = m_1$, i.e. for a reaction

Figure 8.4. Arrangement for ion kinetic energy spectroscopy. An electron multiplier is located behind the slit and after the electric sector.

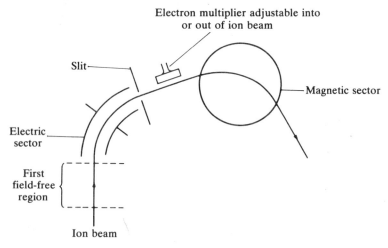

Electron multiplier adjustable into
or out of ion beam

Slit

Magnetic sector

Electric
sector

First
field-free
region

Ion beam

in which the process, $M^{2+} + N \rightarrow M^+ + N^+$ (charge exchange; see section 9.3), where N is a neutral gas, occurs, then $(m_2/m_1)(x/y)E = 2E$; the ions of mass m_2 appear at twice the normal electric sector voltage, E. If the doubly charged ion, M^{2+}, decomposes to two singly charged ions, P^+ and Q^+, then $x = 2y$ and $(m_2/m_1)(x/y)E$ becomes $(P/M)2E$ or $(Q/M)2E$ where P, Q and M are the masses of ions P^+, Q^+ and M^{2+}, i.e. there is a metastable ion peak at each of these voltages. An example of this behaviour can be seen amongst the metastable ions in the spectrum of naphthalene. The doubly charged ion $C_{10}H_6^{2+}$ (126 mass units) fragments to yield two singly charged ions, $C_7H_3^+$ (87 mass units) and $C_3H_3^+$ (39 mass units):

$$C_{10}H_6^{2+} \rightarrow C_7H_3^+ + C_3H_3^+$$

$$126 \qquad 87 \qquad 39 \text{ m.u.}$$

The metastable ions occur at electric sector voltages of $1.38E$ (= $(87/126)2E$) and $0.62E$ (=$(39/126)2E$), respectively. When $x = y$, the required sector voltage for focussing ($(m_2/m_1)E$) is less than E since m_2 is less than m_1. Thus, metastable ions may be observed at electric sector voltages below or above the normal operating voltage E without interference from normal ions. The extra ion detector placed behind the slit (figure 8.4) enables focussing of the metastable ions to be carried out by adjusting the electric sector voltage continuously. If the ion detector is raised out of the ion beam, the ions pass on and into the magnetic sector for mass analysis. The mass analysis is necessary to discover which metastable ions are being investigated. In the example given above of the ion $C_{10}H_6^{2+}$ decomposing to give ions at $C_7H_3^+$ and $C_3H_3^+$, mass analysis of the metastable ions found at a sector voltage of $1.38E$ showed that they had a mass-to-charge ratio of 87, i.e. $m_2 = 87$. Thus, $1.38E = (87/m_1)2E$, from which $m_1 = 126$; mass analysis of the metastable ions focussed at $1.38E$ gave the product ion and this combined with the electric-sector voltage gave the precursor ion. The ion at mass 126 must have been the precursor of the ion at mass 87.

There are other methods of scanning mass spectrometers to detect metastable ions without interference from normal ions which thereby reduce the ambiguity of assigning fragmentation pathways. These techniques are collectively called *linked scanning* and have been automated so that the complex modes of scanning are brought under the control of a computer. The term 'linked scan' refers to the fact that the voltage (E) of the electric sector is varied simultaneously with either the accelerating voltage (V) or the magnetic field (B) so as to maintain a specified relationship between them throughout the scan. A treatment of combinations of B, V and E is available (Boyd & Beynon, 1977) and only the commoner ones are discussed here.

For the first linked-scan method, the magnetic field is kept static whilst the accelerating and electric sector voltages are varied such that E^2/V remains constant (Weston, Jennings, Evans & Elliott, 1976). Let the value of the accelerating voltage V be changed to $(m_1/m_2)^2 V$. The kinetic energy of metastable ions, A^+, is therefore changed from $(m_2/m_1)zV$ to $(m_2/m_1)(m_1/m_2)^2 zV$ $= (m_1/m_2)zV$. For the electric sector to pass these ions, its voltage, E, must be changed to $(m_1/m_2)E$. Therefore, as the accelerating voltage is changed by a factor of $(m_1/m_2)^2$, so the voltage of the electric sector must change simultaneously by (m_1/m_2). During a scan, the values of E and V are linked in the sense that E^2/V is a constant. Precursor ions, M^+, have translational energy $zV = \frac{1}{2}m_1 v^2$, from which their momentum, $m_1 v = \sqrt{(2z\, m_1 V)}$. The initial momentum of metastable A^+ ions is $(m_2/m_1)\sqrt{(2m_1 zV)}$ but, since V has been changed by a factor of $(m_1/m_2)^2$, their new momentum is $(m_2/m_1)\sqrt{(2m_1 (m_1/m_2)^2 zV)} = \sqrt{(2m_1 zV)}$, which is the same as that for M^+ ions. Therefore, during the scan, the magnetic field is kept constant at the value required to transmit M^+ ions. By focussing initially on a selected precursor ion M^+ (often but not always the molecular ion) and then performing this linked scan, metastable ions which are products of fragmentations of M^+ in the first field-free region may be determined unambiguously. If sufficient sample is available, many different ions, M^+, may be examined individually, so that whole fragmentation pathways can be characterized. The method is complementary to the defocussing method which determines precursor ions of a chosen product ion (see above). The E^2/V linked-scan spectrum of molecular ions of n-decane characterized the fragmentations shown in scheme (8.5).

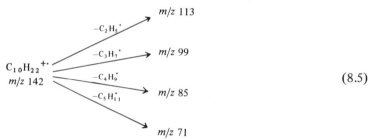

$$(8.5)$$

By performing the analysis also on the ions at m/z 99 and 57, the pathway depicted in scheme (8.6) was determined.

$$C_{10}H_{22}^{+\cdot} \xrightarrow{-C_3H_7^{\cdot}} C_7H_{15}^{+} \xrightarrow{-C_3H_6} C_4H_9^{+} \xrightarrow{-CH_4} C_3H_5^{+}$$

$$(8.6)$$

m/z 142 $\qquad\qquad$ m/z 99 $\qquad\qquad$ m/z 57 $\qquad\qquad$ m/z 41

Both the E^2/V linked-scan method and the defocussing method suffer from the limitations incurred by varying the accelerating voltage: the mass range is limited and the efficiency of the ion source changes during the scan. To overcome these problems, the accelerating voltage is kept constant and linked scanning of the magnetic field and electric sector voltage is performed instead.

When the ratio B/E is kept constant during a scan, the metastable ion spectra obtained are analogous to the E^2/V scan in that all product ions of a chosen precursor are observed (Millington & Smith, 1977; Bruins, Jennings & Evans, 1978). The translational energy of metastable ions, A^+, is $(m_2/m_1)zV$. These ions will be transmitted by the electric sector only if its initial voltage is changed by a factor of m_2/m_1. When the magnetic field is set to focus M^+ ions, then $(m_1v_1{}^2)/r = Bv_1z$ or $B = (m_1v_1)/(zr)$, where v_1 is the velocity of M^+ ions. Ions, M^+, which subsequently decompose in the first field-free region afford metastable ions A^+ with the same velocity, v_1. Hence, to focus metastable A^+ ions, the magnetic field must be changed to $(m_2v_1)/(zr)$, i.e. by a factor of m_2/m_1. Since B and E are changed by the same factor, the magnetic field and electric-sector voltage may be scanned simultaneously such that B/E remains constant and this locates all metastable ions resulting from dissociation of a selected precursor ion.

As expected, the B/E linked-scan spectrum of n-decane is virtually identical to its E^2/V linked-scan spectrum (see above). The isomeric androstanediol derivatives (1) and (2) give very similar conventional mass spectra (table 8.1),

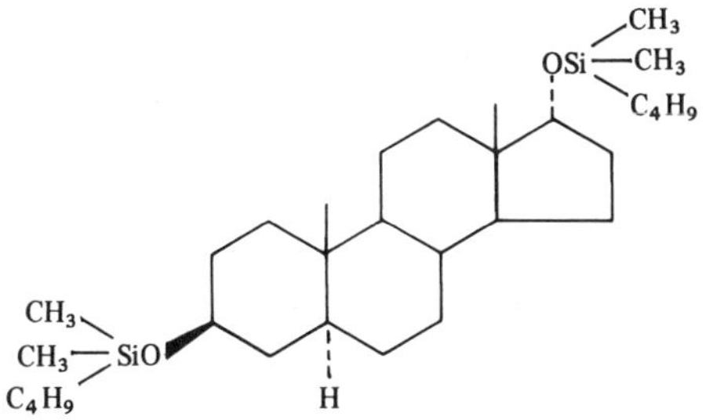

(1)

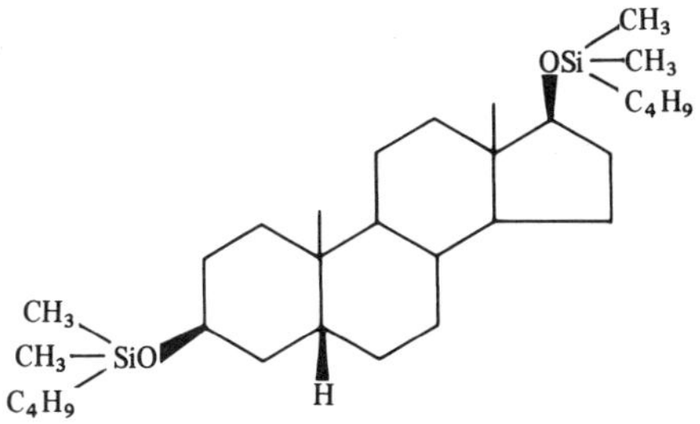

(2)

but the B/E linked-scan spectra of the ions at m/z 463, resulting from loss of $C_4H_9{}^{\cdot}$ from the molecular ions, enable the compounds to be differentiated readily by reference to relative metastable ion abundances at m/z 331 and 255 as seen in table 8.1 (Gaskell, Pike & Millington, 1979).

Table 8.1. *Routine and linked scan mass spectra of steroids (1) and (2)*

	Percentage relative abundances			
	Routine scan		B/E linked scan	
m/z value	Compound (1)	Compound (2)	Compound (1)	Compound (2)
505	1	2	—	—
463	100	100	Precursor ion	
387	8	9	100	100
373	—	—	3	13
345	—	—	1	2
331	6	12	41	84
255	66	56	70	21

A linked scan of the magnetic field and electric-sector voltage in which B^2/E remains constant provides an alternative to the defocussing method. At constant accelerating voltage, the magnetic field is set initially to focus normal A^+ ions with momentum $\sqrt{(2m_2zV)}$. The electric-sector voltage must be changed from E to $(m_2/m_1)E$ to transmit metastable A^+ ions which possess momentum $(m_2/m_1)\sqrt{(2m_1zV)} = (m_2/m_1)^{1/2}\sqrt{(2m_2zV)}$. Therefore, to focus these ions on the detector, B must be changed simultaneously to $(m_2/m_1)^{1/2}B$. Observation of precursor ions for a chosen product ion, A^+, requires scanning such that B^2/E is constant. The technique may be used in the same way as described above for the elucidation of fragmentation pathways of N-phenyl-succinimide. The steroids (1) and (2) both give ions at m/z 255.2118 ($C_{19}H_{27}^+$) in which two silanol groups have been lost. A linked scan such that B^2/E remains constant showed that m/z 463, 387, 373, 345 and 331 were all precursors of $C_{19}H_{27}^+$. From these data, coupled with the B/E linked-scan results shown in table 8.1, several fragmentation pathways were discovered as shown in figure 8.5 (Gaskell, Pike & Millington, 1979). Since the ion at m/z 255 was shown to be a product of m/z 463 and that at m/z 463 a precursor of m/z 255, rapid successive loss of both silanol groups is possible. This illustrates the general principle that the observation of metastable ions does *not* necessarily imply that a single reaction has taken place. When two (or more) successive reactions occur sufficiently rapidly in comparison with the flight time of the ion in the field-free region of interest, a metastable ion for the composite process may be observed. Metastable ions for the individual reactions may also be detected depending on fragmentation rates (figure 8.5).

The final linked-scan method to be described here selectively detects all metastable ions which result from a fragmentation involving elimination of a neutral species with a chosen mass. To do this, the accelerating voltage is held constant and the voltage of the electric sector and the magnetic field are varied such that $(B/E)(1-E)^{1/2}$ remains constant. In an alternative method, which is less satisfactory because the accelerating voltage is changed, the magnetic field is static as a linked scan of accelerating and electric sector voltages is performed (Lacey & Macdonald, 1979). For a single compound, several reactions may occur which give rise to a common neutral fragment. For example, several ions of a derivatized peptide eliminate carbon monoxide and the masses of ions involved in such reactions define the amino-acid sequence

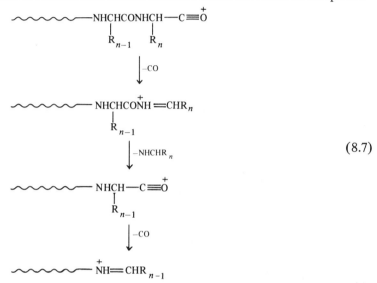

(8.7)

of the peptide (scheme 8.7). Many hydrocarbon ions lose hydrogen as shown by a linked scan of n-decane. The major fragmentations of this type are $C_4H_7^+ \rightarrow C_4H_5^+$, $C_3H_7^+ \rightarrow C_3H_5^+$ and $C_3H_5^+ \rightarrow C_3H_3^+$. Since only the difference

Figure 8.5. Fragmentation pathways of silylated dihydroxysteroids (1) and (2).

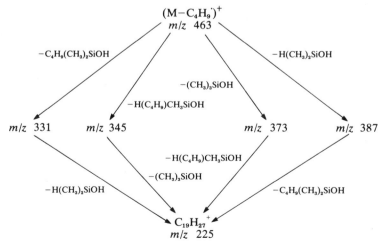

between the masses of precursor and product ions is required, metastable ions resulting from loss of the common neutral species are all detected in one scan. For the analysis of a mixture, components with common structural features may be detected selectively by the so-called *constant neutral* spectrum. Negative-ion chemical ionization of carboxylic acids affords $(M - H)^-$ ions which undergo loss of carbon dioxide. By performing a linked scan for the loss of 44 mass units, carboxylic acids are detected and identified.

In the discussion so far, it has been assumed that during reaction (8.1) the internal energy of M^+ is partitioned between metastable ion, A^+, and neutral species, N, in proportion to their respective masses. However, some *internal* energy may be released as *kinetic* energy during the decomposition and this added kinetic energy imparts an energy spread to the metastable A^+ ions. The effect results in broad peak shapes for metastable ions. Useful kinetic data may be gleaned from analysing the resulting peak shapes (see section 8.1.2) but, in many situations, kinetic data are not required and the breadth of the peaks prevents accurate determination of their position. In particular, when the peaks are so broad that they overlap, the masses of product or precursor ions are difficult to estimate. During linked scanning such that E^2/V or B/E is constant, the mass spectrometer discriminates against extremes in kinetic energy so that all peaks are equally narrow, irrespective of the amount of internal energy converted to kinetic energy during fragmentation. In these scanning modes, the masses of product ions are readily assigned. On the other hand, the defocussing and B^2/E linked-scanning methods preserve energy broadening of the peaks. Therefore, in these modes, kinetic data are available but accurate determination of all precursors of a given product ion is difficult. This situation is illustrated in figure 8.6 which shows for the terpene, limonene, (*a*) its routine mass spectrum, (*b*) the B/E linked-scan spectrum of its molecular ion at m/z 136 and (*c*) the B^2/E linked-scan spectrum of its fragment ion at m/z 93. In spectrum (*b*), metastable ions for losses of methyl (15 mass units), ethyl (29 mass units) and propyl radicals (43 mass units) are observed with good resolution of mass peaks. On the other hand, the B^2/E linked-scan spectrum (*c*) is of much lower resolution, containing only broad peaks. It is observed that, in the first field-free region, ions at m/z 136, 121 and 107 are all precursors of the product ions at m/z 93. Combining the results of these two metastable ion analyses, the following fragmentation pathway can be constructed for limonene:

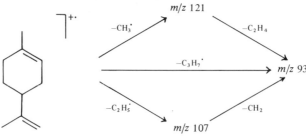

Linked scanning of the electric-sector voltage and magnetic field may be applied to mass spectrometers of conventional or reversed geometries (see below). Automation of the scanning requires a monitor for the strength of the magnetic field and this is usually a Hall-effect probe (section 3.3). This device measures magnetic flux (B) and its output is used to control the scanning of the electric sector voltage to maintain the desired relationship of B to E. Alternatively, control of scanning may be based on a calibration table which correlates scan time with mass during mass calibration (section 3.2).

Figure 8.6. Mass spectra of limonene. (a) Routine mass spectrum recorded on uv-sensitive paper at two different sensitivities, (b) the B/E linked scan spectrum of molecular ions ($C_{10}H_{16}^{+\cdot}$) at m/z 136 and (c) the B^2/E linked scan spectrum of fragment ions ($C_7H_9^+$) at m/z 93. Note that the resolution of spectrum (c) is much lower than that of (b).

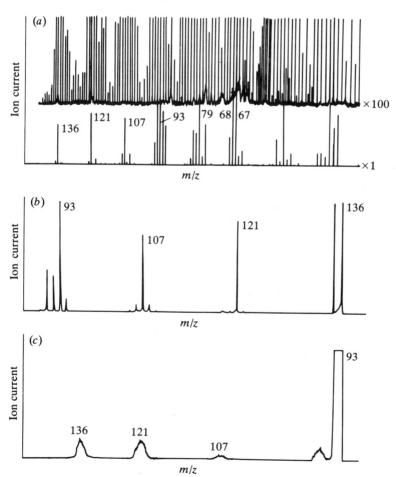

This latter method defines more accurately the magnetic field and leads to more precise mass assignment. With automated systems, it is possible to obtain linked-scan mass spectra in a few seconds, making the technique compatible with GC/MS. Also, a data system may be used to control a conventional scan of a mass spectrometer until a preselected mass peak is detected. At this point, a linked-scan spectrum of that ion is initiated by the computer. When this scan ends, the instrument is programmed to continue the original scan. Hence, a routine mass spectrum and linked-scan spectra of ions of interest in that spectrum may be obtained in one experiment. The alternate use of two different linked-scan modes can achieve accurate mass assignment from the sharp peaks in one mode and analysis of kinetic energy distribution from the broad peaks in the other mode.

Selected metastable ion monitoring in the linked-scan mode is a selective method for detecting and quantifying small amounts of substances (section 6.7).

Characteristics of some selective methods for observing metastable ions are summarized in table 8.2.

8.1.2 Second field-free region

This region extends from the end of the electric sector up to the beginning of the magnetic field. In a single-focussing instrument, it is the space between the ion source and the magnetic field. Because decompositions in the second field-free region occur after the electric sector, there is no discrimination against metastable ions which are therefore mass-analysed in the magnetic field and appear in routine spectra from both single- and double-focussing instruments as broader peaks than those arising from normal ions (section 1.4). Focussing in the magnetic field is affected by both the m/z value of an ion and its translational energy. Therefore, the metastable A^+ ions, which have less translational energy than normal A^+ ions, are deflected differently in the magnetic field. Metastable ions have translational energy $\frac{1}{2}m_2v^2$ = $(m_2/m_1)zV$, so they have the same velocity, $v = \sqrt{(2zv/m_1)}$, as their precursor ions. On passing through the magnetic field, the metastable ions experience a centrifugal force, $Bzv = m_2v^2/r$. Combination of the latter two equations gives the radius of curvature (r) of metastable ions in the magnetic field (equation 8.8).

$$r = ((m_2^2/m_1)(2V/zB^2))^{\frac{1}{2}} \tag{8.8}$$

$$r = ((m_1)(2V/zB^2))^{\frac{1}{2}} \tag{8.9}$$

When equation (8.8) is compared with that derived from equation (2.2) for normal M^+ ions (equation 8.9), it is seen that metastable ions will occur in routine mass spectra at an apparent mass $m^* = m_2^2/m_1$. If a metastable ion is observed at m^* in the routine mass spectrum, it is only necessary to apply

Table 8.2. *Characteristics of scanning modes for observation of metastable ions formed in the first field-free region*

Type of scan	Metastable ions determined	Peak shape	Energy analysis
Defocussing	All precursors of a chosen product ion	Broad	Possible
E^2/V linked scan	All products of a chosen precursor ion	Sharp	Not possible
B/E linked scan	All products of a chosen precursor ion	Sharp	Not possible
B^2/E linked scan	All precursors of a chosen product ion	Broad	Possible
$(B/E)(1-E)^{1/2}$ linked scan	Ejection of common neutral species	Sharp	Not possible

this formula to determine the precursor and product ions from which it arises. For example, the mass spectrum of aniline contains large peaks at m/z 93 and 66, and a large metastable ion peak at m/z 46.8. Since $66^2/93 = 46.8$, it is evident that some of the ions at m/z 66 arise from ions at m/z 93 by ejection of 27 mass units.

If there are many peaks in the mass spectrum, it is time consuming to check manually all the possible precursor and product ions which could give rise to an observed metastable ion, and so tables and nomograms (McLafferty, 1966) are available to speed the checking. Also, it is simple to program a computer to do this checking. It is worth remembering that when a mass spectrum is recorded on uv-sensitive paper which is run through the recorder at a uniform speed, as in figure 11.1 (chapter 11), the m/z values are represented on an exponential scale. This fact, coupled with the formula, m^* $= m_2^2/m_1$, dictates that, on the trace, the distance between the precursor and product mass peaks is the same as that between the product and metastable ion peaks. Therefore, simply by moving a ruler along the trace, potential precursor and product ions are determined. The method is fast and convenient, but not particularly accurate, so that tentative assignments must be checked using the formula. Two factors determine that it is not always possible to assign unambiguously m_1 and m_2. Firstly, because some metastable ion peaks are broad, the precise position of m^* may be difficult to measure. Secondly, because two unknown variables (m_1 and m_2) are needed for the single formula, $m^* = m_2^2/m_1$, there is no unique solution. However, the values of m_1 and m_2 are restricted to the m/z values of observed normal ions, and precursor and product ions are often abundant and therefore obvious. If this is not the case, recourse must be made to other, unambiguous techniques described in this chapter such as linked scanning (see above).

Usually, metastable ion peaks have roughly Gaussian shapes in the mass spectrum, frequently with a 'tail' on the high mass side (figure 8.7a). Sometimes the metastable ions appear as broad, flat peaks extending over several mass units (figure 8.7b, c, d); this arises because, during the actual fragmentation process, some of the internal energy in the decomposing ion appears along the reaction co-ordinate as kinetic energy. Figure 8.7e shows the theoretically expected peak shape which is rarely observed normally because energy spread in the ion beam causes filling of the region between the two sharp peaks. Theoretically expected peak shapes can be observed by using slits to cut down the spread of the ion beam. The energy (T) released in giving rise to the broad metastable ions may be calculated from formula (8.10):

$$T = (dm_1/4m_2^2)^2 \ (zV/\mu) \qquad (8.10)$$

where d is the width of the peak in mass units, m_1 and m_2 are the masses of the precursor and fragment ions, and μ is a reduced term $(m_1 - m_2)/m_2$ (Beynon, Saunders & Williams, 1968; Holmes & Terlouw, 1980).

There is another method of examining metastable ions from the second field-free region. Since metastable ions have less translational energy than normal ions, the two types may be differentiated by having an ion repeller voltage at the detector. Suppose normal ions possess translational energy corresponding to a voltage V, and metastable ions V'. If the repeller voltage (V_r) is greater than V (and therefore V') all ions are turned back and strike a metal plate causing secondary electrons to be emitted (figure 8.8). These

Figure 8.7. Metastable ion peak shapes observed, (a)-(d). Peak (a) is the shape normally found, (b)-(d) are wide, flat peaks, and the theoretically expected shape of one of these flat-topped (broad or dish-shaped) peaks is shown in (e).

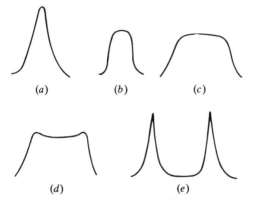

Figure 8.8. Arrangement of scintillator/photomultiplier and ion beam repeller voltage for detection of metastable ions.

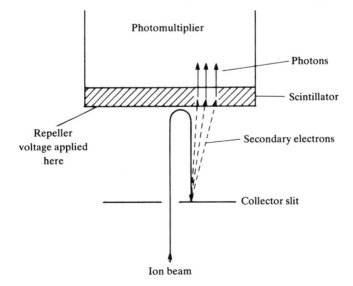

secondary electrons are detected by a scintillator–photomultiplier. If the repeller voltage (V_r) is made less than V but greater than V' then normal ions are not repelled but metastable ones are, again causing emission of secondary electrons. Therefore, the method allows a normal mass spectrum to be obtained at one voltage, $V_r (> V)$, and then, after changing the repeller voltage to $V > V_r > V'$, a second scan records only the metastable ions arising from fragmentations in the second and third field-free regions. Since the potential required just to repel metastable ions may be measured, this system, called the Daly detector, enables analysis of the kinetic energy of such ions. Like the defocussing and linked-scan techniques, it also reveals metastable ion peaks which may be obscured by normal ion peaks in a routine mass spectrum.

8.1.3 Third field-free region

After the end of the magnetic field and before the detector there is a small region (figure 8.1) where decompositions may take place, leading to metastable ions. The region lies after all focussing has occurred so there is no separation of normal and metastable ions which arrive at the collector at the same m/z value, i.e. $m^* = m_1$. However, metastable A^+ ions still possess less translational energy than normal A^+ ions and a detector with a repeller voltage can discriminate between them. The Daly detector (see previous section 8.1.2) is suited to this end and does in fact detect metastable ions formed in either the second or third regions. Metastable ions originating in the third field-free region appear as sharp peaks in this method since no mass separation occurs after the end of the magnetic field.

8.1.4 Other regions

The other places along the flight path of the ions between the source and the detector where metastable ions may be formed lie in magnetic or electric fields and observing them is not simple. At the beginning of the magnetic field (end of second field-free region), the relation $m^* = m_2^2/m_1$ holds but at the end of the magnetic field (beginning of the third field-free region), $m^* = m_1$. Inside the magnetic field, any metastable ions formed will be detected between m_2^2/m_1 and m_1 and this leads to a long tail appearing on the high mass side of the routinely observed metastable ion peak and extending up to the precursor ion peak (figure 8.9). Because metastable ion peaks are, in any case, not large, this diffusion of the peaks into a long tail is only infrequently noticeable at the usual amplification required for recording a normal mass spectrum.

Metastable ions formed in the electric sector are not generally readily observable, but may cause interfering peaks during linked scanning of metastable ions formed in the first field-free region.

8.2 Metastable ions in mass spectrometers of reversed geometry

Double-focussing mass spectrometers of conventional geometry are arranged so that ions pass first through the electric sector and then through the magnetic sector (figure 8.1). Low-resolution and high-resolution mass spectrometry may also be performed with mass spectrometers which have the reverse arrangement of magnetic and electric sectors. A mass spectrometer of reversed geometry is depicted in figure 8.10 and this should be compared and contrasted with figure 8.1. The altered design of figure 8.10 is particularly useful for observing metastable ions formed in the field-free region between

Figure 8.9. Appearance of metastable ion peaks formed within the magnetic sector.

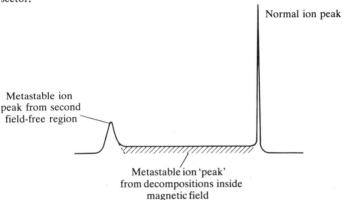

Normal ion peak

Metastable ion peak from second field-free region

Metastable ion 'peak' from decompositions inside magnetic field

Figure 8.10. Path of main ion beam through a mass spectrometer of reversed geometry. The ions pass through the magnetic sector before passing through the electric sector. The field-free region between the sectors is the most important one for studying metastable ions.

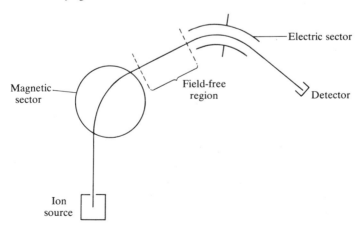

Electric sector

Magnetic sector

Field-free region

Detector

Ion source

the magnetic and electric sectors. The major advantage of the reversed geometry is that the magnetic field may be set to focus only ions of one selected mass into the field-free region. Some of the selected ions may decompose there, giving metastable ions with kinetic energy, $(m^*/m)zV$ where m^* is the mass of the metastable ion and m that of the precursor ion. Therefore, all product ions with different masses, m^*, formed from the chosen precursor, have different kinetic energy and can be separated by the energy-focussing electric sector. With the magnetic field constant to transmit the precursor ion of interest, a single sweep of the voltage of the electric sector identifies all product ions. The technique is called mass-analysed ion kinetic energy spectroscopy (MIKES) since it is in principle the same as ion kinetic energy spectroscopy but obviates the need for a separate mass analysis (section 8.1). The mass peaks obtained are broad due to release of internal energy as kinetic energy during decomposition. Hence, kinetic-energy data are available by analysis of peak shapes, but the limited resolution makes mass measurement difficult. The resolution may be increased somewhat by increasing the accelerating voltage since this increases the kinetic energy of all ions, thereby reducing the effect of small kinetic energy changes caused by fragmentation. To this end, an additional accelerating potential may be applied between the magnetic and electric sectors.

Inasmuch as all product ions of a given precursor are examined, MIKES is equivalent to E^2/V or B/E linked scans with a mass spectrometer of conventional geometry, although the latter techniques give sharp peaks and no kinetic data. Compounds with the same basic skeleton but different side-chains, like steroids or substituted aromatic compounds, are frequently difficult to differentiate from routine mass spectra. With the MIKES technique, it is possible to ignore the ions characteristic of the basic skeleton and hence common to a whole series of compounds, by setting the magnetic field to pass selected ions containing the side-chain. For instance, a series of steroids

$$\xrightarrow{-e} \quad R^+, (R + C_3H_6)^+$$

(3)

(3) with different side-chains (R), when ionized, gives ions R^+ and $(R + C_3H_6)^+$. The metastable ion spectra of either of these types of ion are much more characteristic of differences in structure than are the routine mass spectra. In section 12.2, a further application of MIKES to structural determination is exemplified.

Like the linked-scanning methods discussed in the previous section, MIKES is applied with advantage to the direct analysis of mixtures. The MIKES technique has been suggested as an alternative to combined chromatography/ mass spectrometry. Instead of physically separating components of a mixture by chromatography, the whole mixture is ionized and a single component is examined by setting the magnetic field to transmit ions characteristic of that particular component (frequently the molecular ion). Metastable ion mass spectra then characterize the substance without interference from other components. The method is faster than combined chromatography/mass spectrometry since separation occurs in about 10^{-5} s rather than in minutes or hours for a chromatographic analysis. The sensitivity is as good as that for GC/MS and selected ion monitoring for quantitative analysis is possible with either technique. The MIKES method suffers from several disadvantages. The composition of the ionized mixture may not reflect accurately the composition of the original mixture since some components may suppress the ionization of others. Also, each component generally affords many different ions so that the mass spectrum of the mixture is very complex and several different substances may contribute to the signal at any one m/z value. 'Soft' ionization techniques such as positive-ion and negative-ion chemical ionization, field ionization and field desorption give less fragmentation than does electron impact ionization. A mixture ionized by FD, for instance, is likely to give rise to molecular ions only, so the mass spectrum of the mixture will be relatively simple. A second drawback is that, at any one integer mass, there may be superimposed ions of different elemental composition from different components. The instrument shown in figure 8.10 is not capable of separating such mixtures of ions because they are not energy-focussed before passing through the magnetic field, but instruments which allow separation are available (section 8.4). Despite these complications, many mixtures have been analysed by MIKES. Direct ionization of plant tissues followed by MIKES has allowed the identification of natural alkaloids. Barbiturates in crude drug preparations have also been examined successfully and notable results have been obtained from pyrolysing biological samples in the ion source of a mass spectrometer. For example, the pyrolysis of deoxyribonucleic acids produces a mixture of purine and pyrimidine bases, each of which may be characterized rapidly by MIKES. Any modified or unusual bases may also be identified by the same method.

Mass-analysed ion kinetic energy spectroscopy and related methods have been described extensively in many recent reviews (see, for example, Beynon, Morgan & Brenton, 1979; McLafferty, 1980; Russell, McBay & Mueller, 1980).

8.3.   Triple quadrupole analysers

In ordinary quadrupole mass filters, metastable ions cannot be studied because they are not distinguished from normal product ions. However, when quadrupole mass filters are linked together, fragmentation occurring between analysers gives rise to observable metastable ions. An instrument embodying this principle is shown in block form in figure 8.11, its capabilities for studying metastable ions being roughly the same as those of the reversed-geometry, magnetic-sector mass spectrometer. The first analyser is used to transmit selected ions whilst the second is not used for mass separation; it is a region in which dissociations may occur. The final mass filter is scanned to detect in turn all metastable ions. For instance, for an alcohol or mixture of alcohols, all ions eliminating water may be detected selectively, regardless of their masses, by scanning the instrument such that when the first analyser transmits ions of mass m, the second is set to transmit ions of mass $(m - 18)$. This method is far simpler than the complex scan required to record a constant neutral mass spectrum with a magnetic sector instrument. The properties of quadrupole mass filters that make them particularly compatible with computers and useful for combined chromatography/mass spectrometry and chemical ionization (sections 2.3 and 2.4) also favour the use of triple quadrupole analysers for studying metastable ions. Further advantages over magnetic-sector instruments are outlined below.

8.4. Triple-focussing magnetic-sector mass spectrometers

Mass-analysed ion kinetic energy spectroscopy, as described above, is hampered by the inability to separate ions with the same integer mass but different elemental compositions. If an additional electric sector is placed between the ion source and magnetic sector of an instrument of reserved geometry, high-resolution MIKES is possible. The design of the mass spectrometer is shown in figure 8.12; it can be considered as a conventional high-resolution mass spectrometer (cf. figure 8.1) up to the field-free region of interest, after which is placed an electric sector for analysis of metastable ions by ion kinetic energy spectroscopy. During an analysis of a mixture of N-methylaniline and methylphenols with the instrument shown in figure 8.10, the selected precursor ions at m/z 107 (schemes 8.11 and 8.12) fragmented in the field-free region to give a complex mixture of metastable ions indicating more than one decomposing entity. Operating at a resolution of 15 000, a triple-focussing mass spectrometer separated $C_7H_7O^+$ ions with mass

Figure 8.11. Schematic diagram of the triple quadrupole analyser showing path of main ion beam.

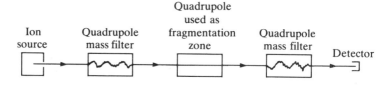

107.0497 and $C_7H_9N^{+\cdot}$ ions with mass 107.0735. The different precursor ions were then characterized independently through their metastable ion mass spectra. In the field-free region after the magnetic sector, $C_7H_7O^+$ ions were observed to dissociate by loss of CO and H_2CO (reaction 8.11) and $C_7H_9N^{+\cdot}$ ions by loss of HCN and CH_3N (reaction 8.12). The decompositions (8.11) and (8.12) are typical of phenols and aromatic amines, respectively (see

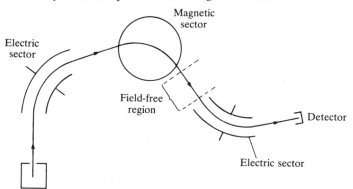

$$(8.11)$$

$$(8.12)$$

chapter 10), and the components were readily distinguished and identified. The increase in specificity attendant upon high resolution makes feasible analysis of complex mixtures. For example, the steroid glycoside digitoxin may be determined in blood or urine by direct analysis.

8.5   Collisional activation

When ions possessing high translational energy collide with neutral atoms or molecules, a small fraction of their translational energy is converted into internal energy. The ensuing increase in ro-vibronic energy may cause

Figure 8.12. Path of main ion beam through a triple-focussing mass spectrometer. Only the most important field-free region is indicated.

decomposition. Therefore, in studies of metastable ions, it is beneficial to introduce a gas (often helium or argon) into the region where metastable ions are formed. The gas is 'leaked' into a small *collision cell* which can be maintained at a relatively high pressure whilst the regions around it are still at very low pressure. The collision cell is placed in the flight path of precursor ions and in the field-free region of interest. Collision-induced dissociation of precursor ions within the collision cell may be usefully employed in any of the techniques discussed in the previous sections. As much as 10 eV of energy may be added to the internal energy of an ion during collision. Since electron impact ionization imparts similar amounts of energy to molecular ions, the metastable ion mass spectra of collisionally activated molecular ions are often similar to routine EI spectra. Also, the large amount of energy transferred on collision makes the subsequent fragmentation virtually independent of the original internal energy of the precursor ion. Therefore, ion abundances observed in collisional activation mass spectra are reproducible and largely independent of the mode of ionization.

Metastable ions observed in the absence of collisional activation increase in abundance if a high pressure is created in the appropriate field-free region. Also, the number of different types of metastable ions will increase as some fragmentation pathways are only observed when collisional activation is used. If the pressure in the collision cell is increased too much, ions will be scattered on collision and few of them will reach the detector. Therefore, there is an optimum pressure, usually 10^{-6}-10^{-4} torr, for best sensitivity. Because an ion undergoes several ion/molecule collisions during flight through the high pressure region, consecutive decompositions may occur and metastable ions will be seen for the composite process.

Ions generated in a quadrupole mass filter are not accelerated by a high voltage so they possess much less translational energy than ions accelerated in a magnetic sector instrument. Collisional activation of ions of relatively low translational energy is more efficient than that of high-energy ions, so the triple quadrupole analyser can be very sensitive. Nitrophenol may be determined at a level of 10 pg with a reversed-geometry, magnetic-sector mass spectrometer whilst 120 fg can be detected with the triple quadrupole analyser. The reason for this difference in efficiency with translational energy is not entirely clear, but a number of factors is probably responsible. For example, there are more competing reactions available to ions of higher translational energy when they undergo collisions. Positively and singly charged precursor ions may afford unwanted doubly or negatively charged product ions without a change in mass, as well as the required singly charged product ions of lower mass. On the other hand, observation of so-called charge permutation processes, such as those shown in scheme (8.13) where B is a molecule of the neutral collision gas, can lead to valuable analytical information for elucidating ion structures.

$$M^+ + B \rightarrow M^{2+} + B + e$$

$$M^+ + B \rightarrow M^- + B^{2+} \tag{8.13}$$

The power of collisional activation is such that other techniques, such as mass-analysed ion kinetic energy spectroscopy, are rarely used without it. Collisional activation is highly useful when used in conjunction with a 'soft' ionization technique (chapter 7). Field desorption and negative ion chemical ionization particularly are likely to yield only molecular or quasi-molecular ions. This makes it possible to assign molecular weight, but little information concerning the structure of an unknown compound is gained. In such cases, once the molecular or quasi-molecular ions are recognized, fragmentation may be induced by collision. In one experiment, data with the combined advantages of soft ionization, EI-like fragmentation and metastable ion mass spectra are available. Electron impact ionization of derivatized peptides yields useful ions indicative of structure, but molecular ions are of low abundance or absent. If, instead, peptides are ionized by positive-ion or negative-ion chemical ionization, quasi-molecular ions, $(M + H)^+$ or $(M - H)^-$, are observed and may be induced to fragment by collision. The resulting metastable ions, recorded by MIKES with a mass spectrometer of reversed geometry or by B/E linked scanning with a conventional instrument, provide the information required for complete structural analysis. The same approach has been used for characterization of saccharides and polynucleotides ionized by field desorption.

Collisional activation is useful also in studies of ion structures. Collisional activation mass spectra of $C_4H_6D_2{}^{35}Cl^+$ ions from the two differently labelled 1-chlorohexanes (4) and (5) are identical, suggesting that both neutral compounds afford a common ion. The formation of a chloronium ion, often

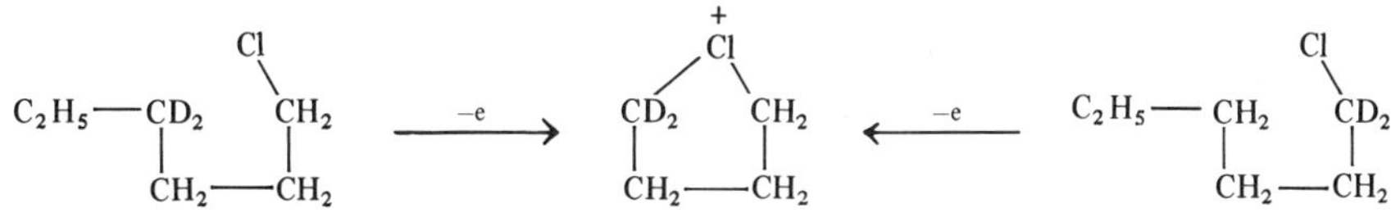

(4)   (5)

postulated for reactions in solution chemistry, would be consistent with these results. Unfortunately, ions can rearrange during collision with another body so that such studies cannot give unequivocal evidence of ion structure.

8.6 Summary and miscellany

Normal fragment ions in a routine mass spectrum are products of reactions, some of which may be complex. The precursor ions of any given product ion are assigned by educated conjecture since normal ions provide no information as to their origin. The observation of metastable ions removes much of the guesswork since metastable ions define reaction pathways. In an unknown complex mass spectrum, the additional information gained by examining metastable ions is invaluable for spectral interpretation. Consider an unknown molecule, M, giving molecular ions, $M^{+\cdot}$ and fragment ions at m/z values less than that of $M^{+\cdot}$ by 15 and 42 mass units. The $(M - 15)^+$ ion will be due almost certainly to elimination of a methyl radical from $M^{+\cdot}$ but the character of the $(M - 42)^+$ ion is less easy to rationalize. It could be due to loss of 42 mass units (e.g. C_2H_2O, C_3H_6) from $M^{+\cdot}$ or of 27 mass units (e.g. HCN) from $(M - 15)^+$. The structural inferences in each case are rather different, a loss of ketene (CH_2CO) being characteristic of the acetyl group and propene of a propyl group, whilst elimination of hydrogen cyanide is indicative of an aromatic amine or nitrile. Therefore, it is important to distinguish between $M^{+\cdot} \to (M - 42)^{+\cdot}$ and $(M - 15)^+ \to (M - 42)^+$. If there were a metastable ion for the latter reaction, then this would be consistent with elimination of HCN and there would be no basis for postulating an acetyl or propyl group. If such a metastable ion were not observed, it does *not* necessarily rule out the reaction $(M - 15)^+ \to (M - 42)^+$. Conversely, if a metastable ion is observed for $M^{+\cdot} \to (M - 42)^{+\cdot}$, it usually, but not always, implies that a single reaction has occurred. Rapid successive loss of $CH_3^\cdot$ and HCN in the same field-free region could afford a metastable ion for the composite process. When metastable ion data are combined with accurate mass measurement of the normal ions, an unambiguous scheme can often be obtained (Rose, 1981).

For compounds with subtle differences in structure, such as stereoisomers, and positional and geometric isomers, the vast majority of peaks in their routine mass spectra may be common and obscure any small differences that would have distinguished them. Selective detection of metastable ions is more sensitive to fine differences in structure because it allows the mass spectromist to ignore the common ions and observe in isolation the behaviour of those ions most likely to differentiate similar structures. This principle has been discussed with reference to the analysis of isomeric carotenoid compounds by linked scanning (Rose, 1982).

The introduction of a collision gas into the field-free region of interest increases the abundance of metastable ions. Alternatively, ions may be excited by focussing a laser beam into the region. The laser method has the advantage over collisional activation that by changing the wavelength of the photons, exactly known amounts of energy may be used to induce *photodissociation* of precursor ions (section 9.7). As a result, the amount of internal energy that is released as kinetic energy during photodissociation may be determined very accurately by ion kinetic energy spectroscopy.

Care must be taken in assigning the masses of precursor and product ions from metastable ion data, especially when the method of recording mass spectra gives rise to broad peaks. In routine mass spectra, use of the equation $m^* = m_2^2/m_1$ may not lead to unambiguous assignments of m_1 and m_2.

When scanning to detect metastable ions in the absence of normal ions, there may be several complicating factors. During studies of dissociations in one particular region of a mass spectrometer, metastable ions formed in another (often within the electric sector) may be focussed accidently, affording interfering peaks. Metastable ions formed in one field-free region may have sufficient energy to fragment in the next field-free region. Such consecutive reactions, when studied purposely, can provide useful information on fragmentation pathways. Analysis of components of low molecular weight in complex mixtures by mass-analysed ion kinetic energy spectroscopy is hazardous because components of higher molecular weight may give rise to interfering peaks. An advantage of the triple quadrupole analyser is that it is less prone to spurious signals.

Metastable ions are not just useful for characterizing inorganic and organic compounds (particularly natural products). The ability to study the reactions of relatively long-lived (low-energy) ions and the energetics of such processes permits fundamental research on the structures of ions. Further information on the topic of metastable ions may be obtained from more specialized literature (Jennings, 1971; Cooks, Beynon, Caprioli & Lester, 1973; Beynon & Caprioli, 1980; Holmes & Terlouw, 1980).

9 Theory of mass spectrometry

9.1. Introduction

A thorough understanding of mass spectrometric processes is severely hampered through poor knowledge of the structures and electronic states of ions. The fragmentation of ions depends on the excess of internal energy they contain and this is governed mostly by the thermal energy in the molecule immediately prior to ionization, by the energy gained during ionization and by the subsequent environment of the ion which determines whether or not it collides with molecules. Despite inadequate knowledge of ion structures, many generalizations may be made which are useful for understanding mass spectrometric fragmentations of compounds.

9.2. Energy states resulting from ionization

Consider a diatomic molecule M ionized to yield a molecular ion $M^{+\cdot}$. The vibrational energies of both the molecule and the molecular ion may be represented in the usual way by Morse curves as shown in figure 9.1. For the moment, it is only necessary to consider the zeroth vibrational energy level for M because, even at 500 °K, the higher vibrational levels are sparsely populated. However, it is necessary to consider the vibrational energy levels, $\nu = 0, 1, 2, \ldots$ in the ion $M^{+\cdot}$. If ionization is very fast, then by the Franck–Condon principle, there will be a vertical transition between M and $M^{+\cdot}$. The minima of the two Morse curves for M and $M^{+\cdot}$ are unlikely to coincide so that a vertical transition will leave $M^{+\cdot}$ in a vibrational state above the zeroth level. For example, transition a in figure 9.1 gives the ion in the vibrational state $\nu = 4$ and therefore the ion has an excess of internal energy. A vertical transition may occur to a vibrational level in $M^{+\cdot}$ above the dissociation limit and in this case the ion will contain energy in excess of the dissociation energy (D). Accordingly, $M^{+\cdot}$ will fragment as soon as it begins to vibrate. It is also possible that the ground state of the ion $M^{+\cdot}$ is completely dissociative (transition b, figure 9.1) so that decomposition of the ion follows immediately after its formation.

From a number of molecules M, ions $M^{+\cdot}$ are produced with different amounts of internal energy proportional to the transition probabilities between M and $M^{+\cdot}$. Also, some 'hot' molecules (M) are ionized from vibrational levels above the zeroth level and the ions formed will again contain different amounts of internal energy. It is important to remember that this distribution of energies is not a 'Maxwell–Boltzmann' distribution since the latter applies to collisionally equilibrated species, and in mass spectrometry the ions are not so equilibrated because ion/ion and ion/molecule collisions are rare, except during chemical ionization and collisional activation, which will both be described later. Each ion behaves as a separate entity containing its own specific amount of energy which may be sufficient to induce fragmentation.

In complex molecules, the Morse curves must be replaced by potential energy surfaces, but similar principles to those used in the discussion of diatomic molecules may be applied. For a species that has various vibrational modes, each mode will be represented by a potential energy surface. If the

Figure 9.1. Morse curves for a diatomic molecule (M) ionized to either a non-dissociative ($M_n^{+\cdot}$) or dissociative state ($M_d^{+\cdot}$) of the ground-state ion by vertical transitions a and b. D is the dissociation energy.

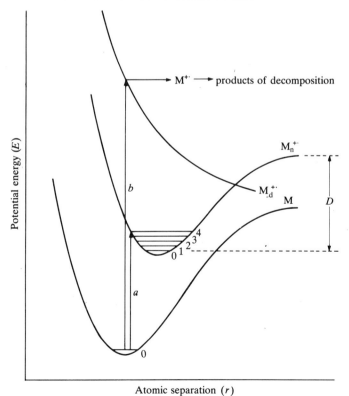

Atomic separation (r)

surfaces cross, a radiationless transfer of energy can occur. Consider the two vibrational states $M_1^{+\cdot}$ and $M_2^{+\cdot}$ of the ion $M^{+\cdot}$ in figure 9.2 which cross at some point X.

Ionization by vertical transition from the $\nu = 0$ level of the molecule to the $\nu = 2$ level of the ion state $M_1^{+\cdot}$ can occur, but the two states $M_1^{+\cdot}$ and $M_2^{+\cdot}$ coincide at X and crossing from state $M_1^{+\cdot}$ to state $M_2^{+\cdot}$ may take place. In state $M_2^{+\cdot}$, the internal energy transferred from $M_1^{+\cdot}$ may be above the dissociation energy so that the ion fragments. In complex molecules there are many such energy surfaces by which transfer of energy may occur and so, after ionization, equilibration of the excess of internal energy in the ion can take place rapidly by these surface crossing processes. The presence of closely spaced rotational energy levels associated with the vibrational levels means that, in polyatomic molecules, surface crossing becomes easy and some vibrational energy is converted into rotational energy and vice versa. However,

Figure 9.2. Ionizing transition from molecule M to ion state $M_1^{+\cdot}$ followed by radiationless transition at X to ion state $M_2^{+\cdot}$.

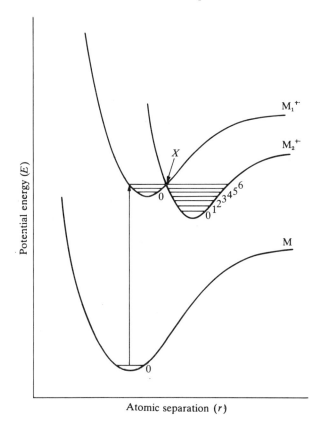

Atomic separation (r)

despite such energy transfer and equilibration into all the oscillators in the
ion, there may be insufficient excess of energy to cause fragmentation and
this results in a stable ion. Various complications to this basic approach to
ionization and fragmentation must be considered.

The only ionization discussed so far corresponds to removal of the most
loosely bound electron to give a vibrationally excited ion in its electronic
ground state. At ionizing energies greater than the first ionization energy it is
possible to remove electrons from lower orbitals to give electronically excited
ions. Figure 9.3 shows the electron configuration of a ground-state ion
(removal of an electronic from orbital ψ_1) and an excited-state ion (removal
of an electron from ψ_3). Figure 9.4 represents the formation of these ions
from the molecule by vertical ionizing transitions. In figure 9.4, the curve $M^{+\cdot}$
represents the ground-state ion and $\overset{*}{M}{}^{+\cdot}$ the electronically excited ion. The
excited ion may lose or redistribute its energy by two main processes: (i)
internal electron redistribution with concomitant emission of radiation to
give a ground-state ion and (ii) radiationless potential surface crossings, as in
figure 9.2, whereby the electronic energy in the excited-state ion can be

Figure 9.3. Orbital levels showing ionization to ground-state molecular ion
($M^{+\cdot}$) and an electronically excited-state ion ($\overset{*}{M}{}^{+\cdot}$).

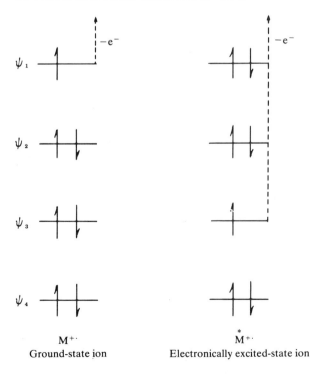

M$^{+\cdot}$ $\overset{*}{M}{}^{+\cdot}$
Ground-state ion   Electronically excited-state ion

transformed into vibrational energy in the ground-state ion. Thus, either by radiation of energy or by radiationless transfer of energy, the excited-state ion can revert eventually to a ground-state ion with an excess of vibrational energy. There are many possible excited states of ions which may be reached on ionization. Energy conversion by surface crossing does not necessarily lead simply to a ground-state ion since, if there is sufficient energy to cause decomposition from any excited state, then decomposition will occur from that state. Because the time for a bond vibration (*c.* 10^{-13} s) is short compared with the maximum lifetime of an excited ion (*c.* 10^{-8} s), many bond vibrations occur with the possibility of radiationless state crossing within this lifetime. The lifetime of an excited-state ion (*c.* 10^{-8} s) is considerably shorter than the time (*c.* 10^{-5}–10^{-6} s) an ion spends in the ion source and in flight to the detector. Therefore, there is ample time for the excess of electronic energy in

Figure 9.4. Ionizing transitions from the molecule M to the ground-state ion $M^{+\cdot}$ and to an excited-state ion $\overset{*}{M}{}^{+\cdot}$, corresponding to the electron configurations shown in figure 9.3.

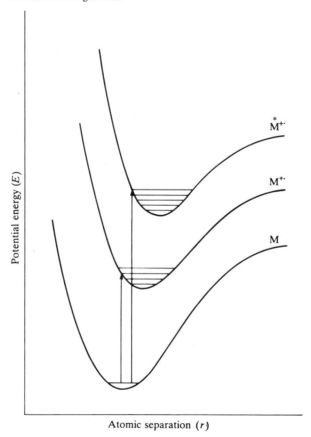

Atomic separation (*r*)

an ion to be converted into an excess of vibrational energy in a lower electronically excited state. As a general working hypothesis in mass spectrometry, it is usually supposed that this less excited state from which most reactions occur is the ground state of the ion. The process may be represented schematically as follows for molecules M interacting with ionizing radiation and fragmenting to yield ions found in the mass spectrum:

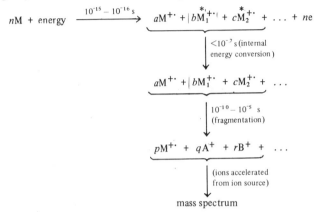

$\overset{*}{M_1}{}^{+\cdot}, \overset{*}{M_2}{}^{+\cdot}, \ldots$ represent electronically excited molecular ions; $M^{+\cdot}$, $M_1^{+\cdot}$, $M_2^{+\cdot}$, $\ldots$ vibrationally excited molecular ions; and A^+, B^+, $\ldots$ fragment ions. The above scheme must be modified slightly for field ionization in which the ions spend a very much shorter time (10^{-11}–10^{-9} s) in the ion source.

Removal of one electron from a molecule leaves an ion in the doublet state (one unpaired spin, so that $2S + 1 = 2$ where S is total spin). During ionization by electron impact, only the total spin of the system (electron plus molecule) needs to be conserved so that an ion may also be formed in its quadruplet state (three unpaired spins). Just as triplet states have longer lifetimes than singlet states, it is likely that quadruplet states have longer lifetimes than doublet states. Hence, fragmentation may take place from relatively long-lived excited-state ions.

Broadly, a picture emerges of ionization producing molecular ions in different energy states, rapid internal energy conversion to produce ions with individual amounts of excess of energy, and fragmentation taking place at different rates depending on the electronic state of and excess of vibrational energy in each ion. Within each ion, the excess of internal energy is considered to be distributed statistically amongst all the vibrational and rotational modes. The distribution of internal energies amongst the ions may be represented as in figure 9.5. Some ions have insufficient energy to fragment (figure 9.5) and the proportion of them increases as the energy of the ionizing radiation decreases to the ionization energy. For this reason, reducing the electron beam energy in an electron impact source reduces the amount of fragmentation observed in the mass spectrum. It is possible to consider a mean or average effect when discussing the total behaviour of many such ions, but this situation of isolated species reacting independently of one another is quite unlike the mean effect resulting from collisional processes in solution or normal gas-phase chemistry. In solution chemistry, collision processes leave molecules with sufficient energy to react or decompose. In mass spectrometry, the ion initially produced does not normally collide with other species and usually decomposes endothermally if sufficient energy is transferred during ionization. As remarked earlier, these considerations do not hold during chemical ionization and collisional activation.

9.3.   Formation of ions

9.3.1   Electron-impact ionization

Electrons accelerated through a potential of V volts have a de Broglie wavelength of $\sqrt{(1.5/V)}$ nm, e.g. 75-V electrons have a wavelength of 0.14 nm corresponding to short-wavelength radiation. During the approach of an electron to a molecule, the impacting electron waves and the electric field of the molecule mutually distort one another. The distorted electron wave can be considered to be composed of many different sine waves and some of these component waves will be of the correct frequency (energy) to interact

Figure 9.5. Distribution of internal energies in ions. The true shape of the curve would be governed by the cross-sections for all the various ionizing transitions. The hatched portion indicates the numbers of ions with insufficient energy to fragment.

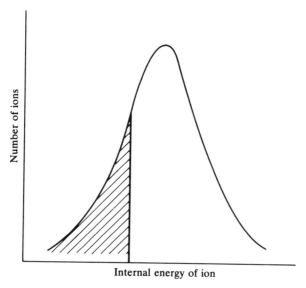

Internal energy of ion

with molecular electrons. Thus, the impact of an electron may lead to electronic excitation in the molecule by promoting an electron from a lower to a higher orbital (this is the same effect as in ultraviolet spectroscopy). Similarly, a molecular electron may be promoted to an outer orbital (compare this with vacuum ultraviolet Rydberg bands) or an electron may be ejected from the molecule altogether to leave a positive ion. Direct attachment of an electron to a molecule to give a stable radical-anion is of low probability. The translational energy of an electron attaching itself to a molecule must be taken up as excess of internal energy in the new radical-anion. Usually, this excess of energy leads either to the electron being 'shaken' off again or to fragmentation of the radical-anion. To form stable radical-anions by direct electron attachment, it is necessary to have sufficient gas pressure for a 'third-body' collision to occur and remove suffient excess of energy that the radical-anion does not fragment. Formation of negative ions is discussed in greater detail in section 7.3. Other electronic effects in the molecule due to electron impact are also observed but there is insufficient space to deal with them here.

The common term 'electron impact' is misleading because an electron is so small in molecular terms that it would have difficulty 'hitting' any part of a molecule it met. It is better to think of the electron as passing close to or even through a molecule rather than of any 'impact' taking place. The effects of the electron on the molecule are felt as it approaches and there is some distortion of the molecule during the transition to an ion, i.e. the ionizing transitions are not strictly vertical and allow more states to be reached than by photoionization. Also, relative to photoionization, more spin states of the ion are accessible by electron-impact ionization. Therefore, in comparison with both field ionization and photoionization, ionization by electrons results in a greater excess of energy in the molecular ions and fragmentation is enhanced. By reducing the electron energy well below the conventional 70 eV, much less excess of internal energy is left in the ion following ionization so that fragmentation is greatly reduced and electron impact spectra become more comparable with photoionization and field ionization spectra.

9.3.2.   Photoionization

The wavelength of radiation required for photoionization is about 100 nm which is much greater than normal molecular dimensions and means that a molecule is subjected to a uniform electric field. In quantum mechanical terms, only the effect of the radiation on the molecule need be considered, unlike the case of electron impact when the wavelength of the radiation and molecular dimensions are similar and mutual quantum effects must be considered. Because the electric field of the photons is essentially uniform over the whole molecule, almost no molecular distortion occurs during ionization. The ionizing transitions are very fast (10^{-16} s) and are vertical in accord with the Franck-Condon principle. Also, because of spin conservation from molecule to ion, fewer states are accessible than by electron-impact ionization. In comparison with the latter, somewhat less vibrational energy is deposited in a molecular ion after photon impact and, like field ionization spectra, photo-ionization spectra yield abundant molecular ions and relatively few fragment ions. Provided the energy ($h\nu$) of the incident photons is known, the excess of energy ($\overset{*}{E}$) deposited in the ion during ionization is given exactly by the expression $\overset{*}{E} = h\nu - I$, where I is the ionization energy. Thus, it is possible to vary the excess of energy in the molecular ion in a controlled way by closely controlling the energy of the incident photons. This technique is used for investigations of the kinetics of ion decomposition.

9.3.3. *Chemical ionization*

This involves the collision of two bodies, an ion and a molecule, which are massive compared with an electron. The collisional interaction is slower, allowing equilibration of energy between the two colliding species. The ions are formed by electron-impact ionization of a *reactant gas* at relatively high pressures in the ion source. If the reactant gas ions, A^+, collide with molecules, B, of the sample under investigation, a chemical reaction can occur in which B is ionized – hence the term 'chemical ionization'. Usually, the reaction entails addition or subtraction of charged entities to or from molecules B to give 'quasi-molecular ions'. For example, if the reactant gas contains H_3^+ ions, reaction can occur with B as shown in equation (9.1), in which proton transfer occurs.

$$H_3^+ + B \to H_2 + (B + H)^+ \tag{9.1}$$

Chemical ionization is described in greater detail in section 7.2. By suitable choice of reactant gas, negative ions can be produced efficiently (section 7.3).

9.3.4. *Field ionization*

If a sharp edge or fine wire is maintained at a high electrical potential, the small radius of curvature at the edge or along the wire produces an intense electric field. In such a field, the electric field (molecular orbitals) of a molecule is distorted and the potential barrier to transfer of an electron from the molecule to the edge or wire is reduced considerably. Under these conditions, quantum tunnelling of an electron occurs, converting the molecule into a radical–cation which is promptly repelled by the high positive electric potential. By use of a negative potential, negative ions can be formed similarly. Field ionization is discussed more fully in section 7.5.

9.3.5. *Charge exchange*

During collision between two molecules or an ion and a molecule, charge exchange (charge transfer) may occur as illustrated by equation (9.2) in which neutral caesium atoms and phosphorus trichloride molecules afford positive and negative ions, respectively.

$$Cs + PCl_3 \rightarrow Cs^+ + PCl_3^- \qquad (9.2)$$

Charge exchange can be utilized by forming ions in the ion source of a mass spectrometer and directing them through a region containing neutral gas. Ion/molecule collisions occur and, in suitable cases, charge transfer produces new ions as exemplified in equation (9.3) in which initially formed I^- ions yield $ICl^{-\cdot}$ ions after collision with neutral ICl.

$$I^- + ICl \rightarrow I^\cdot + ICl^{-\cdot} \qquad (9.3)$$

Ions formed by these methods are frequently difficult to produce by direct ionization.

9.3.6. Ion structures

It is essential to remember that although the structure of a molecule put into a mass spectrometer may, at least in principle, be determined by other physical methods (e.g. X-ray analysis), after ionization it is not necessarily correct to assume that the ion has the same structure as the orginal molecule. Immediately after the ionization step, the original molecule continues to vibrate as an ion and this vibration, together with the different content and distribution of electrons in the ion compared with the orginal molecule, may lead rapidly to a structure quite different from that of the original molecule. There is ample experimental evidence to suggest that frequently rearrangement does occur rapidly after ionization, hydrogen rearrangement being particularly easy.

9.4. Theories of fragmentation rates

A quantitative treatment of mass spectrometric fragmentation has been attempted using mathematical expressions which statistically distribute the available excess of internal energy in the newly formed ion amongst all its oscillators and rotators. Use of this quasi-equilibrium theory (QET) allows the rate of decomposition of the ion through cleavage of any one particular bond to be calculated for any given excess of internal energy (Rosenstock, Wallenstein, Wahrhaftig & Eyring, 1952; Vestal, Wahrhaftig & Johnston, 1962). An almost identical approach is used in RRKM theory (Ramsberger, Rice, Kassel and Marcus; Marcus, 1952), first applied to calculate chemical reaction rates from the distribution of energy amongst the oscillators and rotators in the transition state of a chemical reaction. Such calculations are particularly applicable to ion/molecule collisions where the collision complex can be approximated to a transition state. As a further elaboration, consideration of phase space, the six-dimensional space necessary to describe the positions (x, y, z) and momenta (mv_x, mv_y, mv_z) of the colliding species and/or fragments from decomposition, gives a total description of the distribution of energy into various reaction (decomposition) channels

(Light, 1967). For all but very simple molecules, these mathematical approaches are too time-consuming to be evaluated without the aid of a computer.

These theories have been used successfully to predict the mass spectra of some alkanes and other simple compounds. Rates of fragmentation calculated from RRKM theory with allowance for phase space are used to compare predicted and observed reaction rates in ion/molecule reactions (Safron, Weinstein, Herschblach & Tully, 1972; Marcus, 1975; Klots, 1976; Chesnavich & Bowers, 1978). Some of the success achieved may only be apparent because, in applying the theories, it is necessary to assume that the structure of an ion is known. Frequently, the structure of a molecular ion is equated with that of the molecule from which it came. Similarly, 'best guesses' are made about the structures of fragment ions. Because of such assumptions, there is reservation about the predicted rates of fragmentation even when these agree with observed rates; where there is disagreement, it is uncertain whether this is due to unwarranted assumptions about the application of the theory or about the ion structures. Nevertheless, as knowledge of ion structures improves, comparison of predicted and observed rates of fragmentation is leading to a better understanding of the finer details of fragmentation processes and these refined treatments are becoming more and more important.

There is a simplified version of the QET equations which is useful for discussing mass spectra in a semi-quantitative fashion and can be used with complex molecules without the need of a computer. The simple expression giving the rate of decomposition (k) of an ion is shown in equation (9.4):

$$k = \nu \, ((E - E_0)/E)^{N-1} \tag{9.4}$$

where ν is a frequency factor, E the excess of internal energy in the ion, E_0 the energy of activation for the decomposition, and N the number of oscillators in the ion. Effectively, the frequency factor is a measure of the entropy factor in the decomposition. For example, for a simple bond cleavage it is put equal to the bond-vibration frequency which may be obtained from an infrared spectrum. For a rearrangment reaction, a particular spatial disposition of atoms is probably required and it may be expected that ν would be at a lower frequency than for simple bond cleavage (Field & Franklin, 1957). The number of oscillators in a molecule containing n atoms is given by the expression $(3n - 6)$; often it is found that not all the oscillators appear to be equally effective, in which case a calculated value for k is arrived at which is too large, and the number of oscillators (N) is usually divided empirically by three for better results. The part of the expression (9.4) containing the energy terms, $(E - E_0)/E$, is a dimensionless quantity and energy can be considered in any convenient units, e.g. electron volts.

The simplified rate equation (9.4) has been used to discuss the fragmentation of molecules in a general sense. As simple examples of its use, the following may be mentioned: the greater the energy transferred during ionization, i.e. as $(E - E_0)/E$ increases, the faster an ion decomposes; for two alternative fragmentations of an ion that have similar activation energies, the rates are proportional to the frequency factors, i.e. $k_1/k_2 = \nu_1/\nu_2$; if the frequency factors are similar, the rates of fragmentation are approximately proportional to the activation energies; comparison of the rate for a simple bond-fission reaction with the rate for fission accompanying rearrangement suggests that the rearrangement process may become more prominent as the internal energy of the ion decreases (section 10.9). It should be noted that these rates (k) refer to individual ions with internal energy (E) and that the rates are not equivalent to the rates found in solution chemistry which are the result of an equilibration process amongst millions of molecules. However, it is possible to consider in a qualitative sense the behaviour of an ion having an 'average' excess of energy and extrapolate its behaviour to all the other ions of the same structure but slightly different internal energies. In this way the simple QET equation (9.4) can be used to illustrate some general principles and it is used in this way in the later discussion of ion-rearrangment processes (section 10.9). The approximate equation (9.4) is very inaccurate at low internal energies (E) and should not be used in these regions.

When the simplified QET equation is used for a process such as $A^+ \rightarrow B^+$, subsequent fragmentation of B^+ is sometimes ignored. Particularly at higher internal energies, these further fragmentations are important and the abundance of ion B^+ (or A^+) cannot be simply related to its rate of formation (or decomposition). As stressed in section 1.1, the abundance of an ion in a mass spectrum depends upon both its rate of formation and its rate of decomposition, and care should be exercised in the use of the QET equation if observed ion abundances are to be related to calculated rates of formation or decomposition. Abundances of ions considered theoretically in this way are unlikely to reflect very closely the observed abundances unless some account is taken also of the distribution of excess of energy amongst the ions. The actual distribution of excess of energy may be such that the use of an 'average' excess, as suggested above, would lead to gross differences between theory and practice. The actual distribution (an *energy deposition function*) is difficult to determine experimentally and often a very simplified function (delta or square) is chosen for calculations using the QET or RRKM equations. Where energy deposition functions approximate the true distribution obtained experimentally (as from photoelectron or photoion spectra), agreement or contrast between predicted and observed rates of fragmentation is more significant.

9.5. Thermochemical arguments

Measurement of the minimum energy required to ionize a molecule (*ionization energy*, *I*) or to cause the appearance of fragment ions (*appearance energy*, A) provides valuable thermochemical data. For a molecule, M, giving first a molecular ion, $M^{+\cdot}$, and then a fragment ion, F^+, with ejection of a neutral species, N, the ionization and appearance energy are related to heats of formation (ΔH_f) as shown in equation (9.5).

$$M \to M^{+\cdot} + e; \qquad\qquad I = \Delta H_f(M^{+\cdot}) + \Delta H_f(M)$$

$$M \to M^{+\cdot} + e \to F^+ + N + e; \qquad A = \Delta H_f(F^+) + \Delta H_f(N) - \Delta H_f(M)$$

$$\therefore A = I + \Delta H_f(F^+) + \Delta H_f(N) - \Delta H_f(M^{+\cdot})$$

$$(9.5)$$

Thus, measurement of I or A can yield values for heats of formation of ions or neutral species, providing the remaining data are available. There are uncertainties about some of the values obtained. These uncertainties stem both from difficulties in the measurements themselves and from relatively poor knowledge of ion structures (Howe, 1973). Conversely, knowledge of ion structures can be improved by using thermochemical data. Consider the simple example shown in equation (9.6) in which two different molecules, M_1 and M_2, fragment to produce ions, F^+, of the same composition.

$$M_1 \to F^+ + N_1 + e^-; \qquad \text{appearance energy, } A_1$$

$$M_2 \to F^+ + N_2 + e^-; \qquad \text{appearance energy, } A_2$$

$$\Delta H_f(F^+) = A_1 + \Delta H_f(M_1) - \Delta H_f(N_1)$$

$$\Delta H_f(F^+) = A_2 + \Delta H_f(M_2) - \Delta H_f(N_2) \qquad\qquad (9.6)$$

By measurement of A_1 and A_2 and insertion of heats of formation for M_1, M_2, N_1 and N_2, the heat of formation of F^+ can be compared for the two reactions. If the value found for $\Delta H_f(F^+)$ is the same for the two fragmentations, it provides some positive evidence that the structure of ion F^+ is the same in each case. Similarly, a gross difference in the values found for $\Delta H_f(F^+)$ would suggest that the ions, F^+, from each fragmentation have different structures, even though they have the same elemental composition.

These types of thermochemical argument can be used to predict the likely course of fragmentation or the structures of ions and tables are available listing the known heats of formation of ions (Rosenstock, Draxl, Steiner & Herron, 1977).

Because there is a need to know the heats of formation of more ions and molecules than have been measured, many schemes have been devised for predicting heats of formation. For molecules particularly, these calculated values are mostly as accurate as the measured values and can be used with confidence. For ions, the situation is not so straightforward although good schemes have been advanced for estimating ionic heats of formation (Jolly

& Gin, 1977; McKelvey, Alexandratos, Streitwieser, Abboud & Hehre, 1976) and *ab initio* molecular orbital calculations can provide surprisingly detailed information (Bentley, 1979). The use of these calculated heats of formation greatly extends the degree of speculation that can be applied to mechanisms of ion fragmentation.

The use of thermochemical arguments is further exemplified in section 9.8.

9.6. Ion lifetimes

Discussion of theoretical aspects of mass spectrometry has so far been concerned with the rates at which ions fragment and with predictions on the course a fragmentation will take based on rate equations and thermochemical arguments. However, it should be noted that a mass spectrometer samples a time 'window' depending on the time between formation and detection of an ion. Thus, the appearance of a mass spectrum depends not only on the behaviour of ions after their formation with various amounts of excess of internal energy but also on the time window sampled. Thus, in field ionization, the time spent by an ion in the ion source after formation is very short (10^{-11}-10^{-9} s) so that very fast decompositions can be examined (the limiting rate of any decomposition is equal to a bond vibration frequency and is usually of the order of 10^{-13} s^{-1}). This time window has led to the development of *field ion kinetics* in which the fragmentation behaviour of ions can be examined soon after formation. For greater detail on this rapidly expanding field of research, the reader is referred to excellent reviews (Beckey, 1961; Derrick, 1977).

On the other hand, ions that contain very small amounts of excess of energy or undergo extensive rearrangement before fragmentation will decompose much more slowly, in say 10^{-7}-10^{-6} s. These long-lived ions (metastable ions) provide structural information on ions with small, but ill-defined, excess of internal energy. The uses and investigation of metastable ions are discussed more fully in chapter 8 (Cooks, Beynon, Caprioli & Lester, 1973). By 'tuning' a mass spectrometer to these different time windows, valuable information on fragmentation or structure of ions can be obtained but the observed mass spectra are generally quite different in appearance.

9.7. Ion beams and collisional activation

The interaction of ions with molecules (ion/molecule collisions) has been introduced in the discussions on chemical ionization (sections 7.2 and 9.3) and collisional activation (section 8.5). Fundamental investigations of ion/molecule collisions are carried out in highly specialized apparatus in which beams of ions of closely controlled total energy are allowed to interact with molecules, sometimes also in energy-selected beams. The collision process may be elastic, when the total momentum of the particles is the same after collision as before, or inelastic, when some of the kinetic energy of the particles is converted into internal energy. Further information of ion beam chemistry can be obtained from many books and reviews but the area is too extensive to be discussed in any reasonable fashion in this book (for leading references see Dubrin & Henchman, 1972; Mahan, 1976; Wexler & Parks, 1979).

One important consequence of inelastic ion/molecule collision lies in the ability to impart extra internal energy to an ion by intentionally encouraging collisions with neutral gas molecules. For example, chemical ionization produces molecular or quasi-molecular ions that have little excess of internal energy and therefore afford few fragment ions (chapter 7). Although a suitably chosen reactant gas can give much increased fragmentation (section 7.2), an alternative approach is to pass the ion beam through a small, confined region of relatively high neutral gas pressure. The resulting ion/molecule collisions impart extra internal energy to some of the ions, causing them to fragment more extensively than would have been the case in the absence of collisions. This process of increasing the internal energy of an ion through collision is known as *collisional activation* or sometimes, *collisionally induced decomposition*. Examples of the use of the technique appear in section 8.5. Care must be exercised that the neutral gas density is not so large as to scatter the ion beam entirely so that no ions arrive at the detector.

Another area in which ion/molecule collisions are investigated to probe ion structures is ion cyclotron resonance spectroscopy, described in section 12.4. The use of ion/molecule collisions to effect charge exchange, thereby producing new ion types, has been described earlier.

In photochemistry, molecules are activated through absorption of photons and this added energy can lead to their decomposition as well as to increased reactivity towards other molecules. In the same way, ions can be activated through absorption of photons but, as the ions are normally in regions of low gas density so that ion/molecule collisions do not occur, the added energy induces fragmentation. Thus, a further method for investigating ion structures is to irradiate ions with photons of various known energies and observe the subsequent fragmentation (*photodissociation*). Because ion gas density is usually very low in a mass spectrometer, it is necessary to use intense light sources to irradiate the ions and ensure a sufficiently observable number of photon-induced fragmentations; tuneable lasers form ideal photon sources for these purposes.

9.8. Qualitative theories

Discussion of mass spectrometric fragmentation for structural analysis must be carried out in terms of the structure of the intact molecule before ionization because the structures of ions are generally unknown. Fragmentation of a molecular ion is described with minimum structural change at each step. For example, the loss of $CH_3O^\cdot$ from methyl benzoate can be described as simple bond cleavage, without knowledge of the actual structures of the ions, simply by considering a minimum of structural change during ionization and decomposition necessary to remove this fragment:

$$[C_6H_5CO\text{-}OCH_3]^{+\cdot} \longrightarrow C_6H_5CO^+ + CH_3O^\cdot$$

Conversely, loss of $CH_3O^\cdot$ from a substance of unknown structure may be taken to imply that such a unit is present without any knowledge of ion structures other than their elemental compositions. Thus, for structural analysis of compounds by mass spectrometry it is only necessary to investigate and collate the mass spectra of many compounds of known structure and to use this knowledge in the analysis in an empirical fashion. This empirical approach is common in organic chemistry and has proved extremely useful in ultraviolet, infrared and nuclear magnetic resonance spectroscopy. Because the actual structures of ions are unknown, it is not possible to discuss adequately the mechanisms of mass spectrometric fragmentations, and still less the intervention of intermediates and transition states. Nevertheless, 'mechanisms' for mass spectrometric fragmentation are frequently published and sometimes 'supported' by evidence from isotope labelling experiments or other approaches. Whether or not such mechanisms are ever proved correct or incorrect has no effect on the present value of the empirical approach to mass spectrometry for structural analysis. Two qualitative theories of mass spectrometry which have proved of considerable value as aids to structure determination are described. The main attribute of these theories is that they classify the many types of mass spectrometric fragmentation into a few categories of behaviour. Knowing these categories, it is easier to discuss the mass spectrometric fragmentation of new compounds of known or unknown structure.

9.8.1.   Stabilities of fragmentation products

For the fragmentation of an ion to take place it must possess an excess of internal energy sufficient to exceed the activation energy for the reaction considered (E_f in figure 9.6). It has been demonstrated for many fragmentation reactions of ions, and is believed to be generally approximately correct, that the activation energy for the reverse reaction (E_r) is almost zero or at least very small. By Hammond's postulate (Hammond, 1955) for reaction kinetics, it may be supposed, for such endothermic fragmentations, that the transition state will resemble the products of reaction and, therefore, that the stabilities of the products will, to an approximation, determine the course of the fragmentation. Consider two alternative decompositions (9.7a, b) of the molecular ion of methyl benzoate:

$$[C_6H_5COOCH_3]^{+\cdot} \Bigg\langle {\begin{array}{l} C_6H_5CO^+ + CH_3O^\cdot \qquad\qquad (9.7a) \\[6pt] C_6H_5^+ + CH_3OCO^\cdot \qquad\qquad (9.7b) \end{array}}$$

Figure 9,6. Energy diagram showing the relationship between the activation energies (E_f, E_r) for the forward and backward reactions in the decomposition of the ion $(M^{+\cdot})$ to fragments $(A^+, B^{\cdot})$. The appearance energy for the process is marked (A).

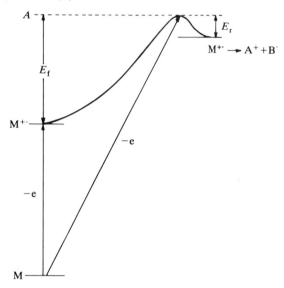

Figure 9.7. Energy diagram showing relationship between competing fragmentation processes from one state of the molecular ion of methyl benzoate. The appearance energies $(A_1$ and $A_2)$ relate to the two reactions in scheme 9.7).

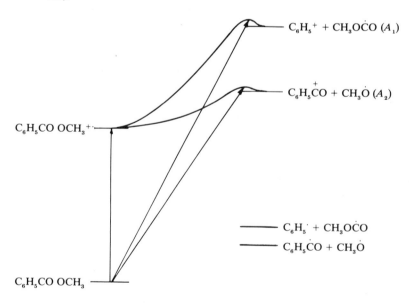

Figure 9.7 shows the relative stabilities of these products with respect to the molecular ion. If the transition state does resemble the products ($E_r \approx 0$) and if the frequency factors (equation 9.4) are similar, then reaction (9.7a) will be favoured over (9.7b) as is found experimentally. For the proper use of this method, it is necessary to take account of the heats of formation of all the products of fragmentation. When the ejected neutral particle is a stable molecular species like CO or CO_2, its large negative heat of formation should not be ignored, since it will lead to enhanced fragmentation by reducing the energy required for the process compared with similar processes in which the neutral fragment ejected does not have a large negative heat of formation. Quite often only the stabilities of the ionic products are considered. For example, the fragmentation of an amine, $RCH_2NR'_2$, beta to nitrogen is often ascribed to 'resonance stabilization' of the positive charge in the fragment ion $[CH_2NR'_2]^+$; the 'stability' of the neutral part R· is ignored (equation 9.8).

$$[R \overset{}{\underset{}{\xi}} CH_2NR'_2]^{+\cdot} \rightarrow R^\cdot + \overset{+}{C}H_2NR'_2 \leftrightarrow CH_2 = \overset{+}{N}R'_2 \qquad (9.8)$$

[For $E_r = 0$, then $E_f = \Delta H_f(R^\cdot) + \Delta H_f(\overset{+}{C}H_2NR'_2) - \Delta H_f(RCH_2NR'_2)$]

The 'wiggly' line is drawn to indicate which bond is considered to break in the process. A charged species is drawn as a structure inside square brackets. An even-electron ion is marked by a (+) sign outside the brackets and an odd-electron species (radical-cation) by a (+·) sign. Many types of mass spectrometric fragmentation have been classified by Biemann (1962) into a few general schemes, some of which are outlined in table 9.1. These classifications

Table 9.1. *Examples of empirically classified fragmentation modes (Biemann, 1962)*[a]

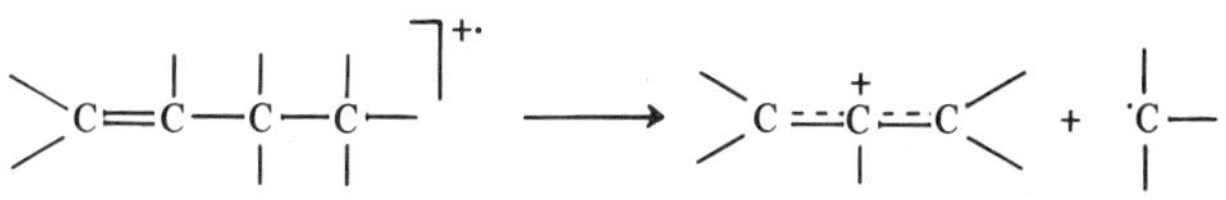

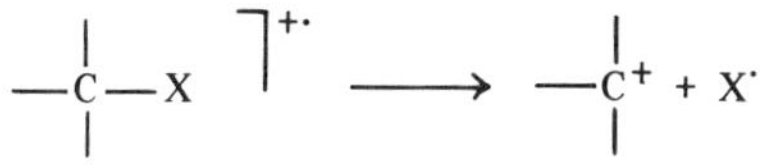

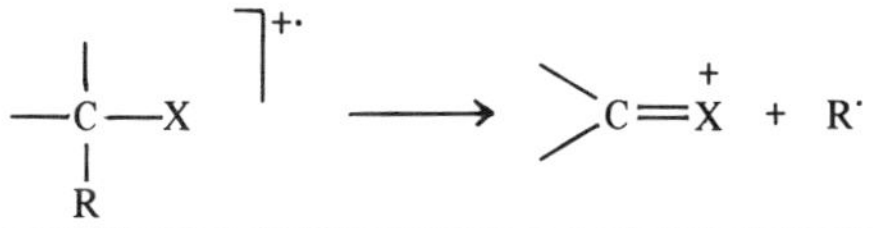

[a]These and other reaction types are classified slightly differently in this book; the Biemann classification represents the first, and still valid, comprehensive effort in this way of approach to reaction paths in mass spectrometry. Type A_3 represents 'allylic' cleavage; Types A_5 and B represent cleavage adjacent to the α-carbon atom.

are empirical, imply actual knowledge of ion structures, and consider fragmentation in terms of the structure of the original molecule. Two examples will suffice to illustrate the argument:

(i) n-Alkane molecular ions decompose extensively to give a series of fragment ions ($C_nH_{2n+1}^+$) separated by 14 mass units:

$$[C_nH_{2n+2}]^{+\cdot} \rightarrow CH_3^+, C_2H_5^+, C_3H_7^+, \ldots$$

However, a branched alkane shows enhanced bond fission adjacent to the branching point and this has been ascribed to the increased stability of secondary or tertiary cations compared with the primary ones generated by n-alkanes:

$$[CH_3-C(CH_3)_2-CH_2-CH_2-CH_3]^{+\cdot} \rightarrow CH_3-\overset{\overset{\textstyle CH_3}{|}}{\underset{\underset{\textstyle CH_3}{|}}{C}}{}^+ + CH_3-CH_2-CH_2^{\cdot}$$

(ii) Bond cleavage of the cinnamyl compound (1) takes place predominantly at the position shown and this is supposed to be due to the stability of the resulting cinnamyl cation (2). The subsequent loss of H_2 from this cation may be ascribed to cyclization with formation of the stabilizing indenyl cation (3):

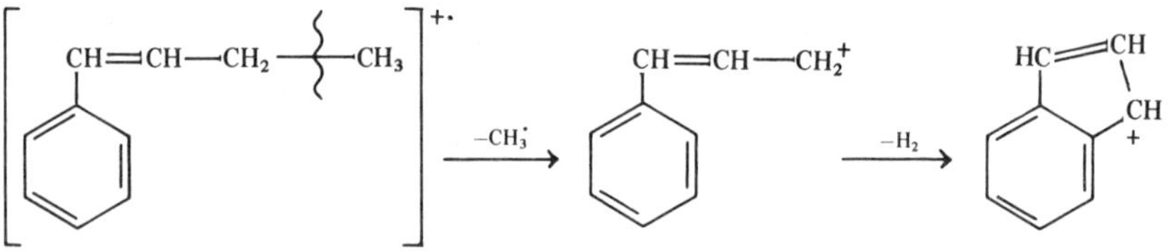

(1)   (2)   (3)

9.8.2. Charge localization

An alternative theory of mass spectrometric fragmentation applied extensively by Djerassi (see, for example, Budzikiewicz, Djerassi & Williams, 1967) supposes that, after ionization, the charge on the molecular ion may be considered to be localized at some particular place in the ion. For an aliphatic amine, $RCH_2NR'_2$, this site of charge localization is the nitrogen atom because one of the 'lone-pair' electrons on nitrogen would be easiest to remove. The charged site is then considered to 'trigger' fragmentation by one- or two-electron shifts.

$$R-CH_2-\ddot{N}R'_2 \xrightarrow{-e} R\overset{\frown}{-}CH_2\overset{\cdot+}{-}\ddot{N}R'_2 \longrightarrow R^{\cdot} + CH_2 = \overset{+}{N}R'_2$$

A double-headed arrow indicates a two-electron shift and a single-headed arrow, or 'fish-hook', a one-electron shift. Charge stabilization by the product ion is therefore a 'bonus' making the reaction more favourable. It has also been argued that it is specifically the presence of the radical site which is the

driving force for the reaction (McLafferty, 1966). To apply the charge localization treatment it is only necessary to decide where the charge is localized
(this may not in fact be at all obvious in a complex molecule). It is unnecessary
to classify fragmentations by type, although this is often done, and therefore the method provides both a valuable *aide-memoire* for collecting large
amounts of information on mass spectrometric fragmentation and a means
of assessing the possible direction of this fragmentation in compounds for
which mass spectra have not been determined. Similarly, using these same
principles for the interpretation of the mass spectra of compounds of unknown structure affords a convenient guide when considering alternative
possible structural features. The losses of an ethyl radical from the epoxide
(4) and of C_2H_2O from acetanilde (5) are used to illustrate the method.

(4)

(5)

9.8.3.   Comments on the two qualitative theories

For a theory to be worthy of the name, it should have a truly predictive capability. This criterion does not apply to either of the qualitative
theories: neither is predictive as both rest heavily on argument by analogy
or comparison with previous experience. From a theoretical point of view,
the first hypothesis, based on Hammond's postulate of the transition state, is
possibly more satisfying although less fashionable. The second theory, which
requires a charge-localized site to trigger fragmentation, would appear to be
opposite to Hammond's postulate. There is no reason to suppose that triggering is necessary for fragmentation along any particular pathway. Electrons in
a molecule are held in molecular orbitals and removal of one electron affects
the whole structure to various degrees. Even 'lone-pair' electrons are not
strictly localized on atoms although this is a good approximation for ground-
state properties of organic molecules (Dewar & Worley, 1969). For example,
the 'lone-pair' electrons on the oxygen atom in methanol have been estimated
to be about 25 per cent delocalized over the molecule. Perhaps these considerations can lead to a fusion of the two qualitative theories of mass spectrometric fragmentation. Removal of an electron from a molecular orbital will
change the various bond vibration frequencies and strengths to different

extents in different parts of the ion. It is only necessary for sufficient vibrational energy to be concentrated along the reaction co-ordinate for fragmentation to occur. If the electron deficiency in the ion is spread out over a 'molecular' orbital it may nevertheless be concentrated at certain parts in the ion (charge localized). Many interpretations of charge localization really use product stability arguments to decide the direction fragmentation will take and, from this, it has appeared that charge localization leads to 'triggering' of fragmentation. Therefore, triggering often appears as a deduction from charge localization theory when, in fact, product stability has already been invoked to explain preferred modes of cleavage. By way of illustrating this point, the fragmentation of compounds of the type C_2H_5X may be considered:

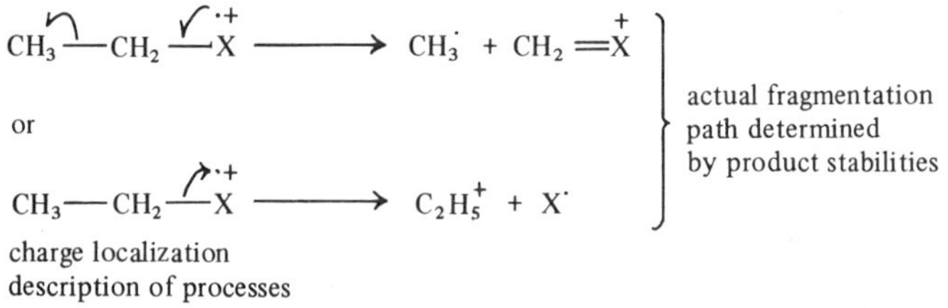

$$CH_3 \!-\! CH_2 \!-\! X \longrightarrow CH_3^{\cdot} + CH_2 \!=\! \overset{+}{X}$$

or

$$CH_3 \!-\! CH_2 \!-\! X \longrightarrow C_2H_5^{+} + X^{\cdot}$$

charge localization
description of processes

actual fragmentation path determined by product stabilities (9.9)

Charge localization alone cannot determine which cleavage in C_2H_5X compounds (scheme 9.9) is more favourable, but product stability arguments can. In fact, although bond strengths have been affected in different ways on ionization, Hammond's postulate may still be applied and the stabilities of products will largely determine the direction of fragmentation in the absence of a large activation energy for the reverse reaction and not forgetting the effects of frequency factors. It may be suggested, therefore, that charge localization can indicate which bonds in a molecule are most likely to be affected on ionization and that product stability arguments will then mostly determine the actual direction of fragmentation. From this point of view, the two theories are complementary and their conventions of the 'wiggly line' and 'fish-hooks' or 'arrows' will accordingly be used somewhat indiscriminately throughout the book as a means of 'electron book-keeping'.

10   Structure elucidation

10.1.   Classifications of mass spectra

The electron-impact mass spectra of many thousands of compounds of known structure have now been determined and are available as libraries of mass spectra. Published volumes of mass spectra, usually taking the form of normalized tables, are available from various sources (section 12.5). If the mass spectrum of a substance of unknown structure has been obtained, it is obviously useful to be able to sort through compilations of published spectra and compare them with it. In this comparative way, it may be possible to identify the unknown compound without needing to rationalize the mass spectrum at all. The task of searching manually through large compilations can be tedious, but preferable to looking through scattered research literature for reports of particular mass spectrometric investigations. Fortunately, searching through large compilations of mass spectra is a task ideally suited to a computer and automatic library searching has already been discussed (section 3.4). Of course, the classification of mass spectra in this way is of limited use if the mass spectrum of the substance to be identified has not been determined previously. In such circumstances, the mass spectrum must be interpreted by rationalization or subjected to computerized spectral interpretation (section 3.4). The remainder of this chapter is concerned largely with interpretation by rationalizing mass spectra.

10.1.1.   Functional-group approach

Classifications of mass spectrometric fragmentations based on the functional-group approach, so successful in organic chemistry, have been equally successful in mass spectrometry. Thus, the fragmentation reactions of aliphatic ketones may be classified together and any fragmentations which appear to be common features of the presence of a ketone group are noted particularly. If an unknown substance is suspected of being a ketone, then its mass spectrum can be examined for features of fragmentation pathways reported for other ketones, and the classifications by functional group make this task much easier. Frequently, the effects of changes in molecular structure on the fragmentation behaviour of the functional group are included in the classifications and provide very useful extra information. For example, a comparison of the effects on fragmentation of a ketone group in ring and straight-chain compounds can help to decide the environment of a ketone group in an unknown substance.

The functional-group approach has its limitations. There may not be sufficient difference between the effects of functional groups to allow a positive identification to be made. Changes in structure can produce large changes in the 'standard' effects of functional groups. The molecule may contain more than one functional group and, if these interact in the fragmentation process, then the effect could be to mask the 'standard' effects of each of the groups. These problems are not exclusive to mass spectrometry but occur frequently in chemistry and illustrate the inherent dangers of relying exclusively on one particular technique for providing an answer to a problem.

There is neither the scope in, nor is it the intention of, this book to provide more than this summary of the functional-group approach to the interpretation of mass spectra. In general, the scheme is very close to similar ones long familiar to chemists in infrared, ultraviolet and nuclear magnetic resonance spectroscopy, and in the classification of reactions of compounds. Accurate mass measurement of the molecular ion in a mass spectrum will give the molecular formula of the compound and this in itself often affords a good indication of the type of functional group present. Several excellent books have been published which classify and discuss mass spectra in detail on the basis of functional groups and the reader is recommended to read them for further information (Budzikiewicz, Djerassi & Williams, 1964*a*, *b*, 1967; McLafferty, 1966; Reed, 1966; Beynon, Saunders & Williams, 1968; Hill, 1972). The behaviour of organic compounds during mass spectrometry, classified according to functional groups, has been well reviewed by Bowie (1971, 1973, 1975, 1977, 1979) and of inorganic compounds by Spalding (1979). These reviews cover not only reactions of positive ions, but also of negative ions.

10.1.2. *Observation of characteristic m/z values*

It is often observed that a certain feature of molecular structure will give rise to a characteristic peak in the mass spectrum such that the presence of an ion at that m/z value may be taken as good evidence for the feature. This approach is not the same as that based on functional group since the latter concentrates on fragmentation pathways whereas one based on characteristic m/z values relies on the identification of one or two particular peaks. Thus, an ion at m/z 43 is often evidence for the presence of the grouping CH_3CO in a molecule; it is of course better to determine the elemental composition of the ion at m/z 43 by accurate mass measurement to support the diagnosis. The presence of an ion at m/z 91 ($C_7H_7^+$) is almost classical in mass spectrometry and usually indicates the presence of either a benzylic residue in the molecule or a structure which readily rearranges to give the $C_7H_7^+$ ion. Similarly, an ion at m/z 30 (CH_4N^+) strongly indicates an amine, particularly a primary one.

Lists of such characteristic peaks have been published and they may be used in much the same way as, for example, nuclear magnetic resonance correlation charts (McLafferty, 1963; Beynon, Saunders & Williams, 1968). Just as peaks near $\delta 7$–8 in a proton magnetic resonance spectrum would suggest the presence of an aromatic structure, so too would the presence of an ion at m/z 91 in the mass spectrum. However, in general, these compilations of characteristic peaks do not have the same usefulness as the correlation charts used in ultraviolet, infrared and nuclear magnetic resonance spectroscopy because, in mass spectrometry, rearrangement reactions accompanying fragmentation are common, i.e. the original molecule is considerably disturbed on ionization. For example, although an ion of composition C_2H_3O at m/z 43 suggests the CH_3CO grouping in a molecule, it is found that many oxygen-containing compounds without this grouping also give abundant ions at this m/z value. The straight-chain ketone $C_2H_5COC_5H_{11}$ has its base peak due to $C_2H_3O^+$ (m/z 43); the use of tables of characteristic peaks would lead to an inference of a methyl ketone, whereas in fact the ion is a product of a complex rearrangement. As a mass spectrometrist gains experience from interpreting mass spectra so the usefulness of characteristic peaks increases.

The method is particularly valuable when closely similar sets of compounds are examined. For example, an ion at m/z 105 ($C_7H_5O^+$) is prominent in mass spectra of molecules containing the grouping C_6H_5CO and, on substitution into the benzene ring, the corresponding ion appears at a different m/z value. Thus, benzophenone yields abundant ions at m/z 105, but 3-chloro-3'-methyl-benzophenone gives ions at m/z 119 ($C_8H_7O^+$, or $CH_3-C_6H_4CO^+$) and m/z 139, 141 ($C_7H_4OCl^+$, or $Cl-C_6H_4CO^+$):

m/z 139,141 m/z 119

In the examination of the mass spectra of closely defined areas of natural product chemistry, characteristic peaks have played a leading role in structure determination. The examination of mass spectra of peptides, benzylisoquinoline alkaloids, indole alkaloids and sterols by workers who have experience with the spectra of these compounds will quickly yield much valuable information. Therefore, where closely related compounds are being examined it is worthwhile searching their mass spectra for peaks that are characteristic of particular common groupings, so that these groupings may be recognized in the mass spectra of similar compounds of unknown structure. For example, benzylisoquinoline alkaloids (1) give a prominent, characteristic ion (2) in their mass spectra which is due to ejection of the benzylic residue; trimethyl-coclaurine ($R^1 = R^2 = R^3 = CH_3$) yields the ion (2) at m/z 206.

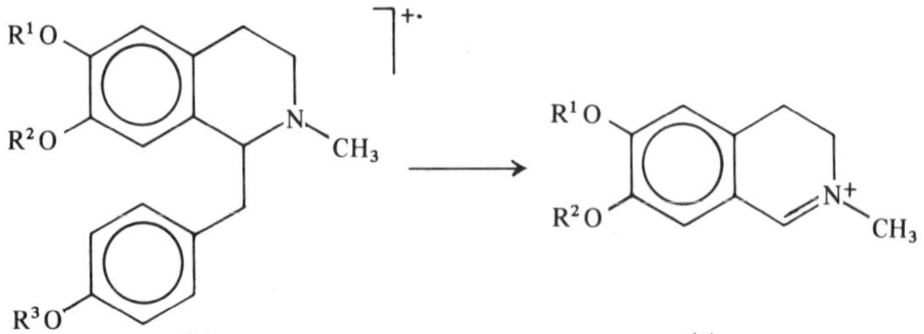

(1)   (2)

10.2. Examination of the mass spectrum

The first step in examining a mass spectrum is usually to decide which peak represents the molecular ion. The spectrum is counted to increasing m/z values until no more ions are observed and this is then the potential molecular ion region. In assigning the molecular ion it is worth recalling that it is accompanied by isotope ions one or more mass units greater $[(M + 1)^{+\cdot}, (M + 2)^{+\cdot},$ etc.]. For chlorine, bromine and sulphur the $(M + 2)^{+\cdot}$ ions are abundant compared with the $(M + 2)^{+\cdot}$ ions of similar compounds without these hetero-atoms.

Figures 10.1*a* and *b* show, respectively, examples of the molecular ion regions for a hydrocarbon with a normal molecular ion region, and a dihalo-compound with peaks up to $(M + 5)^{+\cdot}$. Often ions $(M - 1)^+$ and sometimes $(M - 2)^{+\cdot}, (M - 3)^+$, etc. are observed owing to losses of hydrogen from the molecular ion. Figure 10.1*c* illustrates the molecular ion region of 2-styryl-pyridine which looks very similar to that of the hydrocarbon in figure 10.1*a*. However, the most abundant ion in the nitrogen compound is the $(M - 1)^+$ peak and not the molecular ion. Sometimes these losses of hydrogen are so marked as to lead to the wrong assignment of the molecular ion because the $(M - 1)^+$ ion is so much more abundant than the molecular ion. If such losses of hydrogen are suspected it is worthwhile examining the spectrum at lower ionizing energies or recording the mass spectrum by a 'soft' ionization tech-nique, such as chemical ionization, or observing any metastable ions corres-ponding to losses of hydrogen. Figure 10.1*d* shows the molecular ion region in biphenyl with the marked loss of up to four hydrogen atoms from the molecular ion.

Difficulties are also encountered where molecular ions are of low abun-dance or even non-existent so that it becomes impossible to determine directly the molecular weight of the compound. Sometimes the difficulty can be resolved by examining the fragment ions and deducing the molecular weight. It is more satisfying to re-examine the mass spectrum after ionization with low electron beam energies, or electric fields, or after chemical ionization since these methods often afford prominent molecular or quasi-molecular ions when, under normal electron-impact conditions, only uncertain ones are found.

To be due to molecular ions, a candidate peak must pass several criteria. Firstly, the molecular ion peak must be of largest mass in the spectrum, apart from the natural isotope peaks owing to the presence of ^{13}C, etc. However, it is reasonable to expect some 'background' ions of relatively low abundance, for example, ions derived from vacuum pump oil, from impurities such as plasticizers in samples and from column bleed in combined gas chromatography/mass spectrometry. These impurity ions may extend to high mass and complicate the assignment of the molecular ion peak but, with experience, a mass spectrometrist can usually distinguish such signals without too much trouble. Secondly, in the analysis of organic compounds, the elements most

Figure 10.1. Comparison of the molecular ion regions of (*a*) phenylacetylene, (*b*) 1-bromo-2-chlorobenzene, (*c*) 2-styrylpyridine and (*d*) biphenyl.

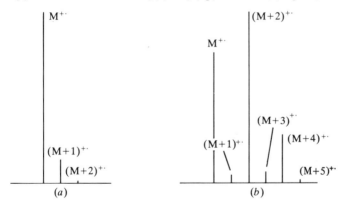

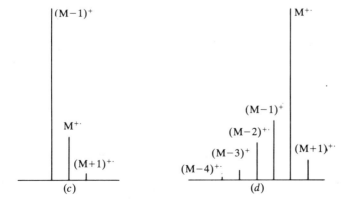

commonly met (C, H, N, O, S, Cl, Br, etc.) very rarely give rise to losses of 5-14 and 21-25 mass units, inclusively. If a candidate peak for the molecular ion shows such losses, it is most unlikely to be correctly assigned. On the other hand, losses of H˙ and $CH_3^˙$ (1 and 15 mass units) are commonly observed for molecular ions of organic compounds. Thirdly, the molecular ions must contain all the elements that are in evidence elsewhere in the spectrum. For instance, if the characteristic isotope ratio for chlorine is seen in some lower mass ions, the molecular ion region must show it too. In high-resolution mass spectra, fragment ions cannot contain greater numbers of any element than the true molecular ion. Lastly, if there are metastable ions which implicate the candidate peak as being due to product ions of some higher mass species, then it cannot be the true molecular ion peak.

If only the molecular weight or molecular formula of the compound is required, then deciding on the molecular ion will be the only examination of the spectrum required. Even so, this can still lead to much valuable information. As has been shown above, the very appearance of the molecular ion region can furnish information. It is useful to remember that an *odd* molecular weight indicates an *odd* number of nitrogen atoms in the compound. Abundant $(M + 2)^{+˙}$ ions would suggest the presence of chlorine or bromine, and less abundant ones the presence of sulphur. An approximate idea of the number of carbon atoms in the molecule can be gained from the relative heights, h and h', of the $M^{+˙}$ and $(M + 1)^{+˙}$ peaks, respectively. Since ^{13}C is present in about 1.1 per cent natural abundance in organic compounds, the approximate number of carbon atoms is given by $100\,h'/1.1h$. For example, in the mass spectrum of naphthalene, which has ten carbon atoms, $h'/h = 10.9/100$ and $(100 \times 10.9)/(1.1 \times 100) = 10$ to the nearest integer. The method is not very accurate for more than ten or twelve carbon atoms and may yield highly erroneous results if the compound contains nitrogen (contribution of ^{15}N to $(M + 1)^{+˙}$ peak) or if there are overlapping peaks due to impurities.

To obtain the molecular formula of a compound, it is necessary to measure the accurate mass of the molecular ion. Knowing the accurate mass, the molecular formula can be obtained from tables or by calculation (Beynon & Williams, 1963; Lederberg, 1964). Table 10.1 shows some of the elemental compositions possible at m/z 100. If the measured accurate mass of the molecular ion of an unknown compound was 100.0635, it can be seen from the table that this mass corresponds very closely to that calculated for $C_4H_8N_2O$. The error between calculated and observed masses is often reported in parts per million (ppm). In the example cited, the difference between calculated and observed values is 0.002 mass units, i.e. the error is two parts in 1 000 637 which is about 2 ppm. Errors of up to ± 10 ppm are usually acceptable but the smaller the error, the more certain the molecular formula. It may be mentioned here that accurate mass measurement is not a substitute for elemental analysis although the two complement each other very well. The mass spectro-

Table 10.1. *Some elemental compositions at nominal mass 100*

Composition	Accurate mass
$C_3H_6N_3O$	100.0511
$C_4H_8N_2O$	100.0637
$C_5H_{10}NO$	100.0762
$C_6H_{14}N$	100.1126
C_7H_{16}	100.1251
C_2F_4	99.9936
$C_2H_4OSi_2$	99.9700

meter yields the exact molecular formula but may give little indication of the purity of a compound. The elemental analysis provides a cross-check on the purity of a compound and on its molecular formula.

Having obtained a molecular formula, it may then be used to calculate the number of double-bond equivalents in the compound. For the commoner elements met in organic chemistry, this may be done very simply by substituting CH_3 for each halogen, CH_2 for each oxygen or sulphur atom, and CH for each nitrogen. For example, carrying out these substitutions on the molecular formula C_6H_6ONCl gives C_9H_{12}; an alkane with nine carbon atoms would have a molecular formula C_9H_{20}. The difference between these formulae, C_9H_{20}–C_9H_{12}, is eight hydrogen atoms and therefore there are four double-bond equivalents in C_6H_6ONCl.

During the initial examination of a mass spectrum, its general appearance is frequently a guide to the nature of the compound under investigation. A spectrum exhibiting many fragment ions increasing in abundance towards low m/z values will suggest a predominantly aliphatic structure whereas a spectrum with few fragment ions, abundant molecular ions and doubly charged ions will suggest an aromatic structure. Figure 10.2 illustrates this point by comparing the mass spectra of naphthalene and 1-pentanal.

The next step in examining the mass spectrum is to note the major fragment ions and attempt to elucidate the main fragmentation pathways.

The major fragment ions may occur at 'characteristic' m/z values (see above) and give immediate information. In the absence of nitrogen atoms, fragment ions occurring at odd m/z values are even-electron species resulting from simple bond fission, whereas fragment ions at even m/z values are odd-electron species produced by multiple bond cleavage, suggesting that rearrangement may have occurred. In this latter case, extra care must be taken in interpreting the spectrum because some of the fragment ions may well arise from rearranged species.

It is best to check supposed fragmentation pathways by looking for metastable ions in the routine spectrum; recall that the position of the metastable ion (m^*) is given by the expression m_2^2/m_1 where m_1 is the mass of the precursor ion and m_2 the mass of the product ion. The mass spectrum of ethyl-benzene ($M^{+\cdot}$ at $m/z = 106$) shows large peaks for $(M - 15)^+$ at m/z 91 and $(M - 41)^+$ at m/z 65. The presence of metastable ions at m/z 78.1 and m/z 46.4 proves that one fragmentation path at least is $M^{+\cdot} \rightarrow (M - 15)^+ \rightarrow (M - 41)^+$ since the calculated values for the metastable ions for these processes are m/z 78.1 ($= 91^2/106$) and m/z 46.4 ($= 65^2/91$). Often metastable ions for suspected fragmentation reactions are not visible in the routine spectrum and it is then necessary to look for them by specialized techniques (see chapter 8).

Having obtained the molecular weight and the main features of the fragmentation of the compound under investigation, it may be possible to make tentative or even definite structural assignments. Comparison of the mass spectrum with those in a library may allow immediate identification of the compound (section 10.1). At this stage, it is advisable to assemble any other available physical data on the compound. Interpretation of a proton magnetic resonance spectrum will benefit from a knowledge of the molecular formula

Figure 10.2. Comparison of mass spectra of (*a*) an aromatic (naphthalene) and (*b*) an aliphatic compound (1-pentanal).

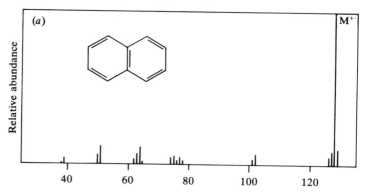

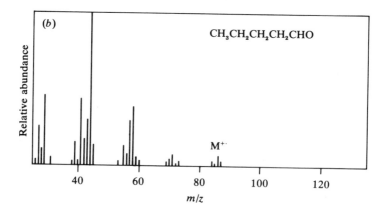

which gives the exact number of protons present and also indicates the presence of particular heteroatoms. The ultraviolet spectrum will yield information on the likely degree of unsaturation so that the number of double-bond equivalents obtained from the molecular formula may indicate whether there are any rings, and if so how many. The infrared spectrum will also often give information on the state of saturation of the molecule and may give very valuable clues to the presence of carbonyl, hydroxyl and other groups. If it is known that such groups are present, it is possible to examine the mass spectrum in conjunction with the known fragmentation behaviour of these functional groups (section 10.1). If the infrared spectrum indicates the presence of a hydroxyl group, then reference to the fragmentation reactions of alcohols in general may reveal similar fragmentations in the spectrum of the unknown substance. It is very much a case of pooling all available information, using it logically, and calling on experience.

It may not be possible to proceed any further with the interpretation without knowledge of the elemental compositions of the fragment ions. Loss of 27 mass units from the molecular ion of C_6H_6ONCl may be ejection of HCN or C_2H_3 · and accurate mass measurement on the fragment ion will determine which it is. A simple mass spectrum containing few fragment ions requires only a little extra work to obtain the accurate masses of these ions and hence their elemental compositions. On the other hand, a complex mass spectrum requires many accurate mass measurements and it should be remembered that any peak at an integer m/z value may represent ions of different compositions. Therefore, at high resolution, some peaks will appear as doublets or triplets. A long list of elemental compositions corresponding to the most abundant ions in a mass spectrum is difficult to sift for information of help in structural work. The problem has been recognized especially following the application of computers to accurate mass measurement (section 3.4) since this technique generates a great deal of information which must be examined in detail. A common method of assembling all the data into some order is through the use of an *element map*. An example of one common type of these maps is given in figure 10.3 and relates to the mass spectrum of a natural product, genipin. The ions are arranged in an array of increasing complexity of elemental compositions from left to right, and increasing m/z values from top to bottom. Thus, the molecular ion is found on the right of the map; doublet peaks are marked (d). The sizes of the peaks are not given in figure 10.3, but this information is frequently included also on element maps. Other methods of element mapping have been proposed.

Inspection of the map in figure 10.3 shows, for example, that the ion at m/z 129 cannot arise from m/z 147 (difference of 18 mass units, which could have been due to H_2O) because the former has one more oxygen atom than the latter. The minimum number of carbon atoms associated with all five oxygen atoms is eleven; with three oxygen atoms it is four (m/z 103). The column headed C_xH_yO in the element map shows that nine carbon atoms are associated with one oxygen (m/z 131). The formula for genipin (3) shows which atoms these must be. Thus, if the structure of genipin had been unknown significant structural features could be gleaned from the map.

Figure 10.3. Element map of genipin. The ions are arranged in order of increasing complexity from left to right and top to bottom. Doublet peaks are marked (d).

m/z	C_xH_y	m/z	C_xH_yO	m/z	$C_xH_yO_2$	m/z	$C_xH_yO_3$	m/z	$C_xH_yO_4$	m/z	$C_xH_yO_5$
				60	$C_2H_4O_2$						
				61	$C_2H_5O_2$						
		59	C_3H_7O	71	$C_3H_3O_2$(d)						
		70	C_4H_6O	83	$C_4H_3O_2$(d)	103	$C_4H_7O_3$(d)				
69	C_5H_9	81	C_5H_5O	100	$C_5H_8O_2$						
71	C_5H_{11}(d)	82	C_5H_6O								
		83	C_5H_7O(d)								
77	C_6H_5	94	C_6H_6O	109	$C_6H_5O_2$(d)	128	$C_6H_8O_3$				
78	C_6H_6	95	C_6H_7O			129	$C_6H_9O_3$				
79	C_6H_7	96	C_6H_8O								
80	C_6H_8										
90	C_7H_6	105	C_7H_5O	124	$C_7H_8O_2$	139	$C_7H_7O_3$				
91	C_7H_7	106	C_7H_6O								
92	C_7H_8	107	C_7H_7O								
93	C_7H_9	108	C_7H_8O								
		109	C_7H_9O(d)								
102	C_8H_6	119	C_8H_7O	135	$C_8H_7O_2$						
103	C_8H_7(d)	120	C_8H_8O	137	$C_8H_9O_2$						
		121	C_8H_9O								
		130	C_9H_6O	147	$C_9H_7O_2$	165	$C_9H_9O_3$				
		131	C_9H_7O	148	$C_9H_8O_2$	166	$C_9H_{10}O_3$				
				149	$C_9H_9O_2$	167	$C_9H_{11}O_3$				
				151	$C_9H_{11}O_2$						
				152	$C_9H_{12}O_2$						
				158	$C_{10}H_6O_2$	176	$C_{10}H_8O_3$	194	$C_{10}H_{10}O_4$		
				159	$C_{10}H_7O_2$			195	$C_{10}H_{11}O_4$		
				161	$C_{10}H_9O_2$	179	$C_{10}H_{11}O_3$				
				162	$C_{10}H_{10}O_2$	180	$C_{10}H_{12}O_3$				
				163	$C_{10}H_{11}O_2$						
						190	$C_{11}H_{10}O_3$	208	$C_{11}H_{12}O_4$	226	$C_{11}H_{14}O_5$

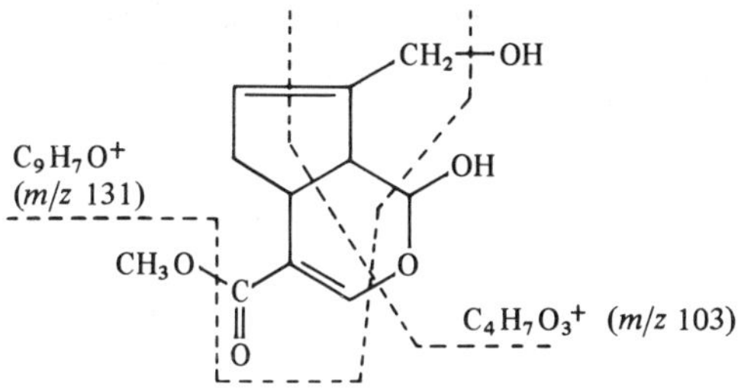

(3)

These maps provide a convenient means of presenting all the information relating to the elemental compositions of the ions in a spectrum but usually do not give data for metastable ions. With the advent of automatic recording of metastable ions in the first and second field-free regions (chapter 8), the use of *metastable-ion maps* in conjunction with element maps is likely to increase.

10.3.   Modification of mass spectra through instrumental parameters

After initial examination of an electron-impact mass spectrum, it may be desirable to modify it by changing the ionizing energy, the method of ionization or the temperature of the ion source. A complex spectrum can be simplified by using radiation of lower energy for ionization. A change in the appearance of a spectrum with time or the presence of unexplained peaks may be due to thermolysis of the sample, or to the sample being a mixture. Modification of mass spectra by the use of derivatives is discussed separately (see chapter 5).

10.3.1.   Ionizing energy

Field ionization and photoionization sources usually operate at a single ionizing energy, for example the 21.21eV helium line, but electron-impact sources have continuously variable electron-beam energies. Electrons from a hot filament are accelerated through a potential of V which ranges from about 5 to 100 V. By convention, *standard* mass spectra are obtained with 70-V electrons because at this energy the yield of ions is near maximum and almost constant over a small voltage range. A typical graph showing ion yield as a function of electron energy is given in figure 10.4. A maximum yield of ions is often obtained at electron energies of 50 rather than 70 V but the former energy is not used for standard work because, close to the maximum, a small variation in voltage may lead to a pronounced change in the mass spectrum. There is little significance attached to the conventional figure of 70 V for the ionizing electrons and a mass spectrum can be obtained at any energy above the ionization energy of the compound examined. For greatest sensitivity it is better to measure spectra at 50–70 V.

The ionization energies of most compounds lie between 7 and 11 eV and, from the ionization energy up to an electron-beam energy of about 20 V, not only does the yield of ions increase but also the *appearance* of the mass spectrum changes appreciably. At, and just above the ionization energy, only molecular ions are formed providing the molecular ion is not in a dissociative state. As the electron-beam energy is increased more vibrational energy is produced in the molecular ion by internal energy conversion processes outlined earlier (section 9.2). The molecular ion then fragments and fragmentation becomes increasingly complex as the initial excess of internal energy of the molecular ion increases with increase in electron-beam energy. The increasing electron energy means that higher and higher excited states of an ion (for example, $\overset{*}{M}_1^{+\cdot}$ in figure 10.5) can be reached by vertical transitions from the molecule M. At some point, vertical transitions to highly excited states of the ion ($\overset{*}{M}_2^{+\cdot}$, for example) are not possible and no more energy can be transferred from the impacting electron. This point is not reached abruptly and maximum transfer of energy from the impacting electron is reached relatively slowly. Thus, as the energy of the impacting electron is increased, the energy which is imparted to a molecule on ionization increases rapidly at first and then much more slowly after about 20 eV. At the same time, the probability of ionization continues to increase so that the actual *yield* of ions continues to increase from about 20 to about 50 eV although the maximum transfer of energy occurs at around 20 eV. Therefore, although more and more ions are formed up to 50–70 eV, little extra fragmentation is observed after 20 eV. In fact, experimental evidence suggests that only about

Figure 10.4. Typical curve illustrating variation in the yield of ions from a compound with increasing electron-beam voltage.

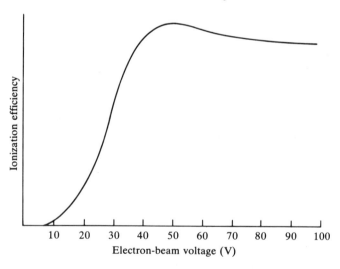

Electron-beam voltage (V)

5–6 eV of energy above the ionization energy of a compound are transferred by 70-V electrons. The effect on the appearance of the mass spectrum is as follows. At the ionization energy, molecular ions are formed; up to about 20 eV the mass spectrum becomes increasingly complex as more and more energy is transferred during ionization, leading to increased fragmentation; after 20 eV and up to 50–70 eV the appearance of the mass spectrum changes little but the total yield of ions increases, leading to greater sensitivity in the mass spectrometer. Figure 10.6 illustrates changes in the mass spectrum of ethyl thiolbenzoate with increasing electron beam energy.

Figure 10.5. Morse curves showing ionizing transitions to excited states $\overset{*}{M}{}_1^{+\cdot}$ and $\overset{*}{M}{}_2^{+\cdot}$ for a molecule M.

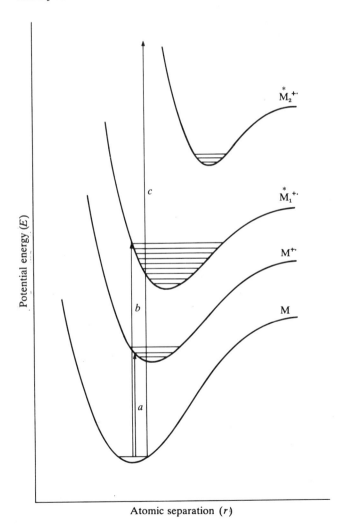

Conversely, a mass spectrum can be simplified considerably by measuring it at electron-beam energies below 20 V and simplification of a complex mass spectrum in this way may lead to a better understanding of the fragmentation paths. Figure 10.7 shows the effect on the mass spectrum of genipin of reducing the electron-beam energy from 70 to about 10 V. The spectrum is greatly simplified and two ions at m/z 78 and 96 of particular importance for its interpretation 'emerge' from the clustered fragment ions at low m/z values (typical of aliphatic compounds) in the higher energy spectrum.

The simplifying effect of reducing the electron-beam energy may be used in conjunction with reduction of the temperature of the ion source (see below) to afford even simpler electron impact mass spectra.

10.3.2. The method of ionization

The appearance of a mass spectrum depends markedly on the method of ionization (see figures 7.1, 7.2, 7.3 and 7.4), so the use of another

Figure 10.6. Changes in the mass spectrum of ethyl thiolbenzoate with increasing electron-beam energy.

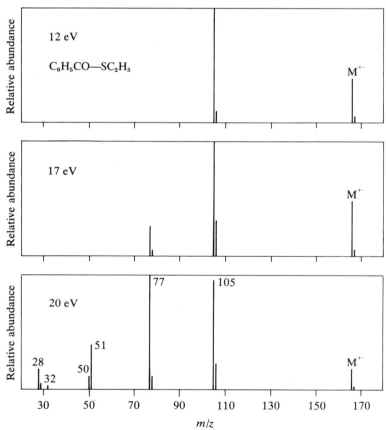

method of ionization often leads to collaborative or complementary information. If the electron-impact mass spectrum indicates that an unknown sample
is a polyhalogenated compound, then negative ion chemical ionization is likely
to yield useful information. For a natural product which is too labile thermally
to give a useful electron-impact, chemical-ionization or field-ionization mass
spectrum, the various techniques discussed in section 12.3 or field desorption
will provide better analytical information. Additional methods of ionization
and their applications are discussed elsewhere (chapter 7 and section 12.3).

10.3.3.   Temperature

The thermal energy imparted to a sample by high temperatures in
the inlet system or ion source may have profound effects on the mass spectrum. The effects observed may result from actual changes in molecular structure or may be caused by the excess of vibrational and rotational energy gained
by the molecule before it is ionized.

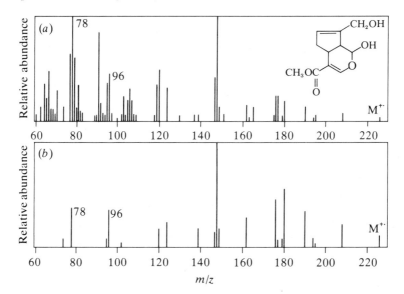

Figure 10.7. Mass spectra of genipin at (*a*) 70 and (*b*) 10 eV. The lower energy
spectrum is much simpler and peaks at *m/z* 78 and 96 remain prominent.

It is often necessary to heat a sample to achieve a great enough vapour pressure in the ion source for mass spectrometry. On heating, a compound may react uni- or bimolecularly to give new compounds. The mass spectrum of the sulphonylhydrazide (4) was complicated by the presence of the azine (5) formed from it thermally (reaction 10.1). Effects of this kind are more often found with heated inlet systems operating at relatively high pressures (10^{-2} torr) particularly if they contain metallic parts which can catalyse reactions. Normally, these effects of temperature on molecular structure are far less severe in direct inlet systems which operate at a lower pressure (10^{-6} torr) and therefore require less heat for vaporization. A direct insertion probe which may be heated or cooled independently is a good device for getting a sample into the ion chamber in a controlled manner. Although bimolecular reaction is much less likely at the low pressures employed, unimolecular reaction can still take place inside the ion chamber.

The temperature of the ion source can affect a mass spectrum markedly other than by actual chemical reaction taking place. On entering the ion source, molecules are not necessarily ionized immediately, but because of the low vapour pressure (long mean-free path) they can collide with its walls many times before ionization. Even if there are no chemical effects at each wall collision, equilibration of thermal energy can occur. The ion sources of electron impact sources often operate at temperatures between 150°C and 250°C so that a molecule can gain a lot of thermal energy before ionization and this energy must be added to that imparted during ionization; that is, the molecular ion contains excesses of energy from the hot source and from electron impact. In practice, it is found that the mass spectra of aromatic compounds with fairly rigid structures show little evidence of this thermal effect but aliphatic structures frequently show greatly enhanced fragmentation because of it. Figure 10.8 compares the partial mass spectrum of a peptide at two temperatures of the ion source and it can be seen that the size of the molecular ion peak is greatly reduced at the higher temperature. Ion sources are available which can be cooled externally, thereby giving some control over the temperature to which a sample is subjected.

The simple thermochemical arguments given above are generally applicable to compounds that are not particularly susceptible to thermal decomposition. For thermally labile compounds (often large natural products), the activation energy for decomposition is lower than that for vaporization and hence at lower temperatures only decomposition occurs. At higher temperatures, volatilization and decomposition are competing processes. Once volatilized, a molecule escapes the region around the direct insertion probe which is dense with sample molecules and where bimolecular interactions can occur. The molecule then enters the low pressure region of the ion source where the chances of decomposition are lower because collisions are so rare. Therefore, if the rate of heating of the sample is very rapid (between ten and several

thousand degrees per second), non-equilibrium conditions obtain and the
sample molecules spend less time at a temperature at which decomposition
is dominant and collisions are common. In effect, the sample is volatilized
into the low pressure region before it has time to decompose in the higher
pressure region. Reactions such as those in scheme (10.1) may also be sup-
pressed under such conditions. Rapid heating normally requires a modified
direct insertion probe and is successful in obtaining electron-impact mass
spectra of thermally labile compounds of low volatility such as large peptides, carbohydrates and even some salts (see section 12.3).

10.4.   Postulation of ion structures

In interpreting a mass spectrum it is convenient to postulate ion
structures to explain the observed fragmentation pathways. The actual structures of ions in a mass spectrometer are generally not known, although the
elemental compositions may be, and hence there is a need to postulate structures. If the compound investigated has a known structure, the mass spectrum
is often interpreted by assuming a minimum of structural change at each fragmentation step. To illustrate this point, part of the fragmentation of acetani-
lide may be considered in three alternative ways (scheme 10.2*a, b, c*). The
first way (*a*) shows ionization and fragmentation in terms of *m/z* values and

Figure 10.8. Partial mass spectrum of the peptide $C_6H_5CH_2OCO-Val-Gly-$
OMe, at (*a*) 160 °C and (*b*) 250 °C. Both spectra are normalized to *m/z* 206.

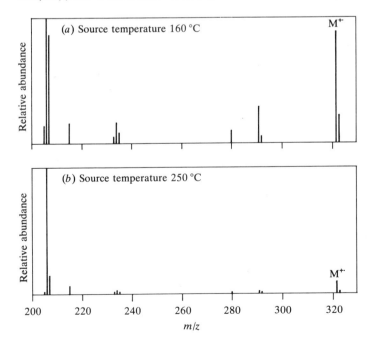

$$\text{C}_6\text{H}_5-\text{NHCOCH}_3 \quad \begin{array}{l} \xrightarrow{(a)} \quad m/z\ 135 \longrightarrow m/z\ 93 \\ \xrightarrow{(b)} \quad \text{C}_8\text{H}_9\text{NO}^{+\cdot} \longrightarrow \text{C}_6\text{H}_7\text{N}^{+\cdot} + \text{C}_2\text{H}_2\text{O} \\ \xrightarrow{(c)} \end{array} \qquad (10.2)$$

$$\left[\text{C}_6\text{H}_5-\text{NHCOCH}_3\right]^{+\cdot} \longrightarrow \left[\text{C}_6\text{H}_5-\text{NH}_2\right]^{+\cdot} + \text{CH}_2{=}\text{C}{=}\text{O}$$

gives little information. The second (*b*) gives the same pathway in terms of elemental compositions of the ions; this affords much more information since the compositions of the ions and of the unit lost at the fragmentation step are known. The third way (*c*) adds postulate to experimental fact by assuming or postulating the structures of the molecular and fragment ions as well as the neutral species to be as shown, i.e. with minimum change from the original molecule. These structures may or may not eventually prove to be correct, but this is quite immaterial to the convenience of describing fragmentation pathways in this way. An added advantage of postulation of ion structures appears when the mass spectra of groups of compounds are compared for similar fragmentation behaviour since this becomes much more obvious from structural relationships than from elemental composition relationships. It is easier to see the similarity in behaviour between diethylmethylamine (6) and 1-pentylamine (7) from the postulated ion structures (scheme 10.3*a*) than from the corresponding elemental compositions (scheme 10.3*b*).

$$\left.\begin{array}{l} \underset{(6)}{\text{CH}_3\text{CH}_2\text{N}(\text{CH}_3)\text{CH}_2\text{CH}_3}\Big]^{+\cdot} \longrightarrow \text{CH}_3^{\cdot} + \text{CH}_2{=}\overset{+}{\text{N}}(\text{CH}_3)\text{CH}_2\text{CH}_3 \\[4pt] \underset{(7)}{\text{CH}_3\text{CH}_2\text{CH}_2\text{CH}_2\text{CH}_2\text{NH}_2}\Big]^{+\cdot} \longrightarrow \text{CH}_3\text{CH}_2\text{CH}_2\text{CH}_2^{\cdot} + \text{CH}_2{=}\overset{+}{\text{N}}\text{H}_2 \end{array}\right\}(a)$$

$$\left.\begin{array}{l} \text{C}_5\text{H}_{13}\text{N}^{+\cdot} \longrightarrow \text{CH}_3^{\cdot} + \text{C}_4\text{H}_{10}\text{N}^+ \\[4pt] \text{C}_5\text{H}_{13}\text{N}^{+\cdot} \longrightarrow \text{C}_5\text{H}_{11}^{\cdot} + \text{CH}_4\text{N}^+ \end{array}\right\}(b)$$

$$(10.3)$$

The postulation of ion structure does cause difficulty in discussing thermochemical aspects of mass spectrometry. If it is necessary to postulate or assume an ion structure, then that postulate or assumption is built into any thermochemical result achieved and therefore introduces uncertainty. Discussion of fragmentation mechanism is also clouded by those assumptions. It is one thing to postulate ion structures as an aid to describing fragmentation but it is another thing altogether to discuss detailed aspects of fragmentation mechanisms in terms of assumed structures!

The usefulness of postulating ion structures to describe fragmentation pathways also applies in a reverse sense for structural work. Knowing a fragmentation pathway one can begin to postulate structures which might behave in this way and hence ideas can be formed regarding the structure of an unknown compound. For example, knowing that a molecular ion ejects $C_2H_3O_2^-$ (59 mass units), one can postulate a CO_2CH_3 (methyl ester) grouping in the original molecule without knowing the actual structures of the ions.

10.5. Fragmentation of hydrocarbons

The carbon–carbon bonds in alkanes are weaker than the carbon–hydrogen bonds. In the molecular ion of an alkane, assuming it to have a similar structure to that of the molecule before ionization, the strengths of C–C and C–H bonds are greatly reduced. Table 10.2 compares the C–C and C–H bond strengths in ethane and the ethane ion. If all the C–C bonds in the

Table 10.2. *Bond strengths in ethane and the ethane radical-cation*

	D_{C-C}	D_{C-H}
C_2H_6	83 (347)	96 (402)
$C_2H_6^{+\cdot}$	44 (184)	26 (109)

Bond strength (D) is given in kcal/mole and, in parentheses, in kJ/mole.

molecular ion of an alkane have similar strengths it might be expected that alkane ions would cleave them almost indiscriminately. Figure 10.9 shows the mass spectrum of n-decane in which can be seen bunches of ions approximately 14 mass units apart extending down from the molecular ion. The differences of 14 mass units are not due to elimination of units of CH_2 (an energetically unfavourable process) but rather to fragmentation at different places in the molecular and fragment ions (scheme 10.4). For an alkane, fragment ions appear at m/z values corresponding to the compositions C_nH_{2n+1}. This simple view of the mass spectra of alkanes is complicated by rearrangement processes and excess of internal energy in the ions. The excess of energy

Figure 10.9. Typical electron impact mass spectrum of an alkane, n-decane.

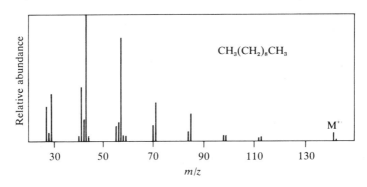

causes rapid fission of the C–C bonds in all ions resulting in a clustering of
ions of small mass near m/z 29, 43, 57 and a rapid falling off in the sizes of
the peaks at higher m/z values up to the molecular ion which is generally of

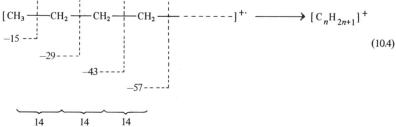

$$(10.4)$$

very low abundance. The excess of energy imparted to the molecular ion may
be reduced by measuring the mass spectrum at low electron-beam energies
and by using a cooled ion source. Under these conditions the mass spectrum
is much more uniform with all fragment ions of similar abundance and without the bunching at low m/z values. Alternatively, chemical ionization could
be employed because, in the chemical ionization mass spectra of alkanes, the
bunching at low m/z values is not so pronounced and quasi-molecular ions,
$(M-H)^+$, usually account for the base peak of the spectrum. The chemical

Figure 10.10. Typical chemical-ionization mass spectrum of an alkane, n-hexa-
triacontane. Isobutane was used as the reactant gas.

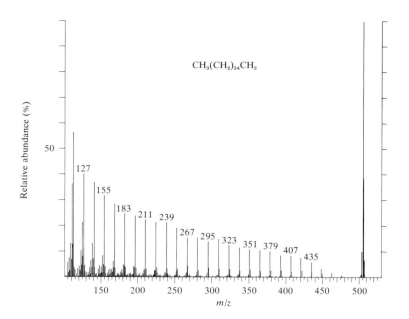

ionization spectrum of hexatriacontane, $C_{36}H_{74}$, obtained by using isobutane as a reactant gas illustrates these features (figure 10.10). The fragment ions, $C_nH_{2n+1}^+$, are due to losses of C_2H_4, C_3H_6, C_4H_8, etc., from quasi-molecular and fragment ions.

Groups of peaks rather than single peaks, 14 mass units apart, are found in alkane mass spectra because elimination of H_2 from various fragment ions can take place. Thus, the ions at m/z values corresponding to C_nH_{2n+1} are accompanied by less abundant ions at C_nH_{2n-1}. Similar losses of hydrocarbon entities yields ions, $C_nH_{2n}^{+\cdot}$.

The supposition that product stability governs the mode of fragmentation (section 9.8) receives some support from the mass spectra of branched-chain alkanes which show enhanced bond breaking at the branching points. The increased fragmentation at these points has been ascribed to the increasing stability of primary, secondary and tertiary carbonium ions. Figure 10.11 shows the mass spectrum of 3,3-dimethylhexane compared with that of its

Figure 10.11. Comparison of mass spectra of (*a*) a straight-chain hydrocarbon, n-octane and (*b*) its branched-chain isomer, 3,3-dimethylhexane. Note the enhanced abundance of ions at m/z 71, 85 and 99 in spectrum (*b*).

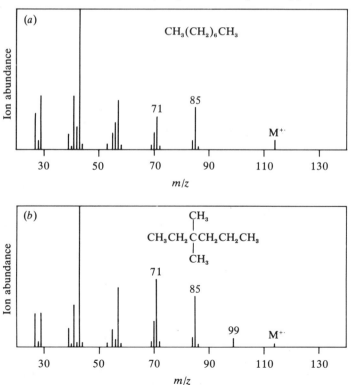

isomer, n-octane. There are increased abundances of ions at m/z 71, 85 and 99 corresponding to bond scission adjacent to the tertiary carbon centre in the branched-chain isomer (scheme 10.5). The phenomenon is useful for determining branching points in alkanes.

m/z 71 ----------┐
$\quad\quad\quad\quad$ CH₃ ┊ $\quad\quad\quad\quad$ m/z 99

$$CH_3CH_2 \overset{\overset{\displaystyle CH_3}{|}}{\underset{\underset{\displaystyle CH_3}{|}}{\text{—}C\text{—}}} CH_2CH_2CH_3 \qquad (10.5)$$

$\quad\quad\quad\quad\quad\quad$ L---------- m/z 85

Cycloalkanes give spectra very similar to those of linear alkanes except that the molecular ions are more abundant. Once a cycloalkane ring has broken, the residual ion behaves like the ions of linear alkanes, again giving bunches of ions 14 mass units apart; for cycloalkanes, principal fragment ions occur at m/z values corresponding to $C_nH_{2n+1-2r}^+$ where r is the number of rings.

The introduction of a double-bond into an alkane lowers the ionization energy but does not alter the appearance of the mass spectrum greatly apart from an apparently preferred cleavage at allylic positions as shown in equation (10.6).

$$\begin{bmatrix} \text{—}\overset{|}{\underset{|}{C}}\text{—}\overset{|}{\underset{|}{C}}\text{—}\overset{|}{C}\text{=}\overset{|}{C}\text{—} \end{bmatrix}^{+\cdot} \longrightarrow \text{—}\overset{|}{\underset{|}{C}}{}^{\cdot} \; + \; \overset{}{\underset{}{}}C\text{—}\overset{\underset{}{\cdots+\cdots}}{C}\text{—}C \qquad (10.6)$$

This latter cleavage has been ascribed to resonance stabilization of the resulting ion. A cycloalkene with the double bond in a six-membered ring exhibits the mass spectrometric 'analogue' of the retro-Diels–Alder reaction. The molecular ion fragments in accord with scheme (10.7), the charge appearing on either fragment.

$$\left[\text{⬡} \right]^{+\cdot} \longrightarrow \; \text{⎡⎤}^{+\cdot} \; + \; \| \qquad (10.7)$$

Thermodynamics suggests that the charge will appear on the fragment with the lower ionization energy. If the ionization energies of the fragments are widely different then one of them will take the charge almost exclusively, but if the ionization energies are similar then either fragment may take the charge and appear as an ion in the mass spectrum. Figure 10.12 illustrates in the mass spectrum of cyclohexene the retro-Diels–Alder process and the larger molecular ion peak found with cyclic compounds. There is evidence that, after electron-impact ionization, migration of the double bond occurs in alkenes. The net practical result is that alkene isomers frequently show very similar if not identical mass spectra. Figure 10.13 shows that the isomeric butenes yield very similar mass spectra. Since it can be difficult to reproduce spectra exactly, especially between different laboratories, an identification of any one of the butenes from its mass spectrum alone would be questionable. It is generally not easy to locate a double bond by mass spectrometry of the olefin itself but it may be located by 'fixing' the bond in a derivative (section 5.3, scheme 5.7). Alternatively, location of double bonds in alkenes may be brought about by chemical ionization with methyl vinyl ether as reactant gas (section 7.2).

With increasing numbers of double bonds, alkenes give larger molecular ion peaks and show less of the 'random' bond cleavage characteristic of alkanes and simple alkenes. Terpenoid molecules frequently fragment by loss of C_5-units or multiples of these. The long-chain terpenoid alcohol, solanesol, breaks down in this way in the mass spectrometer, the favoured positions of bond breaking being allylic (scheme 10.8). The elimination of 43 mass units ($C_3H_7 \cdot$) is commonly observed in terpenoid compounds containing isopropyl, iso-propenyl and isopropylidene groups; in the latter two cases, double-bond migration must occur.

It was mentioned earlier that mass spectra of aromatic molecules are characterized by few fragment ions and abundant molecular ions. Benzenoid

Figure 10.12. Mass spectrum of cyclohexene showing large peaks for the molecular ion and the ion at m/z 54 due to a retro-Diels–Alder reaction

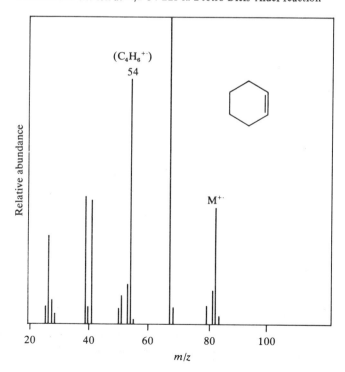

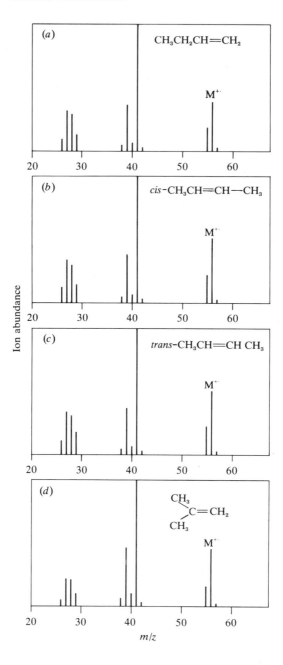

Figure 10.13. Mass spectra of the alkene isomers (*a*) but-1-ene, (*b*) *cis*-but-2-ene, (*c*) *trans*-but-2-ene and (*d*) 2-methylprop-1-ene.

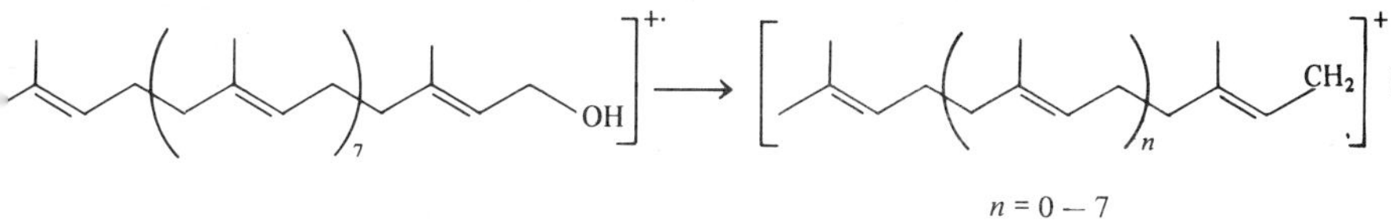

$$\left[\text{—}\text{—}\left(\text{—}\right)_7\text{—}\text{—}\text{—}\text{—}_{OH}\right]^{+\cdot} \longrightarrow \left[\text{—}\text{—}\left(\text{—}\right)_n\text{—}\text{—}\text{—}_{CH_2}\right]^{+} \quad (10.8)$$

$$n = 0 - 7$$

hydrocarbons often afford ions at m/z 91 ($C_7H_7^+$) which can be very abundant, particularly where benzylic groups are present (reaction 10.9). The $C_7H_7^+$ ion is generally considered to have the structure of the tropylium ion.

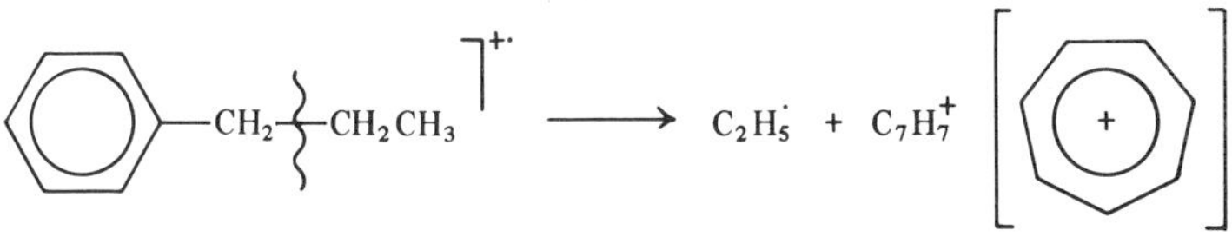

$$\left[\bigcirc\text{—}CH_2\overset{}{\underset{}{\text{—}}}CH_2CH_3\right]^{+\cdot} \longrightarrow C_2H_5^{\cdot} + C_7H_7^+ \left[\bigcirc + \right] \quad (10.9)$$

It is of value to examine the mass spectrum of a suspected aromatic compound for an ion at m/z 91 but it should not be assumed that its presence necessarily implies a $C_6H_5CH_2$ unit in the original molecule. Ions at m/z 91 are frequently observed in the mass spectra of compounds which do not possess a benzyl group; t-butyl-benzene, 1,4-diphenylbutadiene, and camphene (structures (8), (9) and (10)) each give a large peak at m/z 91 which must be due to rearrangement.

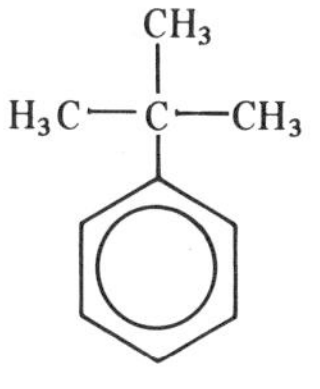

(8) 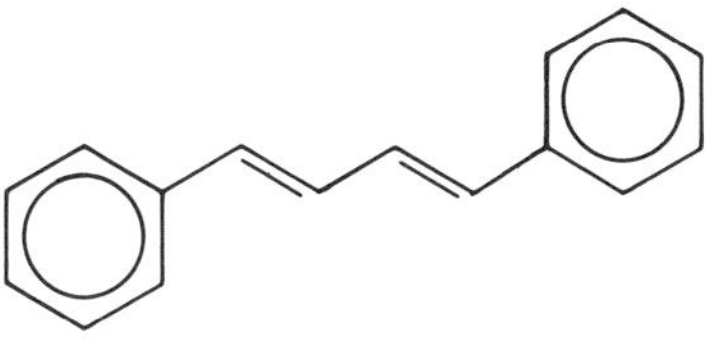   (9)    (10)

However, an ion at m/z 91 of composition C_7H_7 is usually a good guide to the aromatic or benzenoid nature of a molecule. Rearrangement processes in the mass spectra of aromatic compounds are often more obvious and widespread than with aliphatic compounds. The presence of heteroatoms frequently makes these reactions more apparent than in hydrocarbons. The consecutive losses of two CO entities from anthraquinone is an example of one such rearrangement (scheme 10.10; see section 10.9 for other examples).

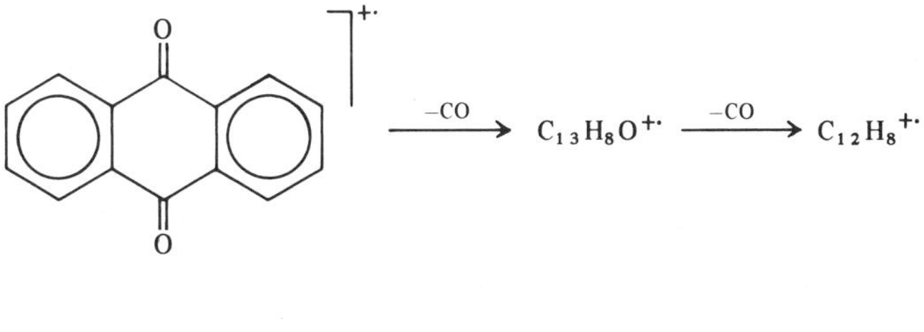

$$\left[\bigcirc\bigcirc\bigcirc\right]^{+\cdot} \xrightarrow{-CO} C_{13}H_8O^{+\cdot} \xrightarrow{-CO} C_{12}H_8^{+\cdot} \quad (10.10)$$

10.6.   Primary fragmentations of aliphatic heteroatomic compounds

The effects of heteroatoms on the mass spectra of organic molecules may be loosely grouped by regarding the heteroatom either as a simple 'saturated' substituent in an alkane or as an 'unsaturated' substituent. Saturated substituents include the halogens, and hydroxyl, amino and sulphydryl groups and are characterized by having lone-pair electrons. Unsaturated substituents include carbonyl and nitrile groups and have π- and lone-pair electrons. Double bonds and aromatic rings may also be included in this category.

Compounds with saturated heteroatomic substituents may be designated R–X, the group X being any one of the common ones listed in table 10.3.

Table 10.3. *Common 'saturated' substituents* (X) *in* R–X *compounds* (R = alkyl)

X	X
F	SH
Cl	SR
Br	NH_2
I	NHR
OH	NR_2
OR	$MetR_n$ (Met = metal atom)

Some of the more important features of the mass-spectrometric fragmentation of these compounds can be grouped under a few headings:

(i)   Simple bond cleavage adjacent to the heteroatom is observed with the charge remaining on the hydrocarbon fragment:

$$R \overset{\rceil^{+\cdot}}{\underset{}{\Big\{}-X} \longrightarrow R^+ + X^\cdot$$

The mass spectra of 1-chloro- and 1-bromopropane (figure 10.14) show prominent losses of halogen. Note the change in the isotope pattern between the molecular ion and the first fragment ion containing no halogen. The loss of halogen is so obvious from the isotope patterns and the number of mass units lost (35, 37 for Cl; 79, 81 for Br) that accurate mass measurement is unnecessary.

The carbon–halogen bond strengths in the molecules CH_3–X (X = halogen; table 10.4) decrease from the very strong C–F bond to the relatively weak C–I bond. However, in the ions CH_3–$X^{+\cdot}$ the bond strengths are greatly reduced for X = F, Cl and Br, but for the iodo-compound the bond strength in the ion is greater than in the molecule. The C–Cl and C–Br bonds in the ions are relatively weak and loss of Cl and Br from the molecular ions of organic chlorides and bromides is generally observed. Although the C–F bond strength is similar to that of C–Cl and C–Br in the ion, simple loss of fluorine from organic fluorides is not commonly observed because of the ease of competing reactions (see below and table 10.5).

Table 10.4. *Relative bond strengths in some* CH_3X *molecules and molecular ions for the processes* $CH_3X \to CH_3{}^{\cdot} + X^{\cdot}$ *and* $CH_3X^{+\cdot} \to CH_3{}^+ + X^{\cdot}$

X	D_{CH_3-X}	$D_{CH_3-X^{+\cdot}}$
F	118 (494)	48 (201)
Cl	81 (339)	49 (205)
Br	67 (280)	51 (213)
I	53 (222)	61 (255)
OH	90 (377)	67 (280)
OCH₃	77 (322)	74 (310)
SH	67 (280)	77 (322)
SCH₃	75 (314)	102 (427)
NH₂	80 (335)	91 (381)

Bond strength (D) is given in kcal/mole and, in parentheses, in kJ/mole.

Table 10.4 gives the bond strengths for several CH_3-X compounds. The order of bond strengths in the molecule are $F > OH > Cl > NH_2 > SMe > SH$, $Br > I$ but in the molecular ion this order changes to $SMe > NH_2 > SH > OMe > OH > I > Br > Cl > F$ (see also Reed, 1962). Note particularly how the C–I, C–S and C–N bond strengths increase in going from the molecule to the molecular ion, possibly due to the fact that iodine, sulphur and nitrogen can stabilize a positive charge much better than fluorine, chlorine, bromine and oxygen. Unlike the other halogens, iodo-compounds frequently fragment with the charge retained on iodine to give an ion at m/z 127.

Elimination of OH˙, SH˙ and NH_2˙ from alcohols, thiols and amines is observed to a slight extent but, particularly with amines, other processes are more important. Cleavage of metal–carbon bonds in simple organometallic compounds is frequently so easy as to lead to low abundances of the mole-

Figure 10.14. Mass spectra of (*a*) 1-chloropropane and (*b*) 1-bromopropane. In each spectrum, elimination of the halogen changes the isotope distribution considerably. Peaks above m/z 45 in the spectrum of 1-chloropropane have been scaled up five times.

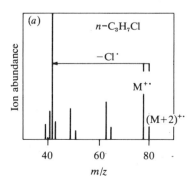
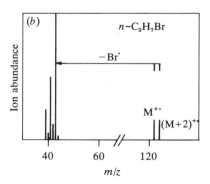

cular ions. One of the most abundant ions in the mass spectrum of tetraethyl-germanium is produced by elimination of an ethyl radical from the molecular ion (figure 10.15):

$$[(C_2H_5)_4Ge]^{+\cdot} \xrightarrow{-C_2H_5^{\cdot}} [(C_2H_5)_3Ge]^+ \xrightarrow{-C_2H_4} [(C_2H_5)_2GeH]^+ \xrightarrow{-C_2H_4} [C_2H_5GeH_2]$$

m/z 190 $\quad\quad\quad$ m/z 161 $\quad\quad\quad$ m/z 133 $\quad\quad\quad$ m/z 105

(ii)   Simple fission adjacent to the heteroatom with simultaneous hydrogen transfer is commonly observed for R–X compounds. Simple cleavage of a molecular ion gives an even-electron species, but cleavage with hydrogen transfer involves more than one bond breaking and gives odd-electron ions.

$$\text{H}\text{——}\text{A}\underset{}{\overset{}{\Big\langle}}\text{—X}\Big]^{+\cdot} \longrightarrow \text{X}^{\cdot} + \text{H}\text{——}\text{A}^+ \quad \text{(Even-electron species)}$$

$$\text{H}\overset{}{\underset{}{\Big\langle}}\text{—A}\overset{}{\underset{}{\Big\langle}}\text{—X}\Big]^{+\cdot} \longrightarrow \text{HX} + \text{A}^{+\cdot} \quad\quad \text{(Odd-electron species)}$$

Whereas loss of F$^{\cdot}$ from fluoro-compounds is not generally observed, the loss of HF is more common perhaps because of its particularly favourable negative heat of formation. The elimination of HCl from chloro-compounds is more favourable than the elimination of HBr from bromo-compounds.

Although the ejection of OH$^{\cdot}$ from alcohols can be observed in their mass spectra, the elimination of H_2O is usually more common, especially when a suitable hydrogen donor site is available. Deuterium labelling has shown that loss of H_2O from alkyl alcohols predominantly involves a hydrogen on a γ- or δ-carbon atom (scheme 10.11) unless steric constraints dictate otherwise.

Figure 10.15. Partial mass spectrum of tetraethylgermanium showing large abundance of $(M - C_2H_5^{\cdot})^+$ ions at m/z 161.

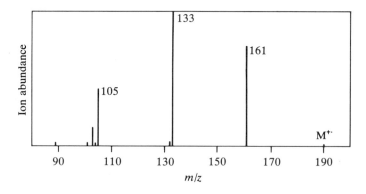

$$\left.\begin{array}{c} \text{C--C--C--C} \\ | \quad\quad\quad | \\ \text{HO} \cdots\cdots\cdots \text{H} \end{array}\right]^{+\cdot} \longrightarrow (M-H_2O)^{+\cdot} \qquad (10.11)$$

A similar process is observed in thiols, which eject H_2S, but amines do not readily eliminate NH_3 possibly due to the stability of the ammonium ion making an alternative fragmentation route more favourable (scheme 10.12; and see category (iii) below).

$$\left.\begin{array}{c} \text{C--C--C--C} \\ | \quad\quad\quad | \\ \text{H}_2\text{N} \cdots\cdots\cdots \text{H} \end{array}\right]^{+\cdot} \longrightarrow \begin{array}{c} \text{C--C--C--C}^{\cdot} \\ | \\ \text{H}_3\text{N}^+ \end{array} \xrightarrow{\quad/\!\!/\quad} (M-NH_3)^{+\cdot} \qquad (10.12)$$

Because of the ease with which H_2O can be ejected from alcohols after ionization, the molecular ions of alcohols are usually of low abundance. In such cases, field ionization and chemical ionization methods are extremely advantageous since these techniques normally yield abundant molecular or quasi-molecular ions. Thus, by comparing electron-impact spectra with those obtained by either field or chemical ionization, details of the molecular ion region become easier to interpret. Alternatively, alcohols may be derivatized to compounds likely to give molecular ions by electron-impact ionization; alkylation, trimethylsilylation or oxidation may serve this purpose. The transfer of hydrogen to hydroxyl during the elimination of II_2O from alcohols seems to involve mainly a six-centre transition state. Deuterium labelling of halogen compounds suggests that a five-membered transition state is most likely during the elimination of hydrohalide. The elimination of ROH from ethers, i.e. cleavage with hydrogen transfer (scheme 10.13), is not abundant perhaps because of the greater ease of alternative reactions as indicated below.

$$\left.\begin{array}{c} \text{C--C--C--C} \\ | \quad\quad\quad | \\ \text{RO} \cdots\cdots\cdots \text{H} \end{array}\right]^{+\cdot} \longrightarrow (M-ROH)^{+\cdot} \quad \text{(low abundance)} \qquad (10.13)$$

The corresponding reaction in organometallic compounds affords (metal + H)$^{+\cdot}$ ions and these are generally found in their mass spectra.

(iii)   Simple cleavage of groups, other than X, from the α-position (α-cleavage) is a common reaction of saturated substituent compounds (scheme 10.14). The ejected radical species may be hydrogen or an alkyl radical.

$$\left.\text{R}\!-\!\!\!\!\begin{array}{c} | \\ \text{C} \xrightarrow{\alpha} \text{X} \\ | \end{array}\right]^{+\cdot} \longrightarrow \text{R}^{\cdot} + \quad \begin{array}{c} \\ \!\!\!>\!\!\overset{+}{\text{C}}\!-\!\text{X} \end{array} \longleftrightarrow \begin{array}{c} \\ \!\!\!>\!\!\text{C}\!=\!\overset{+}{\text{X}} \end{array} \qquad (10.14)$$

It is thought that stabilization of the positive charge by the group X makes the process favourable. This type of fragmentation is more prominent in those compounds having a substituent (X) which can more readily stabilize the charge, as with the NR'R'' group. It has been shown by deuterium labelling that elimination of hydrogen from methanol takes place from the carbon atom and not oxygen (scheme 10.15).

$$H\text{---}\overset{)}{\underset{\langle}{\big|}}\text{---}CH_2\text{---}OH \overset{\rceil\,+\cdot}{} \longrightarrow H^{\bullet} + \overset{+}{C}H_2\text{---}OH \longleftrightarrow CH_2\overset{+}{=\!\!=}\overset{+}{O}H \qquad (10.15)$$

The same process can be observed in ethers, although elimination of an alkyl radical is normally more prominent. The radical may be ejected from either side of the oxygen atom in ethers (scheme 10.16); the larger group (R or R′) appears to be ejected more favourably.

$$R^{\bullet} + \overset{+}{C}H_2\text{---}O\text{---}CH_2R' \longleftrightarrow CH_2\overset{+}{=\!\!=}O\text{---}CH_2R'$$

$$R\text{---}CH_2\text{---}O\text{---}CH_2\text{---}R'\,\rceil\,{}^{+\cdot} \qquad (10.16)$$

$$R'^{\cdot} + RCH_2\text{---}O\text{---}\overset{+}{C}H_2 \longleftrightarrow RCH_2\text{---}\overset{+}{O}\!\!=\!\!CH_2$$

For example, the loss of a propyl radical from ethyl n-butyl ether is more in evidence than the corresponding loss of a methyl radical (figure 10.16). However, examination of the mass spectra of ethers at reduced electron impact energies suggests that the favourable loss of the larger alkyl fragment may be more apparent than real. At low electron energies, subsequent decomposition of the initially produced fragment ions is reduced and the results then suggest that the smaller alkyl fragment is lost more readily. It is a timely reminder of the importance of the initial internal energy in the molecular ion in determining subsequent fragmentation rates (section 1.1 and 9.1) and so determining the abundances of fragment ions.

Similar α-cleavage reactions are found in thioethers and are most marked

Figure 10.16. Mass spectrum of ethyl 1-butyl ether obtained at an electron beam energy of 70 eV. The peak at m/z 59, corresponding to loss of propyl radical from the molecular ion, is much larger than that at m/z 87, corresponding to loss of methyl radical.

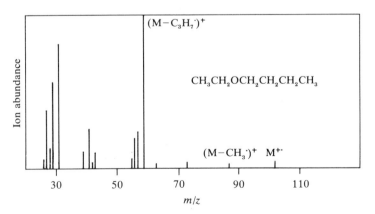

$$
\begin{bmatrix} CH_3-\overset{\curvearrowleft}{\underset{\underset{H}{|}}{C}}H-\overset{\curvearrowright}{|}X \end{bmatrix}^{+\cdot} \begin{cases} \nearrow & C_2H_5{}^+ + X^\cdot \\ \longrightarrow & CH_3CH{=}\overset{+}{X} + H^\cdot \\ \searrow & CH_3{}^\cdot + CH_2{=}\overset{+}{X} \end{cases} \begin{array}{l} \alpha\text{-cleavage} \\ \text{reactions considered} \\ \text{in table 10.5} \end{array}
$$

Table 10.5. *Bond dissociation energies in ions,* $C_2H_5X^{+\cdot}$

Ion from	D_{C-C}	D_{C-X}	D_{C-H}
C_2H_6	44 (184)	26 (109)	26 (109)
C_2H_5F	17 (71)	29 (121)	11 (46)
C_2H_5Br	39 (163)	30 (125)	45 (188)
C_2H_5OH	24 (100)	49 (205)	24 (100)
C_2H_5SH	52 (217)	56 (234)	60 (251)
$C_2H_5NH_2$	31 (130)	66 (276)	29 (121)
C_2H_5CN	48 (201)	36 (150)	56 (234)
$C_2H_5OCH_3$	30 (125)	53 (222)	35 (146)
$C_2H_5SCH_3$	59 (247)	74 (309)	68 (284)

Bond dissociation energy (D) is given in kcal/mole and, in parentheses, in kJ/mole.

in amines in which the nitrogen atom can stabilize the positive charge as an immonium species (scheme 10.17).

$$
[CH_3CH_2N(CH_3)_2]^{+\cdot} \dashrightarrow CH_3^\cdot + \overset{+}{C}H_2{-}N(CH_3)_2 \longleftrightarrow CH_2 = \overset{+}{N}(CH_3)_2 \quad (10.17)
$$

In an amine $R_SCH_2N(R)CH_2R_L$ with small and large alkyl groups R_S and R_L, respectively, the relative importance of ejection of $R_S^\cdot$ over $R_L^\cdot$ increases with decreasing energy of the impacting electrons, as with ethers.

The two types of simple bond cleavage discussed in (i) and (iii) both involve fission adjacent to the α-carbon atom and are competitive processes. In table 10.5 are listed the dissociation energies for α-cleavage of C–C, C–H and C–X bonds in ions of the type $C_2H_5X^{+\cdot}$ (X = H, F, Br, OH, NH_2, SH, OCH_3, SCH_3 and CN). For two competing simple bond fissions in the ground-state ion, the simplified quasi-equilibrium theory suggests that their relative rates of fission (k_1 and k_2) for any internal energy (E) are given by

$$
k_1/k_2 = (\nu_1/\nu_2)[(1 - (E_1/E))/(1 - (E_2/E))]^{N-1}
$$

where ν_1, ν_2 are frequency factors, N the number of oscillators and E_1, E_2 the energies of activation (section 9.4). Approximating this expression through the binomial expansion and putting $\nu_1 \approx \nu_2$ (for simple fission processes, the frequency factors may be equated with the bond vibration frequencies which, for the reactions considered here, are not likely to be sufficiently different to upset the main argument developed by assuming their equality) gives,

$$
\ln (k_1/k_2) \approx ((N - 1)/E)(E_2 - E_1)
$$

i.e. for two simple cleavage processes from a molecular ion, the relative reaction rates leading to initial fragment ions are proportional to the difference between the energies of activation for the two processes. In the absence of significant activation energies for the reverse reactions, the energies E_1 and E_2 may be equated (section 9.8) to the ion bond dissociation energies listed in table 10.5. Hence, inspection of this table should yield at least a qualitative guide to the expected mass spectrometric fragmentation of compounds C_2H_5X, i.e. the bond strength difference when equated with $E_2 - E_1$ should give a semi-quantitative measure of ion abundances in their mass spectra. For comparison, the mass spectra of these compounds are recorded in figure 10.17. Striking relationships may be observed immediately and some of these are annotated here:

(*a*)   From table 10.5 it can be seen that all the bond dissociation energies are low for $C_2H_5F^{+\cdot}$ and that C–H bond cleavage is particularly easy. It might be expected therefore that, with all bond dissociation energies being low, fluoroethane would not show abundant molecular ions and that the $(M - 1)^+$ peak would be most prominent. The mass spectrum shows that this is the case and confirms that loss of F' is an unimportant process in keeping with the C–F bond being the strongest of a set of weak ones.

(*b*)   By way of contrast to fluoroethane, in bromoethane all the bonds are moderately strong with the C–X (X = Br) being the weakest, the C–C bond stronger and the C–H bond the strongest. It might be expected, therefore, that abundant molecular ions would be observed and that the most prominent process would be the loss of Br'. The mass spectrum is in keeping with this deduction.

(*c*)   The mass spectrum of ethane thiol, C_2H_5SH, shows a large molecular ion peak with C–C and C–SH bond cleavage of similar importance. Table 10.5 shows that the bonds in the $C_2H_5SH^{+\cdot}$ ion are all rather strong and that C–C and C–SH α-cleavage reactions are almost equally easy.

(*d*)   In ethanol and ethylamine, the bond dissociation energies for C–C and C–H fissions are very much less than for C–X fission. Therefore, these substances should give moderately abundant molecular ions with marked losses of hydrogen and methyl radicals. Examination of their mass spectra bears out this argument.

Table 10.5 may be used to 'predict' the mass spectra of the simple compounds C_2H_5X. Because the compounds are so simple, further decomposition of the initial fragment ions is greatly reduced. The abundance of an ion in a mass spectrum depends upon its rate of formation and upon its rate of decomposition (section 1.1) but, in the case of C_2H_5X compounds, further decomposition of the initial fragment ions is negligible and their rates of formation determine their abundances. The general principles apparent from an examina-

Figure 10.17. Mass spectra of compounds C_2H_5X (X = H, F, Br, OH, NH_2, CN, SH, SCH_3 OCH_3).

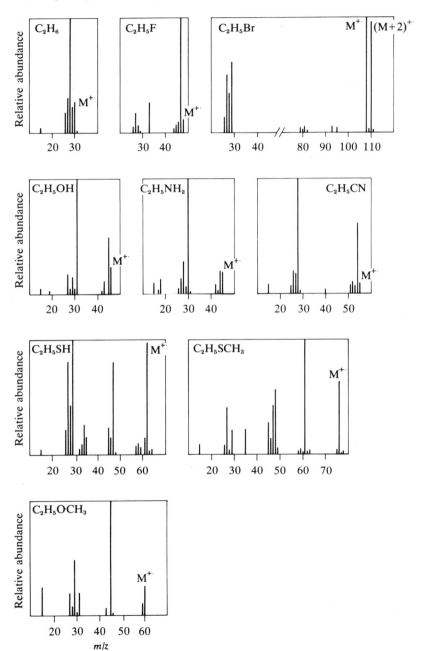

Table 10.6. C–C *bond dissociation energies in n-alkane ions*[a]

Ion from	D_{C-C}
C_2H_6	44 (184)
C_3H_8	30 (126)
C_4H_{10}	21 (88)
C_5H_{12}	15 (63)

[a] Calculated for reactions $C_nH_{2n+1}-CH_3^{+\cdot} \rightarrow C_nH_{2n+1}^+ + CH_3^\cdot$, in kcal/mole and, in parentheses, in kJ/mole.

ation of table 10.5 carry through to a large extent in more complex molecules and serve to provide a thermochemical understanding of many fragmentation processes. In the remainder of this section on fragmentation reactions, reference to the bond dissociation energies given in table 10.5 will prove of value.

It is interesting to note that the C–C bond dissociation energy in ethane is comparatively large and greater than for C–H bond fission; these relative strengths are reflected in the appearance of the mass spectrum which exhibits a facile loss of hydrogen from the molecular ion. However, as the alkyl chain length increases in alkanes, C–C bond dissociation energy decreases more rapidly than C–H bond dissociation energy. Table 10.6 illustrates this effect for loss of a methyl radical from n-alkane molecular ions; for n-butane the relative strengths of C–H and C–C bonds are very similar and much less than the values in ethane and most of the other compounds in table 10.5. In long-chain alkyl compounds, $C_nH_{2n+1}X$, this reduction of the C–C bond strength, coupled with the greater statistical probability of the excess of energy in the ion appearing in a C–C bond, may be expected to lead to an increasing importance of general fragmentation of the alkyl chain; such an effect is observed in the mass spectra of these compounds. As the chain length increases, the specific fragmentations associated with the heteroatom, X, decrease in importance and extensive fission of the alkyl chain is observed.

(iv) Bond breaking at positions other than that adjacent to the α-carbon atom is observed in the mass spectra of aliphatic heteroatom compounds. As the proportion of hydrocarbon in aliphatic heteroatom compounds increases, the influence of the heteroatom on the appearance of the mass spectrum decreases; alternative bond breaking processes become more prominent. Thus, in the long-chain compound shown in scheme (10.18), cleavages such as those at the positions γ, δ, ... increase in importance as R increases in length.

In alkyl chlorides and bromides, $C_nH_{2n+1}X$, this simple bond breaking may also lead to particularly abundant ions, $C_nH_{2n}X^+$, for $n = 3, 4, 5$. The propensity for this reaction shown by these compounds has been attributed to the formation of cyclic structures (reaction 10.19) similar to those found in solu-

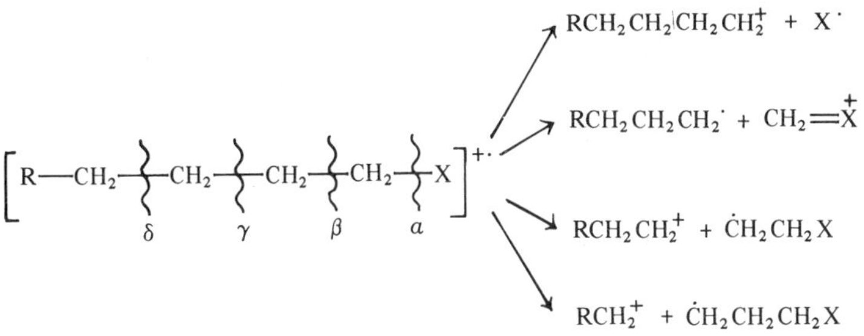

$$\left[R\!-\!CH_2\!-\!CH_2\!-\!CH_2\!-\!CH_2\!-\!X \right]^{+\cdot} \nearrow \begin{array}{l} RCH_2CH_2CH_2CH_2^+ + X^{\cdot} \\[8pt] RCH_2CH_2CH_2^{\cdot} + CH_2\!=\!\overset{+}{X} \\[8pt] RCH_2CH_2^+ + \dot{C}H_2CH_2X \\[8pt] RCH_2^+ + \dot{C}H_2CH_2CH_2X \end{array} \tag{10.18}$$

with the labels δ, γ, β, a on the chain.

tion chemistry. The abundance of ions formed by this process decreases with chain branching in the alkyl moiety and the reaction is not prominent in fluorine or iodine compounds.

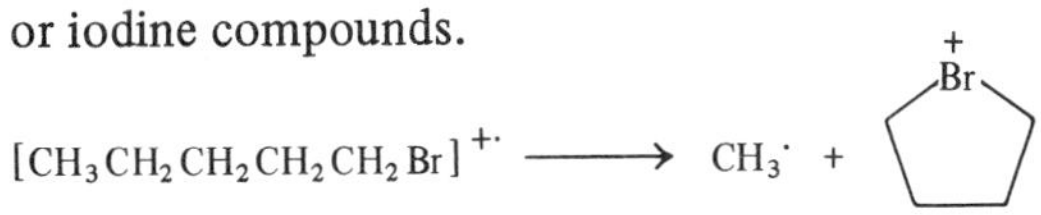

$$[CH_3CH_2CH_2CH_2CH_2Br]^{+\cdot} \longrightarrow CH_3^{\cdot} + \overset{+}{Br} \text{(cyclopentane ring)} \tag{10.19}$$

A similar increased abundance of $C_nH_{2n}X^+$ ions at $n = 3, 4, 5$ is found in amines, but not alcohols or thiols. Unlike the halogen compounds, amines give more abundant ions at $n = 5$ (reaction 10.20), which has been ascribed to a preference for six-membered ring formation (compare losses of HX from R–X compounds where five- and six-membered transition states appear to be involved).

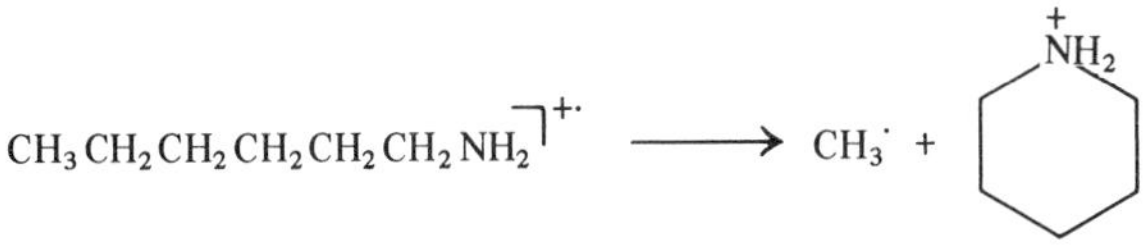

$$CH_3CH_2CH_2CH_2CH_2CH_2NH_2^{+\cdot} \longrightarrow CH_3^{\cdot} + \overset{+}{NH_2} \text{(cyclohexane ring)} \tag{10.20}$$

The commonest 'unsaturated' substituents of alkanes are the carbonyl functional groups $>\!CO$, $-CHO$, $-CO_2H$, $-CO_2R$ and $-CONH_2$, and the sulphoxide, sulphone, phosphate and nitrile groupings.

The mass spectra of ketones are dominated by bond breaking adjacent to the α-carbon atom (α-cleavage) similar to reaction scheme (10.14) shown above for saturated substituents. The comparable breaking of the carbon to heteroatom bond, as in (i) above, is not observed presumably because of the increased strength of $C\!=\!O$ bonds as against $C\!-\!OH$:

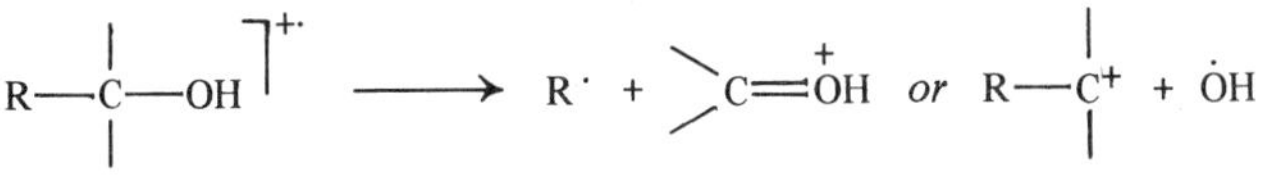

$$\left[R\!-\!\overset{|}{\underset{|}{C}}\!-\!OH \right]^{+\cdot} \longrightarrow R^{\cdot} + \;>\!C\!=\!\overset{+}{O}H \; \text{ or } \; R\!-\!\overset{|}{\underset{|}{C}}{}^+ + \dot{O}H$$

but

$$\left[R\!-\!C\!=\!O \right]^{+\cdot} \longrightarrow R^{\cdot} + \;-\!C\!\equiv\!\overset{+}{O} \; \text{ and not } \; R\!-\!\overset{|}{C}{}^{+\cdot} + O$$

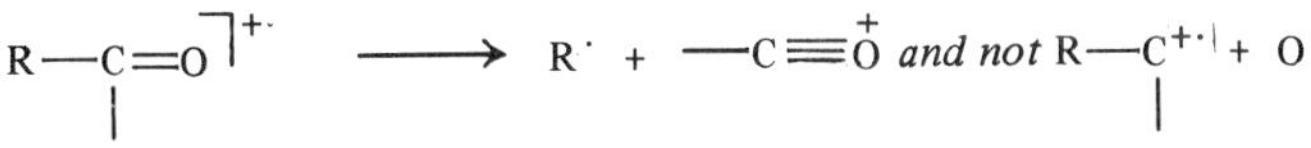

The cleavage adjacent to carbonyl is reminiscent of the photolytic decomposition of ketones and aldehydes. The large degree of resonance stabilization of the positive charge in the fragment ion is said to be responsible for the ease of fission at the α-position. As with α-cleavage of saturated substituent groups, cleavage adjacent to the carbonyl group appears to lead preferentially to the ejection of the largest radical from the molecular ion in 70 eV spectra. The α-cleavage process is frequently followed by elimination of CO from the fragment ion. The mass spectrum of ethyl 1-propyl ketone (figure 10.18) shows the elimination of both ethyl and propyl groups followed by ejection of CO in each case (scheme 10.21). The pathways illustrated in scheme (10.21) may also be rationalized in charge localization terms (scheme 10.22).

$$
\begin{array}{c}
C_2H_5-\underset{\underset{O}{\|}}{C}-C_3H_7 \\
\end{array}^{+\cdot}
\begin{array}{c}
\xrightarrow{-C_3H_7\cdot} C_2H_5CO^+ \xrightarrow{-CO} C_2H_5{}^+ \\
\xrightarrow{-C_2H_5\cdot} C_3H_7CO^+ \xrightarrow{-CO} C_3H_7{}^+
\end{array}
\left.\rule{0pt}{40pt}\right\} \quad (10.21)
$$

$$
\begin{array}{l}
C_2H_5-\underset{C_3H_7}{C}{\overset{\cdot+}{=}}O \xrightarrow{-C_3H_7\cdot} C_2H_5-C{\overset{+}{\equiv}}O \xrightarrow{-CO} C_2H_5{}^+ \\[20pt]
C_2H_5-\underset{C_3H_7}{C}{\overset{\cdot+}{=}}O \xrightarrow{-C_2H_5\cdot} C_3H_7-C{\overset{+}{\equiv}}O \xrightarrow{-CO} C_3H_7{}^+
\end{array}
\left.\rule{0pt}{40pt}\right\} \quad (10.22)
$$

In aldehydes, one of the groups adjacent to the carbonyl is simply hydrogen, and elimination of atomic hydrogen from these compounds is observed:

$$
CH_3CH_2C{\underset{\overset{\|}{O}}{\overset{\displaystyle\diagup H}{}}} \longrightarrow H\cdot + CH_3CH_2C{\overset{+}{\equiv}}O
$$

Figure 10.18. Mass spectra of ethyl 1-propyl ketone showing that ion abundance at m/z 57 corresponding to loss of $C_3H_7\cdot$ is greater than that at m/z 71 corresponding to loss of $C_2H_5\cdot$ at 70 eV.

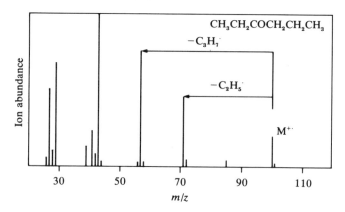

As well as this simple type of bond breaking reaction, ketones and aldehydes exhibit loss of H_2O from their molecular ions. The process has been
examined by deuterium labelling and found to be complex.

The α-cleavage adjacent to carbonyl is still important in the mass spectra
of acids, esters and amides, and these functional groups may be recognized
through the loss of appropriate units from the molecular ions. In each of the
examples shown in scheme (10.23), simple fission at either side of the carbonyl group may occur, but the abundances of the fragment ions so formed
can vary a great deal. Thus, aliphatic acids and amides generally produce only
small peaks in their mass spectra corresponding to the loss of $OH^{\cdot}$ and $NR'_2{}^{\cdot}$
(scheme 10.23a,c), respectively, whereas the ejection of $OR''^{\cdot}$ from esters
(scheme 10.23b) is much more noticeable. However, aromatic acids and
amides do eject $OH^{\cdot}$ and $NR'_2{}^{\cdot}$ from their respective molecular ions. Usually,
cleavage between the alkyl chain and the carbonyl function is more prominent
and $(M - {}^{\cdot}CO_2H)^+$, $(M - {}^{\cdot}CO_2R')^+$ and $(M - {}^{\cdot}CONR'_2)^+$ ions are observed.
Also, the ejected moieties CO_2H, CO_2R' and $CONR'_2$ appear as charged
species at the relevant m/z values. For example, the methyl ester, RCO_2CH_3,
of an aliphatic acid, affords ions at R^+, RCO^+, $CO_2CH_3^+$ and OCH_3^+.

$$
\begin{array}{ll}
R\!-\!\overset{\displaystyle \parallel}{\underset{O}{C}}\!-\!OH\Big]^{+\cdot} \longrightarrow & R^{+\cdot} \text{ and } \overset{+}{C}O_2H \qquad (a) \\[2mm]
& \searrow\; RCO^+ \\[4mm]
R\!-\!\overset{\displaystyle \parallel}{\underset{O}{C}}\!-\!OR'\Big]^{+\cdot} \longrightarrow & {}^{\cdot}R^+ \text{ and } \overset{+}{C}O_2R',\overset{+}{O}R' \quad (b) \\[2mm]
& \searrow\; RCO^+ \\[4mm]
R\!-\!\overset{\displaystyle \parallel}{\underset{O}{C}}\!-\!NR''_2\Big]^{+\cdot} \longrightarrow & R^+ \text{ and } \overset{+}{C}ONR'_2 \qquad (c) \\[2mm]
& \searrow\; RCO^+
\end{array}
\qquad (10.23)
$$

Similar α-cleavages are found in the mass spectra of sulphoxides and sulph-
ones, but not in nitriles. Although metal carbonyls, $Met(CO)_n$, are not really
analogous to organic carbonyl compounds, it is convenient to note here that
α-cleavage of the carbon monoxide ligand is extremely easy in these organometallic compounds. Where several carbonyl ligands are present, fragment
ions are observed corresponding to successive losses of CO:

$$Met(CO)_n^{+\cdot} \longrightarrow Met(CO)_{n-1}^{+\cdot} \longrightarrow Met(CO)_{n-2}^{+\cdot} \longrightarrow \text{etc.} \longrightarrow Met^{+\cdot}$$

The α-cleavage in amides is very important for sequence analysis of amino-
acid residues in peptide chains. Bond breaking adjacent to carbonyl yields
type A and type B ions as shown in scheme (10.24).

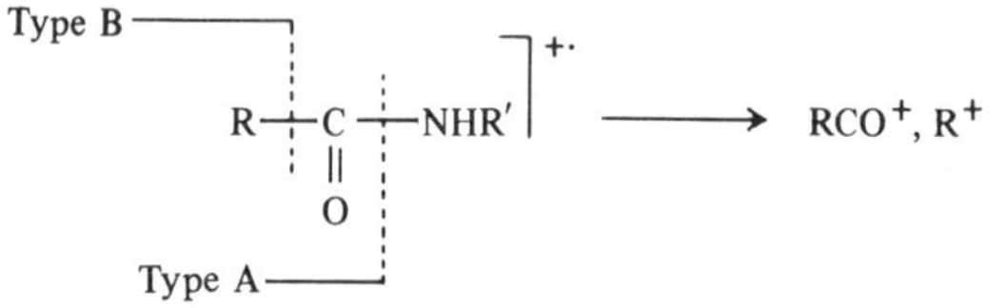

Type B ⎤+·

$$R \;\vert\; C \;\vert\; NHR' \longrightarrow RCO^+, R^+ \qquad (10.24)$$
$$\underset{O}{\overset{\|}{}}$$

Type A ⎦

Type A ions may eject CO to give type B ions. A peptide chain normally cleaves at each CO—NH bond to yield type A fragment ions and by examining these ions, the sequence may be determined from the mass spectrum. It is usual to acetylate the free amino-end of the peptide and to convert the acid terminal to its methyl ester. The tripeptide, Leu–Ala–Gly, methylated and acetylated in this way, gives fragment ions at m/z 156 ($C_8H_{14}NO_2$), 227 ($C_{11}H_{19}N_2O_3$), 284 ($C_{13}H_{22}N_3O_4$) and 315 ($C_{14}H_{25}N_3O_5$; molecular ion) from which the sequence is immediately obtained (scheme 10.25).

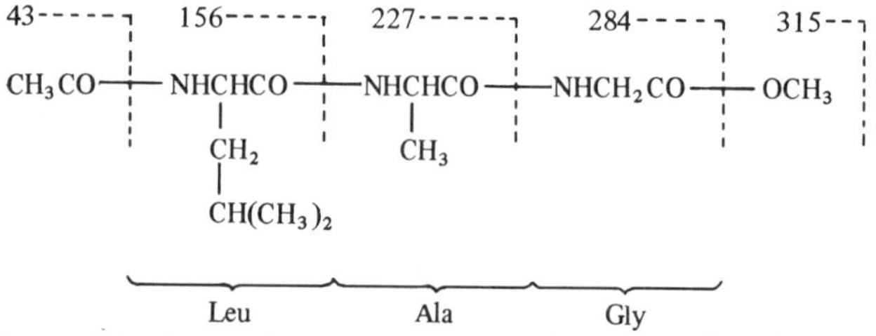

43------   156------   227------   284----   315--

CH₃CO—NHCHCO—NHCHCO—NHCH₂CO—OCH₃

        | |
        CH₂   CH₃
        |
        CH(CH₃)₂

(10.25)

Leu   Ala   Gly

For simple peptides it may be unnecessary to determine the elemental compositions of the fragment ions except as confirmation of the sequence. Larger peptides often give complex mass spectra because of competing side-chain fragmentation, and it may then be necessary to determine the elemental compositions of all ions in the spectrum by accurate mass measurement. From the elemental compositions the sequence may be determined as above and computer programs have been developed to deal with the sorting involved.

A further parallel between compounds with saturated and those with unsaturated substituents is found in the six-centre transfer of hydrogen in carbonyl compounds comparable with the six-centre elimination of H_2O from alcohols. The reaction is often called the McLafferty rearrangement and is exemplified in scheme (10.26) for pentan-2-one. The elimination of H_2O from n-pentanol is shown for comparison. Again, the rearrangement reaction of the odd-electron molecular ion involves more than one bond cleavage and yields an odd-electron fragment ion. In comparing the two reactions of scheme (10.26), notice that, for the carbonyl compounds, the charge generally remains with the heteroatom fragment and a neutral aliphatic moiety is ejected but, for alcohols, the charge is retained on the aliphatic part and the heteroatom is ejected as neutral H_2O.

Just as the elimination of H_2O from alcohols is parallelled by the elimination of H_2S from thiols, the six-centre rearrangement of carbonyl compounds

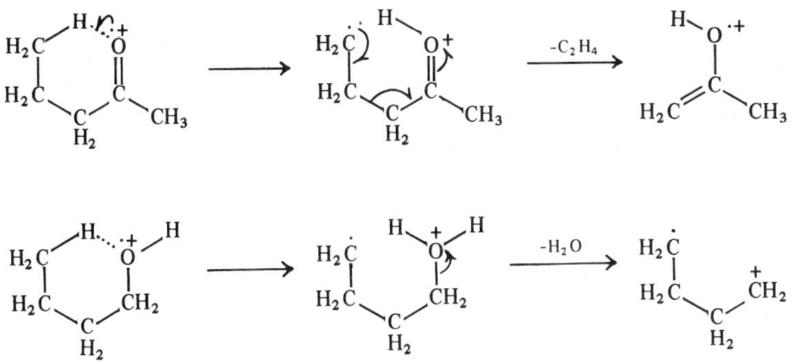

(10.26)

may be regarded as a particular instance of the general and ubiquitous phenomenon shown schematically in reaction (10.27). The atoms X and Y may be
carbon, oxygen, sulphur, phosphorus or nitrogen in various combinations.
Deuterium labelling has demonstrated that hydrogen transfer takes place

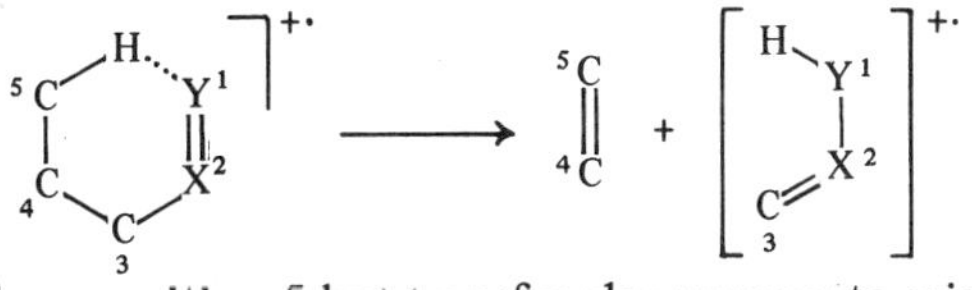

(10.27)

mainly from position 5 but transfer also appears to arise from positions 3 and
4. The rearrangement is observed in the mass spectra of amides and sulphoxides as, for example, in reaction schemes (10.28).

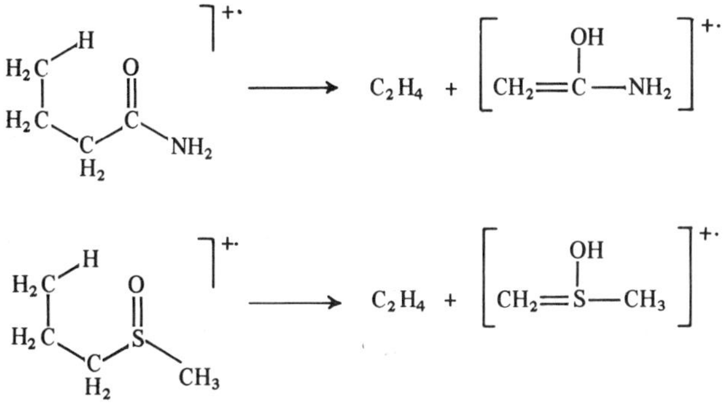

$$C_2H_4 + \left[CH_2{=}\overset{\displaystyle OH}{C}{-}NH_2 \right]^{+\cdot}$$

(10.28)

$$C_2H_4 + \left[CH_2{=}\overset{\displaystyle OH}{S}{-}CH_3 \right]^{+\cdot}$$

Esters that have an alkoxy group longer than methoxy also show the
rearrangement and give a peak corresponding to the ionized free acid:

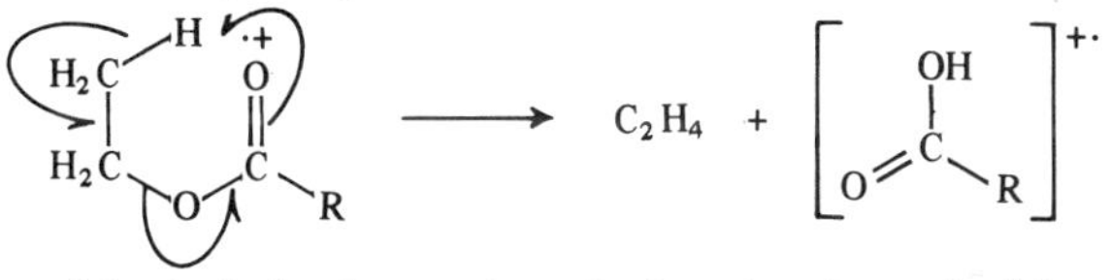

$$C_2H_4 + \left[\overset{\displaystyle OH}{O{=}C}{-}R \right]^{+\cdot}$$

Sometimes this single hydrogen transfer in esters is masked by a double
hydrogen transfer to give what is thought to be the protonated acid. The

1-propyl ester of acetic acid (figure 10.19) gives a large peak at m/z 61 due to this double rearrangement (reaction 10.19). Deuterium labelling suggests that two hydrogen atoms are transferred mainly from the β- and γ-carbon atoms of the alkoxy group. A similar double hydrogen rearrangement with formation of a protonated species is found in the mass spectra of many types of esters including carbonates and those of diacids.

$$CH_3C\begin{smallmatrix}O\\\\OC_3H_7\end{smallmatrix} \quad \right]^{+\cdot} \xrightarrow{\text{Double hydrogen transfer}} \quad CH_3C\begin{smallmatrix}\overset{+}{O}H\\\\OH\end{smallmatrix} \quad + \quad C_3H_5^{\cdot} \quad (10.29)$$

$$m/z \ 102 \qquad\qquad\qquad m/z \ 61$$

The generalized simple cleavage of aliphatic chains in reaction type (iv) of saturated substituents is found in a modified form with unsaturated substituents in alkanes. Thus, esters of general formula, $C_nH_{2n+1}CO_2R$, give rise to fragment ions of compositions, $C_nH_{2n}CO_2R$. However, unlike halide compounds where the most abundant fragment ions have a composition $C_nH_{2n}X$ with $n = 4$, the ester fragment ions are most abundant at the compositions corresponding to $n = 2, 6, 10$, etc. The effect has been explained in terms of transfer of hydrogen and fragmentation as shown in the reaction sequence (10.30).

$$(10.30)$$

Figure 10.19. Mass spectrum of 1-propyl acetate showing peak at m/z 61 due to double hydrogen rearrangement. The peaks above m/z 100 have been scaled up 100 times.

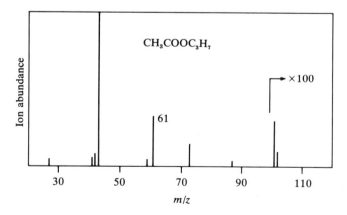

$CH_3COOC_3H_7$

×100

61

Ion abundance

30   50   70   90   110

m/z

10.6.1. Summary

Many simple fragmentation reactions of heteroatom molecular ions may be summarized by reference to scheme (10.31) in which X represents the heteroatom and R^1, R^2 and R^3 are substituents; when X is doubly bonded to the adjacent carbon atom R^3 is not present.

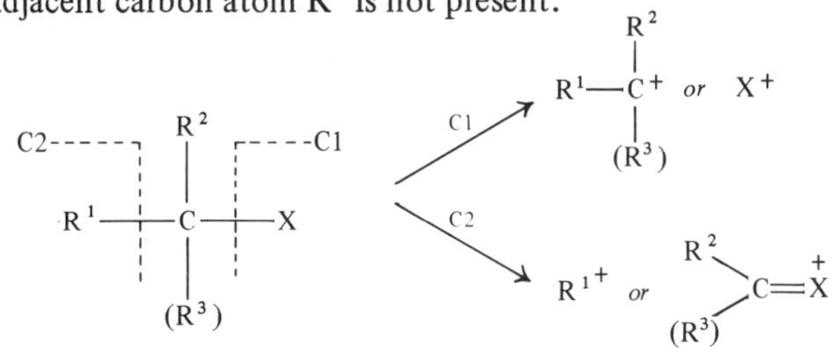

$$(10.31)$$

Type C1 cleavage is breaking of the C–X bond and is observed with saturated heteroatom substituents (X = Cl, Br, I, OH, SH, NH_2). The C1 cleavage is not observed with unsaturated heteroatom substituents. Type C2 cleavage is commonly observed for both saturated and unsaturated heteroatom substituents of alkanes. The extent to which C1 or C2 bond breaking occurs is very dependent on the nature of the heteroatom and the structure of the molecule. Either or both fragments from C1 and C2 reactions may appear as even-electron charged species in the mass spectrum as shown in scheme (10.31).

The other common process observed in aliphatic heteroatom compounds is rearrangement of hydrogen accompanying elimination of a neutral species to give odd-electron fragment ions. Again the process can be represented in a common formalism (scheme 10.32) by two reaction types (RE1 and 2). With saturated substituents (X), the rearrangement-elimination reaction (RE1) is general except for amines, but, for unsaturated substituents, the alternative process (RE2) is general. The processes shown in scheme (10.32) have been drawn with the six-membered ring transition state, but it should be recalled

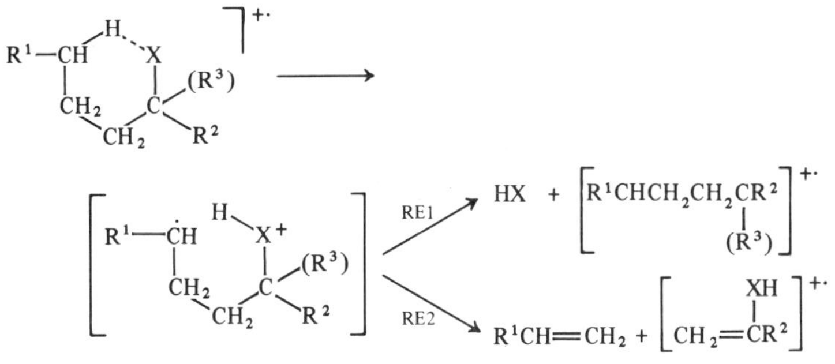

$$(10.32)$$

that the results of deuterium labelling are not unequivocal and some hydrogen transfer occurs through smaller and larger cyclic transition states. For halogen compounds, a similar five-membered state appears to be more important.

A double bond or an aromatic ring can be considered here as an unsaturated substituent in alkanes. The 'allylic' or 'benzylic' types of cleavage adjacent to these unsaturated centres are comparable with, for example, the α-cleavage (C2) of carbonyl compounds if the unsaturated centres take the place of the heteroatom (scheme 10.33).

$$CH_3 \overbrace{} CH{=}O^{\top +\cdot} \longrightarrow CH_3{}^{\cdot} + H\overset{+}{C}{=}O \longleftrightarrow HC{\equiv}\overset{+}{O}$$

(10.33)

$$CH_3 \overbrace{} CH_2{-}CH{=}CH_2{}^{\top +\cdot} \longrightarrow CH_3{}^{\cdot} + \overset{+}{C}H_2{-}CH{=}CH_2 \longleftrightarrow CH_2{=}CH{-}\overset{+}{C}H_2$$

The allyl cation is isoelectronic with the formyl cation in the resonance formalism; in the first case, a positive charge is delocalized over three carbon atoms and, in the other, over one carbon and one oxygen atom. However, the comparison is not a particularly good one because the olefinic double bond itself may be considered isoelectronic with the chloro-function, for example. As with cleavage adjacent to a chlorine atom, cleavage adjacent to the olefin group occurs with hydrogen transfer and is an important process (scheme 10.34).

$$R{-}Cl^{+\cdot} \to HCl + [R - H]^{+\cdot}$$

$$R{-}CH{=}CH_2{}^{+\cdot} \to H_2C{=}CH_2 + [R - H]^{+\cdot}$$

(10.34)

In many ways, double bonds and aromatic rings have characteristics in common with both saturated and unsaturated substituents.

It must be stressed that, so far, only primary fragmentation of the molecular ion has been considered. The effects of subsequent fragmentation may modify the mass spectrum to such an extent that the primary processes appear to have little prominence. For example, if the rate of fragmentation of M^+ to give A^+ is slower than the fragmentation of A^+ to B^+ (scheme 10.35), then the size of the peak corresponding to A^+ in the mass spectrum will be very much

$$M^+ \to A^+ \to B^+$$

(10.35)

reduced (section 1.1). Also, subsequent fragmentation is not so easy to classify and will be discussed later. If only for these two reasons there is little to substitute for experience gained in the interpretation of mass spectra. Attempts to program computers to elucidate structures from mass spectra have met with some success (section 3.4). In this respect, the exceptional memory characteristics of a computer are invaluable for storing accumulated knowledge and experience.

10.7. Primary fragmentations of aromatic heteroatomic compounds

The mass spectra of aromatic compounds may be subdivided into three classes: those in which the aromatic ring is separated from the heteroatom by an aliphatic part (aralkyl compounds), those in which the aromatic ring is attached directly to the heteroatom, and those in which the heteroatom forms part of the aromatic ring. Since the presence of the aromatic ring may modify the mass spectrum considerably compared with what is observed with aliphatic compounds these three classes are considered separately.

(i)   Aralkyl compounds have mass spectra with characteristics common to aliphatic and aromatic compounds separately. The presence of the aromatic ring appears to confer greater stability on the molecular ions which are much more prominent than for simple aliphatic materials. The aromatic ring may itself act as the substituent X in formula (10.31) as mentioned above, and give corresponding fragmentations. Thus cleavage and rearrangement processes C1, C2 and RE2 (scheme 10.36a,b,c) are observed in 1-butylbenzene.

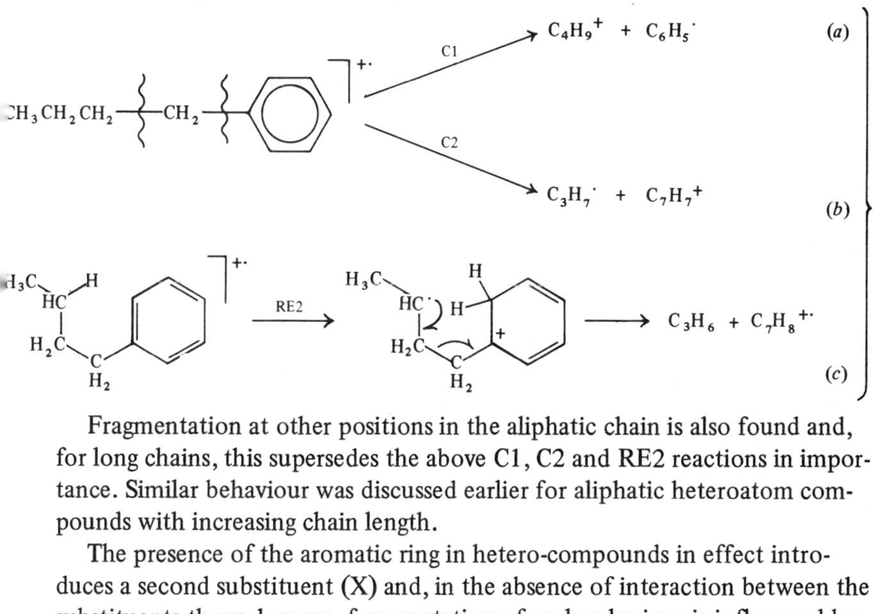

Fragmentation at other positions in the aliphatic chain is also found and, for long chains, this supersedes the above C1, C2 and RE2 reactions in importance. Similar behaviour was discussed earlier for aliphatic heteroatom compounds with increasing chain length.

The presence of the aromatic ring in hetero-compounds in effect introduces a second substituent (X) and, in the absence of interaction between the substituents through space, fragmentation of molecular ions is influenced by each of the substituents. Thus, the decomposition of methyl-3-phenylbutyrate on electron impact gives some of its fragment ions as shown (scheme 10.37).

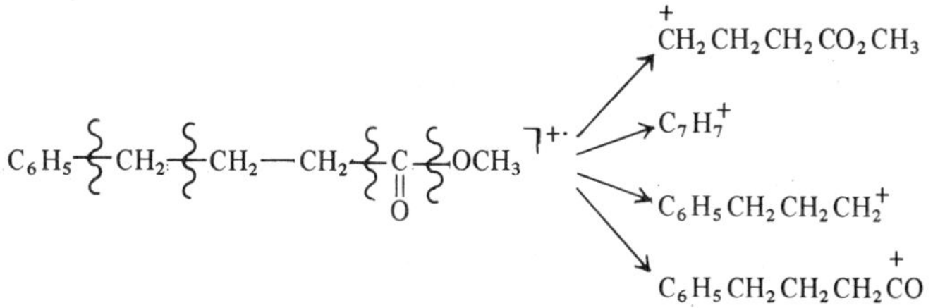

The closer the two substituents to each other, the more they may interact through space or through bonding so as to modify their separate effects.

(ii)   When the substituent heteroatom is attached directly to a benzene ring,

simple C1 cleavage may still occur as with chlorobenzene and nitrobenzene
(see schemes). It is interesting to note that the C—F bond strength is great

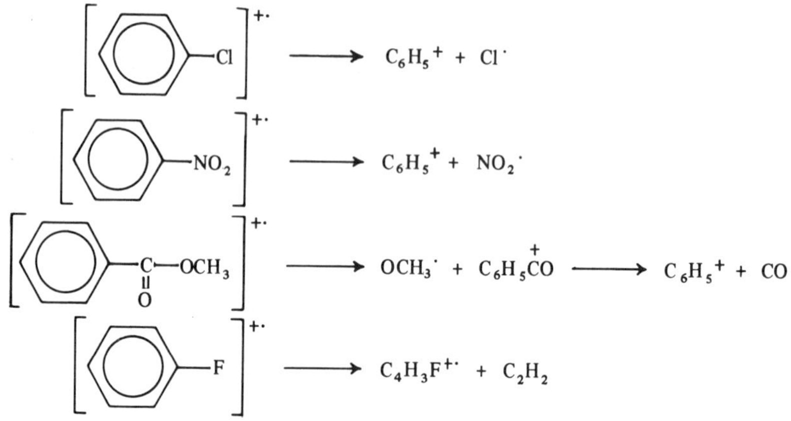

$$[C_6H_5{-}Cl]^{+\cdot} \longrightarrow C_6H_5{}^+ + Cl^\cdot$$

$$[C_6H_5{-}NO_2]^{+\cdot} \longrightarrow C_6H_5{}^+ + NO_2{}^\cdot$$

$$[C_6H_5{-}\underset{O}{\overset{\|}{C}}{-}OCH_3]^{+\cdot} \longrightarrow OCH_3{}^\cdot + C_6H_5\overset{+}{C}O \longrightarrow C_6H_5{}^+ + CO$$

$$[C_6H_5{-}F]^{+\cdot} \longrightarrow C_4H_3F^{+\cdot} + C_2H_2$$

enough to prevent simple cleavage as is the case with aliphatic compounds
and, instead, C_2H_2 is eliminated from the benzene ring. The elimination of
halogen from chlorobenzene may be an ion pair process:

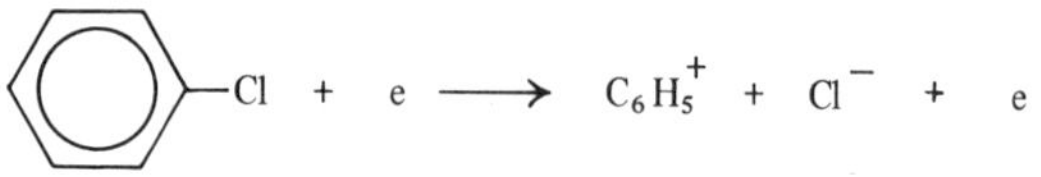

$$C_6H_5{-}Cl + e \longrightarrow C_6\overset{+}{H_5} + Cl^- + e$$

This would be a case of primary cleavage into positive and negative ions on
electron impact and is observed for halogen compounds at low electron-beam
energies. Since mass spectrometers are usually set to monitor only positive
ions, the negative ones are ignored in most mass spectrometry. The existence
of this process does not affect the observed spectrum except in so far as the
energy of the process is different from that in which a chlorine atom is ejected
(section 1.1).

Often, simple C1 cleavage does not occur and rearrangement of the molecular ion takes place before fragmentation. Rearrangment processes are
covered in more detail in section 10.9 but those germane to this description
of fragmentation of aromatic heteroatom compounds are discussed here.

Phenols, phenyl ethers and amines occur commonly in aromatic chemistry
and exhibit complex fragmentation behaviour. Phenol eliminates not OH$^\cdot$, but
CO and CHO$^\cdot$ from the molecular ion. Phenyl methyl ether ejects CH_2O and
CHO$^\cdot$ but not methyl or methoxy radical, and aniline eliminates HCN from
the molecular ion. These processes are quite general (scheme 10.38).

In some circumstances, nitro-compounds do not simply eject NO$_2^\cdot$ but a
complex reaction sequence occurs in which NO$^\cdot$ and CO are eliminated successively (reaction 10.39). This process is observable to a slight extent in the mass
spectrum of nitrobenzene itself but it can increase in importance in other

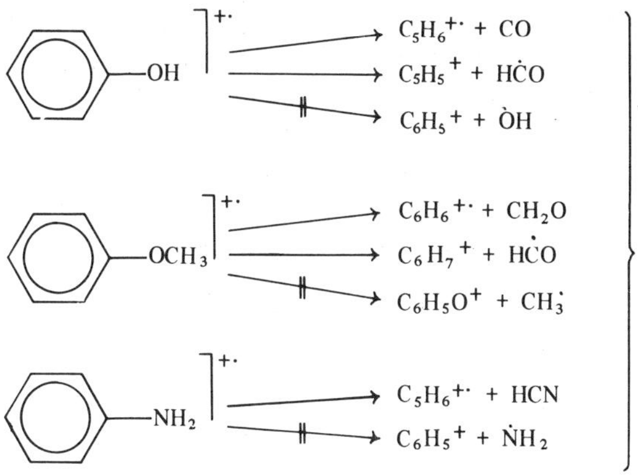

$$
\begin{array}{l}
\text{(phenol)}^{+\cdot} \longrightarrow C_5H_6^{+\cdot} + CO \\
\phantom{\text{(phenol)}^{+\cdot}} \longrightarrow C_5H_5^{+} + H\dot{C}O \\
\phantom{\text{(phenol)}^{+\cdot}} \xrightarrow{\;\;\Vert\;\;} C_6H_5^{+} + \dot{O}H
\end{array}
$$

$$
\begin{array}{l}
\text{(anisole)}^{+\cdot} \longrightarrow C_6H_6^{+\cdot} + CH_2O \\
\phantom{\text{(anisole)}^{+\cdot}} \longrightarrow C_6H_7^{+} + H\dot{C}O \\
\phantom{\text{(anisole)}^{+\cdot}} \xrightarrow{\;\;\Vert\;\;} C_6H_5O^{+} + CH_3^{\cdot}
\end{array}
\Bigg\} \quad (10.38)
$$

$$
\begin{array}{l}
\text{(aniline)}^{+\cdot} \longrightarrow C_5H_6^{+\cdot} + HCN \\
\phantom{\text{(aniline)}^{+\cdot}} \xrightarrow{\;\;\Vert\;\;} C_6H_5^{+} + \dot{N}H_2
\end{array}
$$

$$
\left[C_6H_5\overset{O}{\underset{O^-}{\overset{+}{N}}}\right]^{+\cdot} \longrightarrow \left[C_6H_5-O-N{=}O\right]^{+\cdot} \xrightarrow{-NO^{\cdot}} C_6H_5O^{+} \xrightarrow{-CO} C_5H_5^{+}
$$

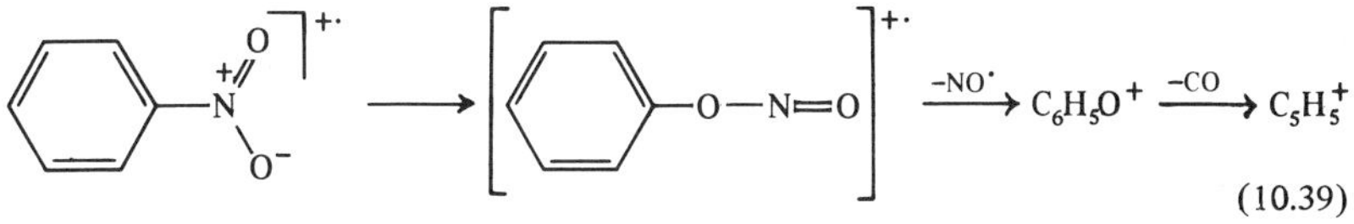

$$(10.39)$$

compounds so as to make the simple elimination of $NO_2^{\cdot}$ inconsequential. Table 10.7 lists the commoner substituents exhibiting such rearrangements and the neutral species ejected.

Table 10.7. *More commonly occurring substituents of aromatic rings which eject neutral particles resulting from rearrangement*

Substituent	Neutral particle(s) ejected after rearrangement[a]
NO_2	NO, CO (NO_2)
NH_2	HCN
$NHCOCH_3$	C_2H_2O, HCN
CN	HCN
F	C_2H_2
OCH_3	CH_2O, CHO, CH_3
OH	CO, CHO
SO_2NH_2	SO_2, HCN
SH	CS, CHS (SH)
SCH_3	CS, CH_2S, SH (CH_3)

[a] Particles shown in parentheses are ejected without rearrangement. Where these are given, it indicates that these particle losses are usually observed along with those resulting from rearrangement; their relative proportions are very variable.

Often there is more than one heteroatomic substituent in an aromatic compound and they may interact to modify the above fragmentation processes considerably. The relative positions of the substituents around the aromatic ring are also important. Whereas the elimination of a methyl radical from the methoxyl group in phenyl methyl ether is a scarcely detectable process, in 2- and 4-methoxyaniline the process is more important than the usual behaviour of the aromatic methoxyl group (loss of CH_2O and CHO'). The ease of elimination of methyl from 2- and 4-methyoxyaniline has been ascribed to a resonance stabilization of the fragment ion. The behaviour of 3-methoxyaniline, in which such resonance interaction is considered impossible, is more 'normal' in that CH_2O and CHO' are eliminated from the molecular ion (reactions 10.40).

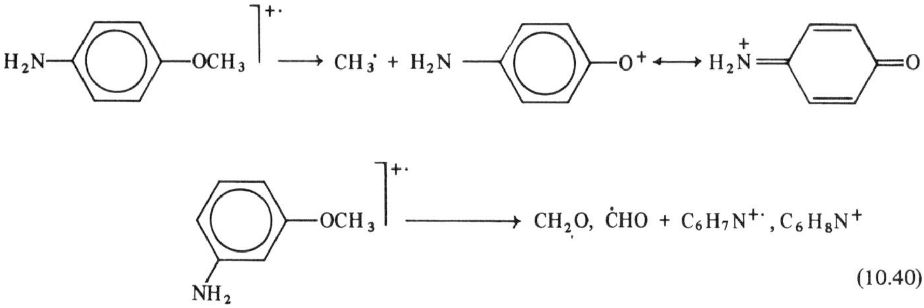

$$(10.40)$$

As in solution chemistry, *ortho*-substituents in benzene rings may behave differently from both the *meta-* and *para*-substituents. The difference in behaviour is often called an *ortho-effect*. The elimination of CH_4O from methyl anthranilate and of OH' from *ortho*-nitrotoluene are examples of this

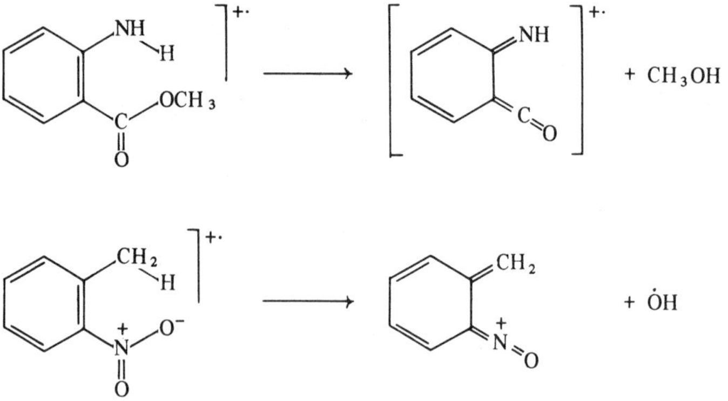

$$(10.41)$$

ortho-effect (reactions 10.41). The process is quite general and may be represented by reaction scheme (10.42) in which A, X, Y and Z may be carbon, oxygen, nitrogen and sulphur.

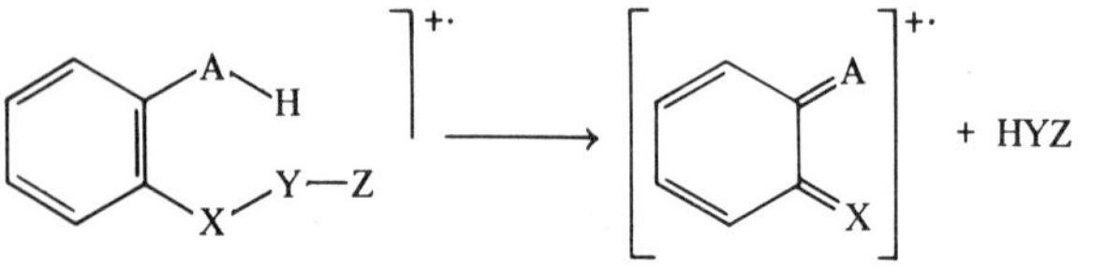

$$+ \text{ HYZ} \qquad (10.42)$$

It is frequently observed that *ortho*-substituents are eliminated more readily than *meta-* or *para*-substituents and a cyclization process is often invoked as an explanation but there is little evidence to support it. This ejection of *ortho*-substituents is also described as an *ortho*-effect. An example is shown in scheme (10.43).

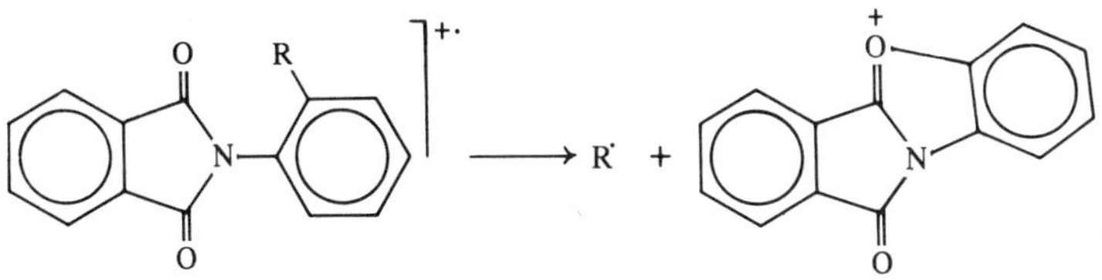

$$R=Cl, NO_2, ...$$

(10.43)

For disubstituted benzenes the order of fragmentation at the substituents may be predicted quite accurately from table 10.8 which gives alongside each substituent the neutral particle commonly ejected during its fragmentation. Generally, the higher a substituent in the table, the more readily fragmentation begins there. Thus, in a chloroacetophenone, elimination of methyl radical is observed before chlorine and in a fluorobromobenzene loss of bromine is observed but not of fluorine (scheme 10.44).

Table 10.8. *Order of ease of fragmentation initiated by the presence of a substituent in a benzene ring*[a]

Substituent	Neutral particle eliminated
$COCH_3$	CH_3
CO_2CH_3	OCH_3
NO_2	NO_2
I	I
OCH_3 ⎤	CH_2O, CHO
Br	Br
OH ⎦	CO, CHO
CH_3	H
Cl	Cl
NH_2	HCN
CN	HCN
F	C_2H_2

[a] Decreasing ease of fragmentation from top to bottom of table. The bracketed substituents are closely similar in ease of fragmentation.

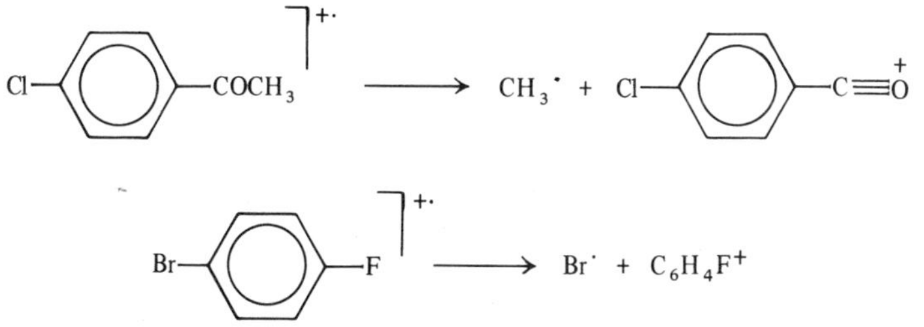

(10.44)

The table is intended as a guide only because strong interaction effects between substituents can modify their behaviour as shown earlier. Benzenes substituted with OCH_3, Br and OH groups require very similar energies for fragmentation so that where these substituents occur together in a benzene compound, table 10.8 serves as only a nominal guide to the order of fragmentation.

The role of substituents in the mass spectra of aromatic compounds has been examined in some detail through the use of Hammett σ-constants (linear free energy relationships), but there is little justification for doing this. The conditions obtaining in electron impact mass spectrometry do not permit the use of the same thermochemical relationships as used for solution chemistry (chapter 9). Lack of knowledge of ion structures and the electronic states leading to fragmentation also provide good grounds for reluctance to draw conclusions from observed correlations with σ-constants.

(iii)   Aromatic compounds like benzene yield relatively stable and hence abundant molecular ions. The most important initial fragmentation of benzene leads to the elimination of C_2H_2 and C_2H_3.

$$C_6H_6^{+\cdot} \begin{cases} \nearrow C_2H_2 + C_4H_4^{+\cdot} \\ \\ \searrow C_2H_3^{\cdot} + C_4H_3^{+} \end{cases}$$

Heteroaromatic compounds also give abundant molecular ions and, apart from side-chain or substituent fragmentations such as those described in sections in (i) and (ii) above for benzene compounds, their fragmentation leads to the ejection of the heteroatom in a neutral species. Whereas benzene ejects C_2H_2 and $C_2H_3^{\cdot}$ from the molecular ion, pyrrole and pyridine eject HCN. Similarly, thiophene ejects $CHS^{\cdot}$ and furan eliminates $CHO^{\cdot}$ from their molecular ions. Fragmentations of the heteroaromatic ring and of its side-chains often com-

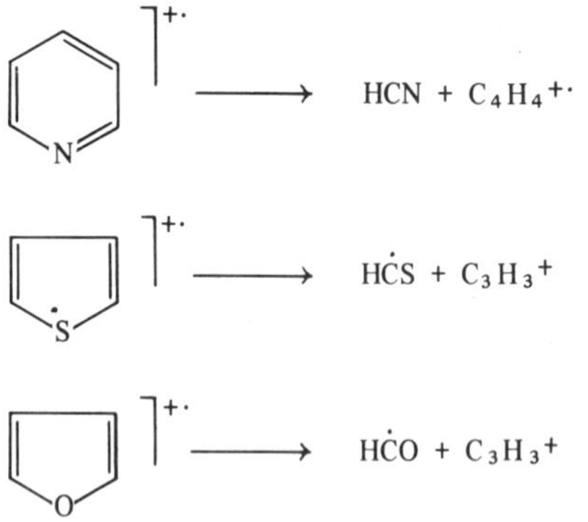

$$\longrightarrow HCN + C_4H_4^{+\cdot}$$

$$\longrightarrow H\overset{\cdot}{C}S + C_3H_3^{+}$$

$$\longrightarrow H\overset{\cdot}{C}O + C_3H_3^{+}$$

pete with each other; the fragmentation of the side-chains may be influenced by their positions in the aromatic nucleus. This last property can be invaluable in deducing the position of a substituent in a heteroaromatic ring. The amount of elimination of $H^{\cdot}$ from 2-, 3- and 4-methylpyridine varies with the position of the substituent, the $(M - 1)^+$ peak being largest in the 3-methyl isomer (figure 10.20). Similarly, the 2-methyl isomer has the largest $(M - 15)^+$ peak corresponding to the loss of a methyl radical. The pyridine-2,3-dicarboxylic acid imide (11) eliminated both CO and CO_2 from the molecular ion to give ions at m/z 134 and 118 with relative abundances of 17 and 4 per cent compared with the base peak in the spectrum (reaction 10.45). The 3,4-isomer (12) afforded ions at the same m/z values but with relative abundances of 1 and 27 per cent.

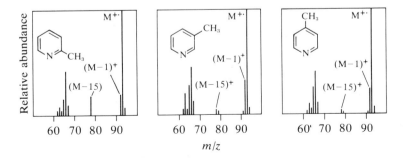

Deuterium labelling strongly suggests that extensive rearrangement of benzene compounds occurs in the molecular ion before fragmentation. It has been suggested that these rearrangements proceed through benzvalene-type

Figure 10.20. Mass spectra of 2-, 3- and 4-methylpyridine showing enhanced size of $(M - 1)^+$ peak in the 3-methyl isomer, and enhanced size of $(M - 15)^+$ peak in the 2-methyl isomer.

intermediates. Sometimes, this sort of rearrangement is apparent from the nature of the fragment ions. For example, the mass spectra of oxazoles contain features very similar to those of isoxazoles and it has been suggested that rearrangement occurs via an azirane (scheme 10.46) as found in the photochemistry of these compounds.

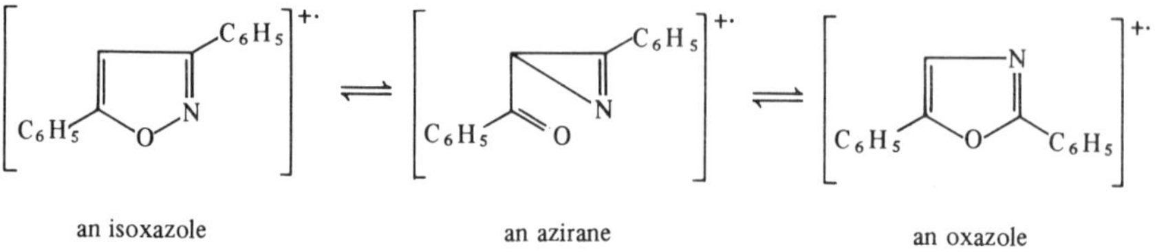

| an isoxazole | an azirane | an oxazole |

(10.46)

Because C2 type side-chain cleavage of heteroaromatic compounds is often prominent in their fragmentation, it has been suggested that the fragment ions are analogous to the supposed tropylium structure found with benzyl compounds, i.e that ring expansion has occurred (reactions 10.47). However, there are no firm experimental details to support this contention and there is even doubt about the correctness of the tropylium structure for the $C_7H_7^+$ ions from benzylic compounds. Nevertheless, in this type of fragmentation behaviour, heteroaromatic compounds are very similar to simple benzene compounds.

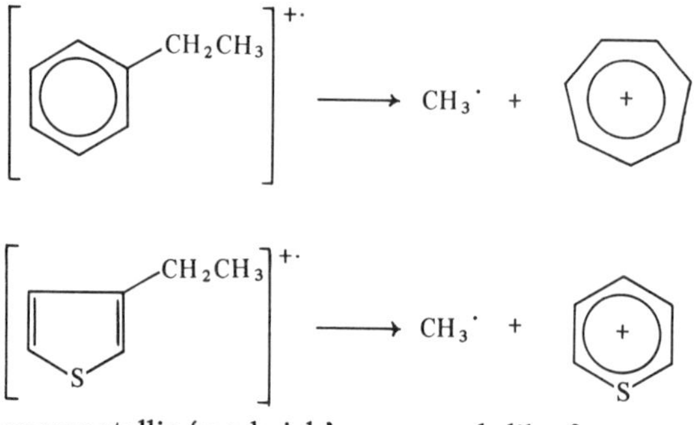

(10.47)

The organometallic 'sandwich' compounds like ferrocene exhibit many characteristics of the fragmentation behaviour of aromatic compounds. Abundant molecular ions are observed and any side-chains present normally fragment like their simple benzene counterparts. Also, just as heteroaromatic compounds decompose by fragmentation of the aromatic ring, so the metallocene molecular ions decompose with destruction of the sandwich structure to leave an ionized metal atom ($Met^{+\cdot}$).

10.7.1. Summary

Aromatic compounds exhibit many of the features found in the mass spectra of aliphatic compounds but these are usually considerably modified because of (*a*) the increased stability of the molecular ions, (*b*), the aro-

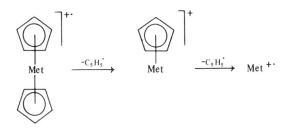

matic ring system itself behaving as a substituent (X) in an alkane, (c) substituents on the aromatic ring changing fragmentation behaviour and (d) heteroatoms in the aromatic ring influencing the fragmentation of the side-chain depending on its position.

10.8.   Subsequent decomposition of primary fragment ions

Most molecular ions formed by electron impact are radical–cations that have one unpaired electron. Radical–cations are known in solution chemistry to be reactive species and the molecular ions formed in mass spectrometry frequently decompose by ejection of a radical to form an even-electron charged species (scheme 10.48a). In even-electron species, the electrons can pair in orbitals to give closed shells; such even-electron cations in solution

$$[CH_3CH_2CH_3]^{+\cdot} \rightarrow CH_3^{\cdot} + C_2H_5^+$$

m/z 44 (even mass number)   radical m/z 29 (odd mass number)
radical–cation   cation
(odd number of electrons)   (even number of electrons)

(10.48a)

$$[C_6H_5OCO-OCH_3]^{+\cdot} \rightarrow CO_2 + C_7H_8O^{+\cdot}$$

m/z 152 (even mass number) m/z 108 (even mass number)
radical–cation   radical–cation
(odd number of electrons) $\downarrow$ $-\dot{C}HO$

$$C_6H_7^+$$
m/z 79 (odd mass number)
cation
(even number of electrons)

(10.48b)

chemistry, the carbenium ions, are generally less highly reactive than odd-electron species like radical–cations.

If the radical-cation does not eject a radical to produce an even-electron ion, rearrangement may occur with subsequent elimination of a non-radical fragment to give another odd-electron ion. The rearranged odd-electron fragment can then decompose by ejection of a radical to yield an even-electron ion as in the example of methyl phenyl carbonate (10.48b). It is found that many of the ions in a mass spectrum occur at odd m/z values (even-electron ions). Ions at even m/z values (odd-electron species) are produced by rearrangement. For odd numbers of nitrogen atoms in an ion this behaviour is reversed, i.e. odd m/z values correspond to odd-electron species and even m/z values to even-electron species.

The 'even-electron rule' states that odd-electron ions (radical–cations) may eliminate either radicals or neutral molecules but even-electron ions (cations) may fragment only by loss of neutral molecules and not of radicals. The situation is summarized in scheme (10.49):

$$A^{+\cdot} \longrightarrow C^+ + N^{1\cdot}$$
$$A^{+\cdot} \longrightarrow D^{+\cdot} + N^2$$
$$B^+ \longrightarrow E^+ + N^3 \qquad\qquad (10.49)$$
$$B^+ \stackrel{||}{\longrightarrow} F^{+\cdot} + N^{4\cdot}$$

A simpler statement of the rule is that successive losses of radicals are forbidden. The rationale of the rule, in charge localization terms, is that in an even-electron species there are no unpaired electrons to 'trigger' radical reactions. Qualitative thermochemical theory argues that product stability largely governs fragmentation pathways and because even-electron species are generally more stable than odd-electron species, loss of a radical from the former to give the latter is energetically unfavourable. For most compounds, the rule is obeyed and may be usefully applied when rationalizing mass spectra. For example, odd-mass ions which contain no nitrogen can fragment to further odd-mass ions but not to even-mass ones. However, the even-electron rule has more than its fair share of exceptions (Karni & Mandelbaum, 1980) and violations appear to be associated with dissociations in which either or both of the odd-electron products (radical and radical-cation) are particularly stable. Polychlorinated, polybrominated and particularly polyiodinated compounds sometimes exhibit successive losses of the relatively stable halogen radicals. Violations are sometimes observed when the loss of two radicals leads to ions for which structures containing high degrees of conjugated unsaturation may be postulated (reaction 10.50). However, arguments based on thermochemical stabilities of assumed structures are not well founded. It is, of course, not necessary to invoke successive loss of two methyl radicals if a

$$(10.50)$$

compound shows $M^{+\cdot}$, $(M-15)^+$ and $(M-30)^{+\cdot}$ ions since losses of $CH_3\cdot$ and C_2H_6 from the molecular ion could explain the results. Unless the elimination of a radical from an even-electron ion is substantiated by observation of meta-stable ions for the reaction $(M-15)^+ \rightarrow (M-30)^{+\cdot}$, for example, it should not be proposed. Generally, the even-electron rule does have a place in mass spectrometry as a useful 'rule of thumb'. When even-electron ions are induced to fragment by collisional activation (section 8.5), the even-electron rule no longer holds, presumably because the attendant increase of internal energy in the ions overcomes the high activation energy for loss of radicals.

It will be appreciated from the foregoing discussion that subsequent decomposition of primary fragment ions often revolve around the behaviour of even-electron species. It may be recalled that one of the first items to be noted in a mass spectrum is which of the major fragment ions occur at odd m/z values and which at even m/z values.

The initial fragmentation reactions of molecular ions can be grouped into a few general schemes, but the decomposition reactions of the fragments themselves are not so easy to classify. The first fragmentations of a molecule in a mass spectrometer are generally those giving most structural information because after several decompositions along any particular fragmentation pathway, the relationship between the ions and the original structure becomes too tenuous to be of much value. Fragment ions (A^+ and B^+) from a molecular ion ($M^{+\cdot}$ in scheme 10.51) may yield valuable information about $M^{+\cdot}$, e.g. loss

$$M^{+\cdot} \rightarrow A^+ \rightarrow B^+ \rightarrow C^+ \rightarrow D^+ \rightarrow \qquad (10.51)$$

of H_2O, $CH_3\cdot$, $CH_3O\cdot$, etc., but by the time fragmentation reaches the ion D^+ much of the original structure of $M^{+\cdot}$ has been destroyed and little structural information may be obtained. However, ions at the low mass end of a spectrum should not be ignored. Important clues to the structure of $M^{+\cdot}$ may still be obtained as with $C_2H_3O^+$ ions at m/z 43 from CH_3CO groups and CH_4N^+ ions at m/z 30 from primary amines. Often such useful low-mass ions arise from primary fragmentation of the molecular ion. The loss of C_2H_3O from a molecular ion can be recognized by the presence of $(M - C_2H_3O\cdot)^+$ ions as well as $C_2H_3O^+$ ions at m/z 43. The later stages of a fragmentation pathway frequently only become clear after the structure of a compound has been elucidated. These later fragment ions then serve as confirmation for a proposed structure rather like the use of the 'finger-print' region in infrared spectroscopy which may be used for confirmatory evidence for an assignment but is not very useful for initial diagnosis.

Aliphatic fragment ions decompose further by elimination of C_2H_4:

$$R{-}CH_2{-}CH_2{}^+ \rightarrow R^+ + C_2H_4$$

Aromatic fragment ions dissociate by ejection of C_2H_2, so that aromatic compounds are often recognized by series of ions m/z 77 and 51 or m/z 91, 65 and 39:

$$C_6H_5^+ \xrightarrow{\;-C_2H_2\;} C_4H_3^+$$
$$m/z\ 77 \qquad\qquad m/z\ 51$$

$$C_7H_7^+ \xrightarrow{\;-C_2H_2\;} C_5H_5^+ \xrightarrow{\;-C_2H_2\;} C_3H_3^+$$
$$m/z\ 91 \qquad m/z\ 65 \qquad m/z\ 39$$

Heteroatomic molecules, after initial fragmentation, often decompose further after hydrogen transfer. The migrating hydrogen atom may arise from a β-, γ- or δ-carbon to the heteroatom. The common C2 cleavage reaction of heteroatomic compounds yields even-electron fragment ions which, after hydrogen transfer, may fragment along two different pathways to give even-electron ions containing the heteroatom. The processes may be represented generally by scheme (10.52) in which X is nitrogen, oxygen or sulphur. Ions at m/z 30 and 58 in the mass spectrum of N-ethylpropylamine (scheme 10.53) are examples of this type of fragmentation.

$$\left[\begin{array}{c} R-CHXCH-R^3 \\ \;\;\;\;\;| \;\;\;\;\; | \\ \;\;\;\;\;R^1 \;\;\; R^2 \end{array}\right]^{+\cdot}$$

$$\xrightarrow{-R^{1\cdot}} R-CH=\overset{+}{\overset{\cdot}{X}}CH-R^3 \longrightarrow [R-H] + CH_2=\overset{+}{X}CH-R^3$$
$$\underset{R^2}{|} \qquad\qquad\qquad \underset{R^2}{|}$$

$$\xrightarrow{-R^{2\cdot}} R-\overset{+}{CHX}=CH-R^3 \longrightarrow [R^3-H] + R\overset{+}{CHX}=CH_2$$
$$\underset{R^1}{|} \qquad\qquad\qquad \underset{R^1}{|}$$

(10.52)

$$C_3H_7-\overset{+\cdot}{N}H-C_2H_5$$

$$\xrightarrow{-H\cdot} C_2H_5-CH=\overset{+}{N}H-C_2H_5 \xrightarrow{H\text{-transfer}} C_2H_4 + CH_2=\overset{+}{N}H-C_2H_5$$
$$\qquad\qquad\qquad\qquad\qquad\qquad\qquad m/z\ 58$$

$$\xrightarrow{-C_2H_5\cdot} CH_2=\overset{+}{N}HC_2H_5 \xrightarrow{H\text{-transfer}} C_2H_4 + CH_2=\overset{+}{N}H_2$$
$$\qquad\qquad\qquad\qquad\qquad\qquad m/z\ 30$$

(10.53)

Hydrogen transfer may also occur with elimination of the heteroatom as the neutral species and is quite common for halides and esters. For example, if X = Br ($-CH(R^2)R^3$ absent in scheme 10.52) then HBr is eliminated as in reaction sequence (10.54):

$$C_2H_5-CH-\overset{+\cdot}{Br} \longrightarrow CH_3^\cdot + C_2H_5-CH=\overset{+}{Br} \xrightarrow{H\text{-transfer}} HBr + C_3H_5^+ \quad (10.54)$$
$$\underset{CH_3}{|}$$

The elimination of a neutral fragment from even-electron ions is particularly marked in the mass spectra of many simple organo-metallic compounds. Ejection of a butyl radical from the molecular ion of tetra-1-butyl tin affords an even electron ion which then eliminates C_4H_8 with hydrogen transfer to give another even-electron ion; this latter ion then decomposes again by elimination of a second C_4H_8 unit to yield a third even-electron ion.

$$(C_4H_9)_4Sn^{+\cdot} \xrightarrow[-C_4H_9^\cdot]{} (C_4H_9)_3Sn^{+\cdot} \xrightarrow[-C_4H_8]{\text{H-transfer}} (C_4H_9)_2SnH^+ \xrightarrow[-C_4H_8]{\text{H-transfer}}$$

$$(C_4H_9)SnH_2^+$$

When rearrangement of a molecular ion accompanies its fragmentation, the product ion may itself fragment like the molecular ion of the postulated rearranged species. The point may be clarified by two examples. Ejection of CO from the molecular ion of pyranocoumarin (13) yields a fragment ion which is conveniently postulated as the furanocoumarin (14). The subsequent decompositions of the ion (14) and of the molecular ion produced by a specimen of the furanocoumarin itself were almost identical. Analogously, the methoxyphenazine (15) eliminated CH_2O from its molecular ion to give a fragment ion postulated as the phenazine molecular ion (16). The subsequent decomposition of the ion (15) was almost identical to that of the molecular

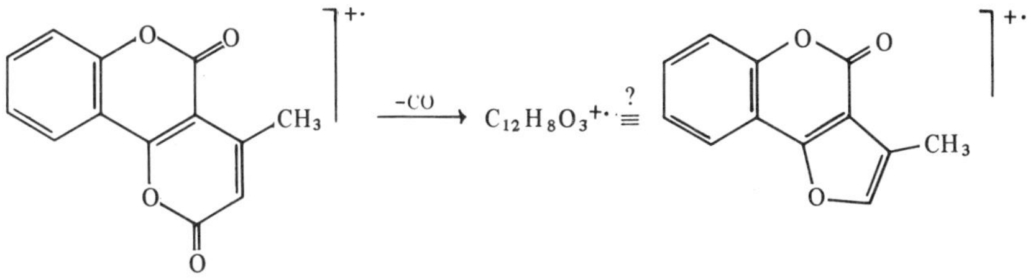

(13) $\xrightarrow{-CO}$ $C_{12}H_8O_3^{+\cdot} \overset{?}{\equiv}$ (14)

Subsequent decompositions are very similar

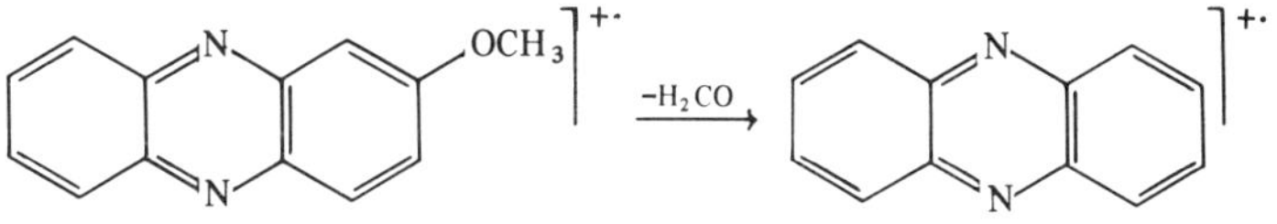

(15) $\xrightarrow{-H_2CO}$ (16)

Subsequent decompositions are similar to those of the molecular ion of phenazine ion of phenazine. The close similarity in behaviour of the respective fragment and molecular ions suggests they *may* have similar structures although, especially for the decomposing ion, these are not necessarily as postulated. There is thermochemical evidence that at least some of the many observed similarities of ion decompositions are accidental and that the ion structures cannot be as postulated.

10.9. Rearrangement accompanying fragmentation

The formation of new bonds between atoms in an ion leads to re-arrangement. Hydrogen migration, which seems particularly facile, has been covered in earlier sections. In some instances, mainly with odd-electron ions, the site from which a hydrogen is transferred is fairly specific. In other cases, generally with even-electron ions, the reaction appears to be far less specific. The randomization of hydrogen in ions is widespread and can make the inter-pretation of deuterium labelling studies difficult or ambiguous.

Skeletal rearrangement is defined as the formation of new bonds between atoms other than hydrogen. The classification of these rearrangements is use-ful for the interpretation of mass spectra because known structural features giving rise to this phenomenon can be compared with the behaviour of com-pounds of unknown but postulated structure. Skeletal rearrangements in mass spectrometry are diverse but attempts have been made to classify them.

One of the earliest and most useful classifications was based on (i) the migration of alkyl or aryl groups and (ii) the nature of the neutral particle ejected after rearrangement (Brown & Djerassi, 1967). These compilations are useful, and for structural organic chemistry the nature of the ejected species may be particularly informative. Typical neutral species eliminated following rearrangement include CO, CO_2, SO_2, CH_2O and HCN, all having favourable negative heats of formation. Two examples are shown in scheme (10.55). The diagnostic value of the ejected neutral particle may or may not be significant.

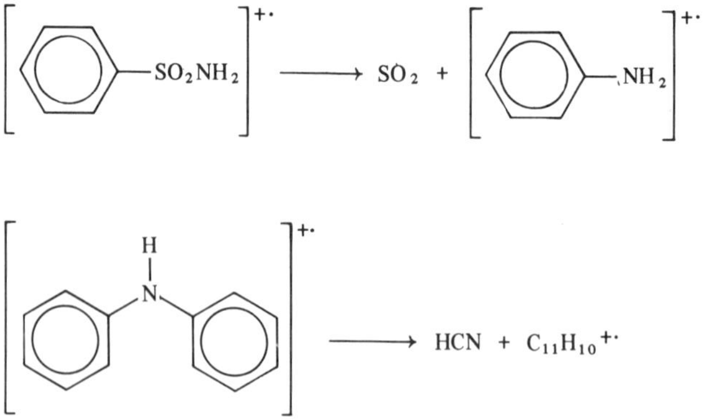

$$\left[\bigcirc\!\!\!-\!SO_2NH_2 \right]^{+\cdot} \longrightarrow SO_2 \ + \ \left[\bigcirc\!\!\!-\!.NH_2 \right]^{+\cdot}$$

(10.55)

$$\left[\underset{\overset{|}{N}}{\overset{H}{\bigcirc}}\bigcirc \right]^{+\cdot} \longrightarrow HCN \ + \ C_{11}H_{10}{}^{+\cdot}$$

Thus, elimination of CO_2 from a molecular ion indicates an unsaturated ester, carbonate or cyclic imide group as likely structural features but loss of CO

occurs from a wide variety of compounds and yields less immediate information. Elimination of CO_2 from unsaturated esters and carbonates is quite
general, although the degree to which it is observed is very variable. The process has been represented by two schemes in which X and Y are doubly or
triply bonded in the unsaturated esters and either R or R′ must be aryl in carbonates (reactions 10.56).

$$\begin{bmatrix} X \\ \| \\ Y \diagdown_{CO_2^{\diagup R}} \end{bmatrix}^{+\cdot} \longrightarrow \begin{bmatrix} X_{\diagdown} \\ \|\diagdown R \\ Y_{\diagup} \end{bmatrix}^{+\cdot} + CO_2$$

(10.56)

$$\begin{bmatrix} R \\ | \\ O \diagdown_{CO_2^{\diagup R'}} \end{bmatrix}^{+\cdot} \longrightarrow \begin{bmatrix} R \\ | \\ O{-}R' \end{bmatrix}^{+\cdot} + CO_2$$

The idea of migration to a positive centre *via* a nominal four-centre process
in even-electron ions is embodied in scheme (10.57), two examples of which

$$\begin{array}{ccc} A^+ & D \\ | & | \\ B & {-} & C \end{array} \longrightarrow \begin{array}{c} A{-}D \\ | \\ B{-}C^+ \end{array} \longrightarrow AD^+ + BC$$

$$\begin{bmatrix} CH_3O & CH_3 \\ | & | \\ O{=}C & {-} & O \end{bmatrix}^{+\cdot} \xrightarrow{-H^{\cdot}} \begin{array}{c} CH_2{=}\overset{+}{O}\nearrow CH_3 \\ | \diagup \\ O{=}C{-}O \end{array} \longrightarrow CH_2{=}\overset{+}{O}{-}CH_3 + CO_2$$

(10.57)

$$\begin{bmatrix} ArSO_2NH & Ph \\ | & | \\ N{=}CR \end{bmatrix}^{+\cdot} \xrightarrow{-ArSO_2^{\cdot}} \begin{array}{c} \overset{+}{NH}\nwarrow Ph \\ | \diagup \\ N{=}CR \end{array} \longrightarrow PhNH + RCN$$

are shown. Whereas these rearrangements appear to be four-centre ones, the
corresponding reaction of trimethylsilyl ethers seems amazingly insensitive to
distance between the reacting centres:

$$\begin{bmatrix} OSi(CH_3)_3 \\ \diagdown \\ (CH_2)_n \\ \diagup \\ OPh \end{bmatrix}^{+\cdot} \xrightarrow{-CH_3^{\cdot}} \begin{array}{c} \overset{+}{O}Si(CH_3)_2 \\ \diagdown \\ (CH_2)_n \\ \diagup \\ OPh \end{array} \longrightarrow \begin{array}{c} Ph\overset{+}{O}Si(CH_3)_2 \\ + \\ (CH_2)_nO \end{array}$$

$n = 2{-}8$

A variety of formally four-centre rearrangements in odd-electron ions can
be collated in one scheme (10.58), very similar to that of (10.57) but migration to a positively charged centre is not implied. The main requirement for
this reaction appears to be that atoms A, B, C and D which may be nitrogen,
oxygen, sulphur or carbon should form contiguous π-centres. It is not sug-

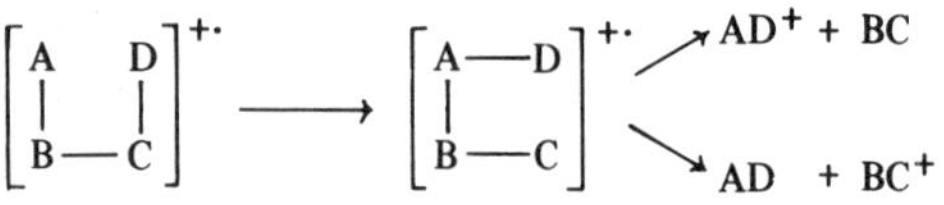

$$\left[\begin{array}{cc} A & D \\ | & | \\ B & \!\!\!-\!\!\! C \end{array}\right]^{+\cdot} \longrightarrow \left[\begin{array}{cc} A\!\!-\!\!D \\ | & | \\ B\!\!-\!\!C \end{array}\right]^{+\cdot} \begin{array}{c} \nearrow AD^+ + BC \\[2mm] \searrow AD + BC^+ \end{array}$$

(10.58)

gested that all these rearrangements actually proceed through a four-membered transition state but when four such π-centres are present in a molecule, then rearrangement ions may be expected in the mass spectrum. Two examples from the mass spectra of trifluoroacetyl amides and cyclic thioimides illustrate the general nature of this rearrangement:

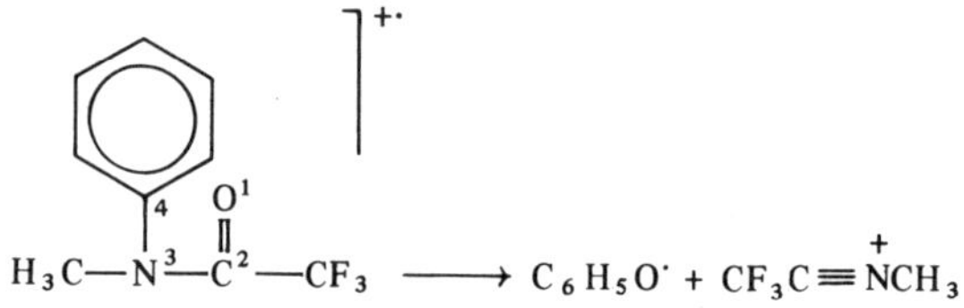

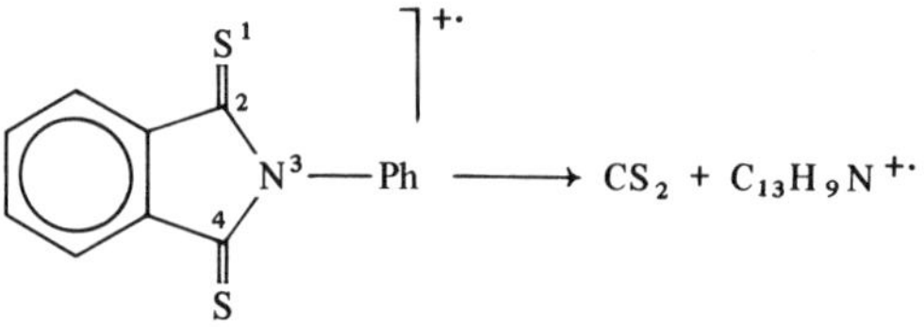

$$H_3C\!\!-\!\!N^3\!\!-\!\!\overset{O^1}{\underset{\|}{C^2}}\!\!-\!\!CF_3 \longrightarrow C_6H_5O^\cdot + CF_3C\!\equiv\!\overset{+}{N}CH_3$$

$$\longrightarrow CS_2 + C_{13}H_9N^{+\cdot}$$

Often rearrangement competes with simple cleavage and the latter may be so easy that peaks representing rearrangement processes are small or not present at all. From this point of view, the closely similar amides (17) and (18)

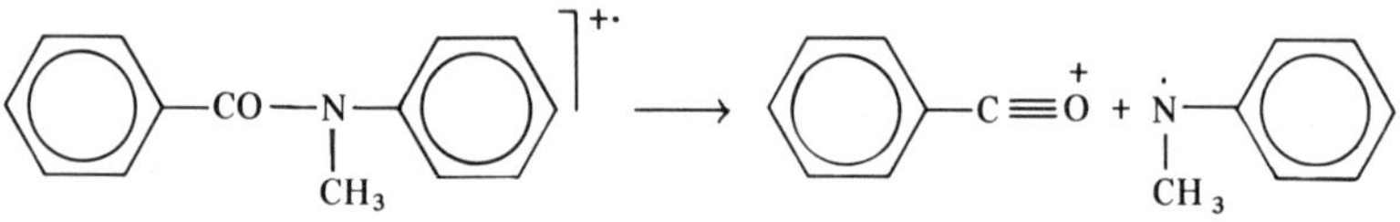

(17)   Simple cleavage

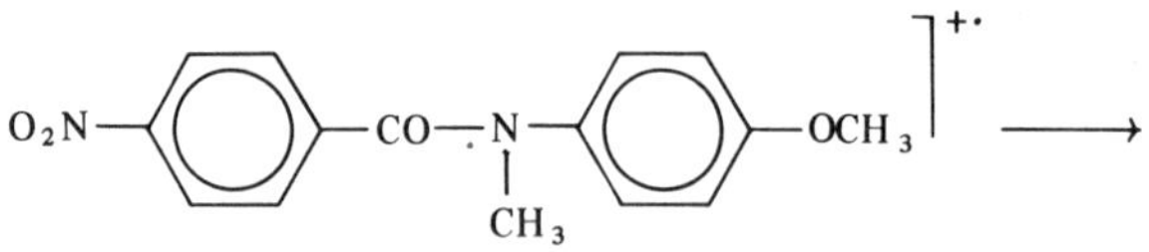

(18)

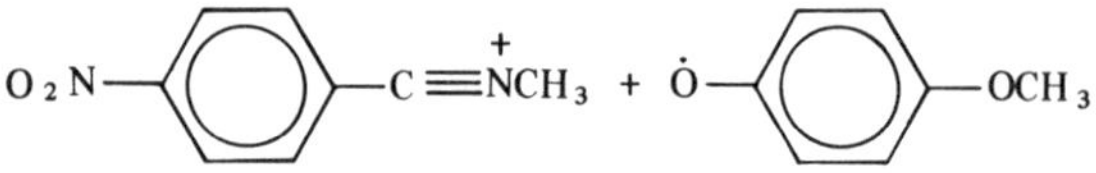

Rearrangement

are interesting since the former shows almost no rearrangement whereas the latter has abundant ions corresponding to substantial rearrangement. Unlike the nominal four-centre rearrangement of even-electron ions, migration of alkyl groups seems uncommon in odd-electron ions.

It was pointed out many years ago (Field & Franklin, 1957) that rearrangement processes would have lower frequency factors (see section 9.8) than simple bond breaking reactions. Subsequent experiments have endorsed this statement. Thus, by gradually reducing the electron-beam energy in electron impact ionization of molecules, it is observed that rearrangement reactions may increase in importance relative to simple cleavage. Using the simplified form of quasi-equilibrium rate equation, the rate of a rearrangement–cleavage reaction (k_R) may be compared with the rate of simple cleavage (k_C) for a molecule with N oscillators. In the equations (10.59), the frequency factors for rearrangement with cleavage and simple cleavage are ν_R and ν_C respectively and the energies of activation for the processes, E_R^0 and E_C^0. The inter-

$$k_R = \nu_R \, (1 - (E_R^0/E))^{N-1}; \quad k_C = \nu_C \, (1 - (E_C^0/E))^{N-1} \qquad (10.59)$$

nal energy in the ion is E. Equations (10.59) combine to give equation (10.60) which, as a first approximation, may be written as in equation (10.61) where A is a constant and $\Delta E = E_C^0 - E_R^0$. When the energy of activation for simple

$$\log (k_R/k_C) = \log (\nu_R/\nu_C) + (N-1) \log (1 - (E_R^0/E))/(1 - (E_C^0/E)) \qquad (10.60)$$

$$\log (k_R/k_C) \approx A + (N-1)(\Delta E/E) \qquad (10.61)$$

cleavage is greater than the energy of activation for rearrangement with cleavage, then $\Delta E/E$ in equation (10.61) is positive and $\log (k_R/k_C)$ increases as E is reduced; i.e. reducing the electron-beam energy, and therefore the energy transferred to the molecular ion, causes the rate of the rearrangement process to increase relative to the rate of the simple bond breaking reaction. Therefore, as the electron-beam energy decreases, the importance of the rearrangement process increases relative to that of simple bond fission. The same conclusion may be reached by comparison of the rates k_R and k_C at different internal energies of an ion (figure 10.21). Using the simplified rate equations (10.59), graphs of $\log k_R$ and $\log k_C$ can be plotted for various internal energies as in figure 10.21a. In this case, at all internal energies (e.g. E_1, E_2), $\log k_C$ is greater than $\log k_R$ and the rate of simple bond fission always exceeds that for rearrangement followed by fission. With energies of activation and frequency factors different from those obtaining in figure 10.21a, the curves for $\log k_R$ and $\log k_C$ may cross (figure 10.21b). Then, at some low excess of internal energy (E_1), the rate of rearrangement with cleavage is faster than the rate of simple bond cleavage but, at higher internal energy (E_2), the order of rates is reversed. Thus, at low electron impact energies, rearrangement is favoured over ordinary bond cleavage and, by reducing the electron impact energy below the normal 70 V, the increased importance of the rearrangement process may be observed by increased relative abundances of ions corresponding to it.

Ions formed by field ionization are detected very soon after formation and it has been suggested that there is too little time for much rearrangement to occur. It was expected that comparison of field and electron impact ionization mass spectra would identify rearrangement processes since they would occur only in the latter. However, there is substantial evidence that rearrangements can occur in field ionization mass spectrometry. The topic is discussed more fully in section 7.5.

10.10. Fragmentation following other methods of ionization

Scattered references to positive-ion and negative-ion chemical ionization, field ionization and field desorption occur in this chapter. In this section the information is collated. Fragmentation of ions is described in empirical terms and rationalization of mass spectra relies upon previous knowledge of the behaviour of large numbers of known compounds in a mass spectrometer. There is ample experience of behaviour following electron impact, but not following the forementioned additional methods of ionization.

Field ionization and field desorption mass spectrometry involve little, if any, fragmentation because molecular or quasi-molecular ions are formed with little excess of internal energy. When molecular ions (M^+) do fragment, the product ions are not dissimilar to the primary fragment ions in electron impact mass spectra although there is more emphasis on direct cleavage of the weaker bonds in ions generated by electric fields. When quasi-molecular ions, $(M + R)^+$ where R = H, Li, Na, etc., are formed upon ionization in an electric

Figure 10.21. Variation of rates of simple bond cleavage (k_C) and of rearrangement with cleavage (k_R) with changes in internal energy of an ion. In (a) the curves do not cross; in (b) the curves do cross.

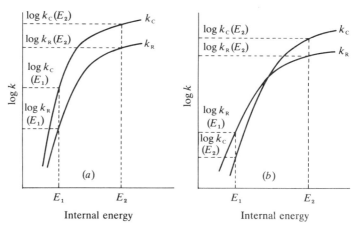

field, their behaviour is similar to quasi-molecular ions produced by chemical ionization (see below). Fragmentation in field ionization and desorption mass spectrometry is sometimes due to thermal decomposition of neutral molecules prior to ionization so that 'fragment ions' are really molecular ions of pyrolysis products. In such a case, the behaviour is governed by the thermochemistry of the neutral molecules, but fragmentation frequently parallels that after electron impact. Collisional activation of otherwise stable ions causes a large increase in their internal energy. The resulting fragmentation is very similar to that in electron impact mass spectrometry in which similar amounts of internal energy are imparted to molecular ions.

Chemical ionization affords quasi-molecular ions, most commonly $(M + H)^+$ ions. Being even-electron species and possessing near thermal energies, these ions are less prone to extensive fragmentation than odd-electron molecular ions formed by electron impact and tend to obey the even-electron rule (see above) inasmuch as they fragment almost exclusively by loss of neutral molecules. Whilst electron impact ionization of esters, $RCOOCH_3$, affords molecular ions which fragment by loss of $CH_3O^{\cdot}$ and $^{\cdot}COOCH_3$, the same esters form $(M + H)^+$ ions on chemical ionization and these quasi-molecular ions eliminate CH_3OH and $HCOOCH_3$. The site of proton attachment in $(M + H)^+$ ions is widely assumed to be the functional group associated with highest proton affinity (a heteroatom). In polyfunctional compounds, the proton may be considered as a bridging atom between two heteroatoms. Cleavages are frequently rationalized on the basis of the assumed, specific site of proton attachment. Thus, a common cleavage is ejection of the functional group with the newly bound proton (scheme 10.62). Since the appearance of chemical ionization mass spectra is highly dependent on the reactant gas used (section 7.2), generalizations are difficult.

$$RX + [H^+] \longrightarrow R\overset{+}{-}X-H \longrightarrow R^+ + HX$$

$$(10.62)$$

$$\underset{\underset{O}{\parallel}}{RC}-OCH_3 + [H^+] \longrightarrow \underset{\underset{O}{\parallel}}{RC}-\overset{+}{\underset{\underset{H}{\mid}}{O}}CH_3 \longrightarrow RC\equiv\overset{+}{O} + CH_3OH$$

A variant of chemical ionization is charge exchange (section 7.2). An electron is transferred from a sample molecule (M) to reactant gas ion (e.g. $Ar^{+\cdot}$, $N_2^{+\cdot}$, etc.) during collision to yield an odd-electron molecular ion ($M^{+\cdot}$). The molecular ions are usually formed with a large excess of internal energy and give mass spectra very similar to those observed by electron impact, although the amount of internal energy in $M^{+\cdot}$ varies depending on the character of the reactant gas.

Even-electron quasi-molecular ions formed by negative-ion chemical ionization are frequently stable so that the only ions observed are those resulting from reactions of the sample molecule with the negatively charged reactant ion. Such reactions are described in section 7.3. Formation of odd-electron molecular anions, $M^{-\cdot}$, by electron capture may be accompanied by fragmentation. For compounds, RX (X = Cl, Br, F, CN and NO_2), fragment ions, X^-, are common; esters and acids, $RCOOR'$ (R' = H, alkyl), give $RCOO^-$ and R^- ions. These simple cleavage reactions are frequently accompanied by hydrogen rearrangements. Skeletal rearrangements observed in negative-ion mass spectrometry are often very complex and *ortho*-effects in polysubstituted aromatic compounds are particularly favoured (reaction scheme 10.63).

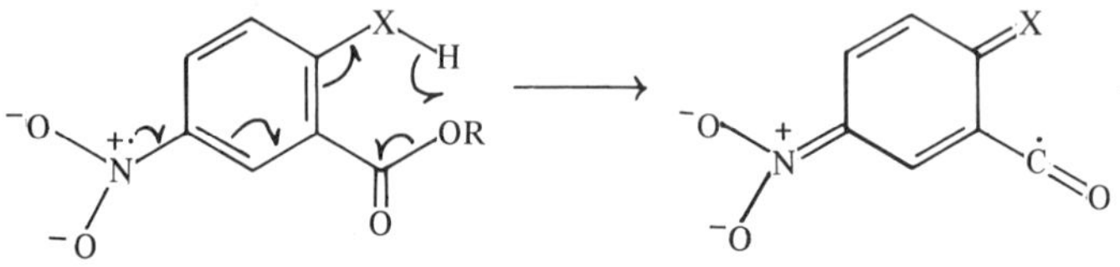

$$+ \; ROH \quad (10.63)$$

(X = O or NH)

The relatively small body of information on the behaviour of negative ions militates against generalization. The interested reader is referred to more specialized literature on negative ions (Bowie & Williams, 1975; Massey, 1976; Bowie, 1975, 1977, 1979).

Since many of these additional methods of ionization are used solely to assign or confirm the molecular weight of an unknown compound or to detect and quantify known compounds, attempts to rationalize the fragmentation observed are not always made and samples are analysed under conditions which minimize fragmentation. Generally, the methods of ionization discussed here and in detail in chapter 7 are more likely to identify the molecular weight and functional groups of an unknown substance but less likely to show fragmentation of its carbon skeleton than does electron-impact ionization.

11 Examples of structure elucidation by mass spectrometry

11.1. Introduction

Before some worked examples are given to illustrate the elucidation of unknown structures by mass spectrometry, some general hints are in order. These salient points have been discussed elsewhere in the book, but the summary here is intended as a helpful guide to the analysis of samples.

A structure is rarely deduced solely from its mass spectrum. What the mass spectrometrist hopes to do by way of structural elucidation is to propose a structure that is consistent with the given spectrum. Unfortunately, it may even be possible to propose several structures that are compatible with the spectrum. Of course, there are degrees of probability of correctness of structural assignment and if the elemental composition (high-resolution mass spectrum) and metastable ions are consistent with the proposed structure and especially if the spectrum of the unknown structure matches well against a standard spectrum from a library of mass spectra, then the correctness of an assignment may be virtually certain. However, there are many instances of different compounds (in particular isomers) giving rise to mass spectra with differences less than the day-to-day variation in the recording of a mass spectrum, so one should be wary of rash structural assignments. Verification of proposed structures by other techniques is recommended. When dealing with very small amounts of material, synthesis, chromatography and uv spectroscopy are widely accepted methods; for larger amounts, ^{1}H and ^{13}C nuclear magnetic resonance and infrared spectroscopy are the techniques most likely to yield useful complementary evidence.

It is very rare that a totally unknown sample has to be analysed because there is almost always other information. If the compound is a product of a laboratory synthesis, then its likely structure is known. If the unknown substance is a natural product, its origin will give many clues as to its identity. The type of experimental work-up often reveals much chemical evidence of structure. In particular, the procedure used to extract the compound will reveal much about its character (e.g. an amine is soluble in aqueous acid, but is extracted into organic solvents from basic aqueous solution). Information about the substance can vary from its highly diagnostic nuclear magnetic resonance or infrared spectrum to much less specific properties such as colour, smell and physical state, all of which should be kept in mind whilst interpreting the mass spectrum.

Initial examination of a mass spectrum can give much information. Even-mass molecular ions indicate a substance with an even number of nitrogens or without nitrogen altogether. When molecular ions have odd mass, the compound has an odd number of nitrogen atoms. Compounds showing few mass peaks, doubly charged ions and abundant molecular ions are often aromatic. On the other hand, aliphatic compounds tend to have their most abundant ions at low mass. Any series of ions 14 mass units apart should be noted for these are usually $C_nH_{2n}X^+$ ions and may reveal the character of the functional group, X. Multiple isotope peaks suggest the presence of metals. The above points are illustrated in the following examples.

During more detailed examination of a mass spectrum, it should be remembered that primary losses of fragments from the molecular ion are likely to provide the most diagnostic information. In the elucidation of fragmentation pathways, stability of ions is an important consideration but the neutral species ejected must not be ignored just because the mass spectrometer does not detect them. The large negative heats of formation of some neutral species, such as CO, CO_2, HCN, N_2, HF and so on, may prove to be the most important factor in making their elimination from ions favourable, rather than the fragment ion having any exceptional stability. Also, even-electron ions (cations) rarely eliminate odd-electron neutral species (radicals) but generally dissociate through loss of neutral molecules to give further even-electron ions. In routine mass spectra recorded from a magnetic sector instrument, metastable ions occur at a mass $m^* = m_2^2/m_1$, where m_1 is the mass of the precursor ion and m_2 that of the product ion. With constant-speed recording the distance on the trace between the product and metastable ion peak is the same as that between the precursor and product ion peak so that the precursor/product/metastable ion relationship is usually readily apparent by simple direct measurement (see figure 11.1, example A). Metastable ions originate from dissociations occurring outside the ion source and are frequently associated with elimination of small neutral molecules like H_2O, HCN and CO from molecular ions. Ejection of a neutral molecule from a molecular ion is necessarily a rearrangement process and, since rearrangements generally have lower rates of reaction than do direct cleavages, they have a greater likelihood of occurring after the ions have left the ion source.

The first three examples given below are simple and are used to illustrate the approach to structure determination. The remaining examples have been chosen to illustrate the application of mass spectrometry to different types of compound and the use of various methods of structure elucidation including accurate mass measurement, additional methods of ionization, computerized library searching and analysis of metastable ions. It is shown that mass spectrometry alone provides information which may or may not be sufficient to propose a complete structure. The complementary role of other techniques is emphasized. For the purposes of this chapter, the method of introducing the sample into the mass spectrometer (direct insertion, chromatographic interface, etc.) is irrelevant. Detailed discussion of an analysis by gas chromatography/mass spectrometry is given in chapter 4 and quantitative aspects of mass spectrometry are exemplified in chapter 6. Specialized metastable ion analyses, in particular the differentiation of isomeric compounds, are covered in chapter 8.

11.2. Example A

The mass spectrum of compound A, recorded on uv-sensitive paper, is reproduced in figure 11.1; only two traces are given for clarity and simplicity. The upper trace is used for counting the spectrum and finding metastable ion peaks and the lower trace is used for structural work. The upper trace was obtained at a sensitivity 100 times greater than the lower trace. The peaks at m/z 28 and 32 (due to N_2 and O_2 in air) were identified by comparison with another, previously calibrated spectrum. First, the spectrum is counted on the upper trace. Notice how the gap between any two peaks slowly decreases as the m/z values increase; a sharp change between two successive gaps indicates the presence of doubly charged ions, e.g. see m/z 28.5 which is the doubly charged ion of mass 57. Counting stops at the last abundant ion, m/z 61, which appears to be the $(M + 1)^{+ \cdot}$ ion (^{13}C isotope) and therefore the molecular ion is at m/z 60. There is a broad metastable ion peak at m/z 29.4. Notice that in figure 11.1 the peak at m/z 42 is equidistant from that of m/z 60 and the metastable ion, suggesting that the metastable ion is due to loss of 18 mass units from the molecular ion. This is confirmed by the fact that $42^2/60 = 29.4$. On the lower trace, many peaks of less abundant ions are not recorded and without the upper trace it would be difficult to count the spectrum accurately.

For convenience here, the mass spectrum is converted now into a normalized line diagram, although usually this would only be done for record purposes after the spectrum had been interpreted. The peak heights on the lower trace are measured (column 2, table 11.1). Because traces intermediate between highest and lowest sensitivity have been omitted from the mass spectrum for the sake of simplicity, some of the peak heights measured on the lower trace cannot be checked (section 1.6) and smaller peak heights are only approximate. The largest peak (base peak) in the spectrum is at m/z 31 and this is put equal to 100 units (column 3, table 11.1); all other peaks are related to this base peak to give their relative abundances and the normalized line diagram (figure 11.2) is then drawn. Peaks of less than 1 per cent of the height of the base peak have been omitted (e.g. m/z 44), as have peaks below m/z 27.

Figure 11.1. Mass spectrum of compound A.

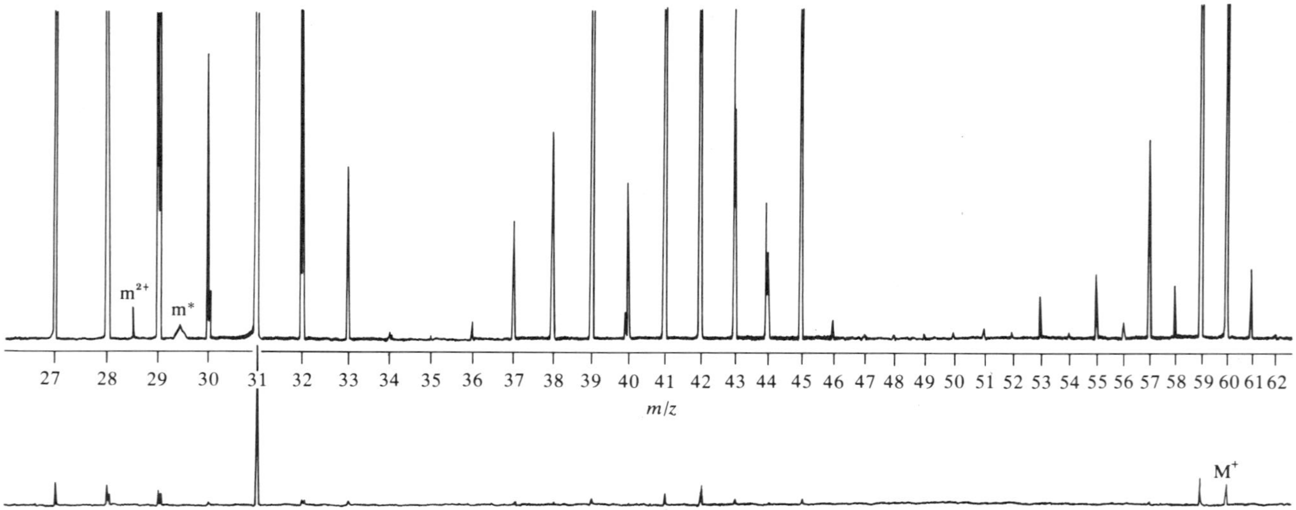

Table 11.1. *Measurements taken from the mass spectrum of a compound* A *(figure 11.1)*

m/z	Peak height (mm)	Relative abundance (%)
60	6	13
59	8	17
45	2	4
43	1.5	3
42	6	13
41	4	8
40	0.5	1
39	2	4
33	1.5	3
32	1.5	3
31	47	100
29	5.5	11
28	6.5	14
27	7	15

Amongst the larger peaks are m/z 59 (loss of 1 mass unit, $H^{\cdot}$, from the molecular ion), m/z 45 (ejection of 15 mass units, $CH_3^{\cdot}$), and m/z 31 (possible 'characteristic' ion of composition CH_3O? - section 10.1). The compound A is of even molecular weight and therefore cannot contain an odd number of nitrogen atoms and the more abundant fragment ions are at odd m/z values (even-electron ions). One significant peak occurs at even mass, m/z 42, and is an odd-electron ion, which means that rearrangement has occurred with loss of 18 mass units (H_2O) from the molecular ion. The metastable ion confirms this elimination of H_2O since $42^2/60 = 29.4$ and a metastable ion is found at

Figure 11.2. Normalized line diagram taken from mass spectrum of compound A in figure 11.1.

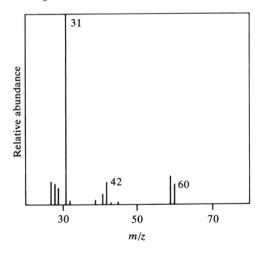

this m/z value. This last piece of information indicates that compound A contains a hydroxyl group and may be an alcohol. From figure 11.1, where the upper trace is recorded at a sensitivity 100 times that of the lower, the ratio of m/z 60 to m/z 61 can be measured to be 100:3.5. Since for each carbon atom of a molecule there is a 1.1 per cent chance that it is ^{13}C, a ratio for $M^{+\cdot}$ to $(M + 1)^{+\cdot}$ of 100:3.3 is expected for a molecule containing three carbon atoms. This is close enough to the observed ratio to indicate that compound A has three carbon atoms. Therefore, it is now known that compound A contains the elements C_3H_2O, amounting to a mass of 54. Since the molecular weight is 60, the total composition must be C_3H_8O, indicative of a saturated alcohol C_3H_7OH. The loss of H$\cdot$ from the molecular ion points to at least one hydrogen on the α-carbon atom (reaction 11.1; C2 cleavage; see section 10.6).

$$\left[-\overset{|}{\underset{\overset{|}{H}}{C}}-OH \right]^{+\cdot} \longrightarrow \ >C=\overset{+}{O}H \ + \ H\cdot \tag{11.1}$$

The loss of H_2O also indicates a hydrogen on a γ- or δ-carbon atom (section 10.6; RE1 reaction). Since a δ-carbon atom implies four carbon atoms of $4 \times 12 = 48$ mass units, there would be insufficient mass remaining to accomodate the oxygen atom in the molecule, concurring with the presence of only three carbon atoms. A γ-hydrogen is ejected with the hydroxyl group on fragmentation of the molecular ion; the remaining hydrogens must fit as shown (scheme 11.2) and compound A is 1-propanol. The ion at m/z 31 confirms the primary alcohol character of A.

$$\left[\begin{array}{c} H_2C\overset{H}{\diagdown} \\ | \quad \ \ OH \\ H_2C\diagdown_{CH_2}\diagup \end{array} \right]^{+\cdot} \longrightarrow \ H_2O \ + \ C_3H_6{}^{+\cdot}$$
$$m/z \ 60 \qquad\qquad\qquad m/z \ 42 \tag{11.2}$$

$$\left[CH_3CH_2 \overset{\zeta}{-} CH_2OH \right]^{+\cdot} \longrightarrow CH_3CH_2\cdot \ + \ CH_2{=}\overset{+}{O}H$$
$$m/z \ 31$$

For comparison, it may be noted that 2-propanol, which does not have a γ-carbon atom, ejects OH$\cdot$ and not H_2O from the molecular ion.

11.3. Example B

Compound B gives two molecular ion peaks (figure 11.3) at m/z 156 and 158 of almost equal height and bromine must be present (the isotopes ^{79}Br and ^{81}Br are of almost equal natural abundance). Taking 79 and 81 mass units from 156 and 158, respectively, leaves 77 in each case. Abundant ions are observed at m/z 77; this is often characteristic of aromatic systems and,

Figure 11.3. Mass spectrum of compound B. Ratio of peaks m/z 159:m/z 158 = 6.2:92.5.

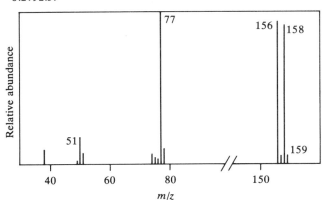

taken together with the general appearance of the mass spectrum (few, but large, peaks present), indicates that the compound is an aromatic bromide. The ratio of m/z 159 to m/z 158 is 6.2:92.5 or, stated in another way, the ^{13}C isotope peak at m/z 159 is 6.7 per cent of the height of the peak at m/z 158. Since each carbon atom contributes approximately 1.1 per cent of ^{13}C isotope, simple division (6.7/1.1 = 6, to nearest integer) gives the number of carbon atoms in compound B as six (section 1.3). Thus, the unknown must have the composition, C_6H_5Br, and is probably bromobenzene. Compound B *is* actually bromobenzene but from the mass spectrum alone it is only possible to suggest this identification as other isomers, such as 'Dewar' bromobenzene, for example, could not be ruled out on the evidence of the mass spectrum alone.

11.4. Example C

The mass spectrum of compound C (figure 11.4) shows a molecular ion at m/z 100 and accurate mass measurement gives 100.0889. Reference to tables or calculation (see table 1.5) affords the molecular composition $C_6H_{12}O$ (C_6 = 72.000; H_{12} = 12.0938; = 15.9949; total = 100.0887; error in measurement = 2 parts in 1 000 887, i.e. <2 ppm, which is acceptable). The composition $C_6H_{12}O$ has one double-bond equivalent (section 10.2) which is accounted for by the carbonyl absorption in the infrared spectrum. Compound C must be an aliphatic aldehyde or ketone. An aldehyde normally shows a loss of hydrogen (C2 cleavage) from the molecular ion but there is no ion at m/z 99 so compound C is probably a ketone $RCOR'$.

Abundant ions at m/z 85 must be due to loss of 15 mass units ($CH_3^{\cdot}$) from the molecular ion and suggest C2 cleavage adjacent to carbonyl (R = CH_3);

Figure 11.4. Mass spectrum of compound C. Accurate mass of molecular ion at m/z 100 is 100.0888. The infrared spectrum shows carbonyl absorption at 1712 cm^{-1}.

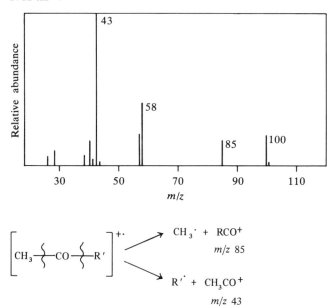

The alternative cleavage adjacent to the carbonyl group would give an ion at m/z 43 as shown and the base peak in the spectrum of compound C occurs at m/z 43. Since the total composition of C is $C_6H_{12}O$, the group R′ must be C_3H_7 and the possible structures of the unknown have been reduced to two, i.e. R′ = 1- or 2-propyl. To distinguish between the two possibilities, the mass spectrum is examined further and the odd-electron ion at m/z 58 noted. This ion arises by the six-centre rearrangement (RE2) of ketones and means that R′ has a γ-carbon atom and so is 1-propyl. Compound C is 2-hexanone.

11.5.   Example D

The mass spectrum of a solid natural product is presented in table 11.2. The base peak at m/z 86 might at first be considered as a candidate for the molecular ion, but this assignment would require a loss of 12 mass units to form the fragment ion at m/z 74. Since such a loss is not likely as it would have to be a carbon atom or remotely a BH˙ radical, it is concluded that the

Table 11.2. *Electron impact spectrum^a of compound D*

m/z	Relative abundance (%)
87	5.6
86	100.0
75	3.6
74	18.1
71	0.3
70	1.6
69	0.7
57	2.8

^{a} All ions over m/z 50 and greater than 0.2 per cent relative abundance are shown. The metastable ion spectrum shows a very small peak for the fragmentation m/z 131→86. Chemical ionization of compound D with isobutane results in abundant ions at m/z 132.

molecular ion peak is greater in mass than m/z 86 and is absent or extremely small. There are no particularly characteristic peaks at low mass and elucidation of the structure from the routine electron-impact mass spectrum alone is not feasible. Linked scanning of the magnetic field (B) and electric-sector voltage (E) such that B^2/E remains constant was used to search for any ions which dissociated just outside the ion source to give m/z 86 (section 8.1). Only an extremely small metastable ion peak corresponding to the dissociation of m/z 131 to m/z 86 was observed. Metastable ions for this fragmentation are not observed in the routine mass spectrum. Since m/z 131 is a precursor to m/z 86, it is possibly, but not proven to be, the molecular ion. To proceed further with the analysis, more information regarding the molecular weight is required. This is best achieved by means of 'soft' methods of ionization (chapter 7).

Chemical ionization of compound D with isobutane resulted in only two significant mass peaks, m/z 132 (100 per cent relative abundance) and 86 (72 per cent relative abundance), which together accounted for 80 per cent of the total ion current. Field ionization and field desorption both afforded abundant ions at m/z 132. It is concluded that the molecular weight of compound D is 131, these additional methods of ionization having yielded $(M + H)^+$ ions (chapter 7). Because the molecular weight is odd, the compound must contain an odd number of nitrogen atoms.

The formation of ions at m/z 86 in the electron-impact spectrum is due to a loss of 45 mass units from the molecular ion, often corresponding to $C_2H_5O^{\cdot}$ from ethyl esters or $^{\cdot}COOH$ from acids. If the substance were an ester, $RCOOC_2H_5$, the ion $R-C{\equiv}O^+$ (m/z 86) would be expected to eject CO to give a signal at m/z 58 for R^+, but this is not observed (table 11.2). Thus, the compound is probably an acid. This deduction is supported by the fact that the compound is a solid with a high melting point; an ester of this mass is likely to be a liquid. Ions at m/z 74 must arise through loss of 57 mass units (usually $C_2H_5CO^{\cdot}$ or $C_4H_9^{\cdot}$) from the molecular ions. The loss of $C_2H_5CO^{\cdot}$ would imply that compound E is an ethyl ketone, but such a compound would give $(M - C_2H_5{}^{\cdot})^+$ ions at m/z 102 and abundant ions at m/z 57 ($C_2H_5CO^+$; C2 cleavage; section 10.6). The former peak is absent and the latter rather small, so that loss of a butyl radical is indicated with m/z 57 being partly due to $C_4H_9{}^+$.

The groups C_4H_9 and COOH account for a mass of 103, leaving 29 mass units still to be explained. As the compound contains an odd number of nitrogen atoms, the outstanding elements are CH_3N. The losses of $C_4H_9{}^{\cdot}$ and $^{\cdot}$COOH are so favourable in the electron-impact spectrum that C2 cleavage is suspected, the stability of the fragment ions being ascribed to formation of immonium species; this requires both groups to be attached to an α-carbon.

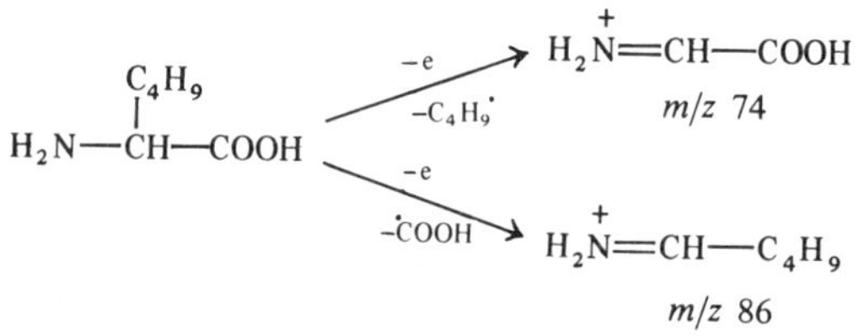

Ions at m/z 74 (relative abundance 18.1 per cent) contain only two carbon atoms, so that the relative abundance of ions at m/z 75 due to the natural ^{13}C content is $18.1 \times 2 \times 1.1/100 = 0.4$ per cent. Ions at m/z 75 are much more abundant than this and require an alternative explanation. It is proposed that the peak at m/z 75 arises mostly from a six-centre rearrangement:

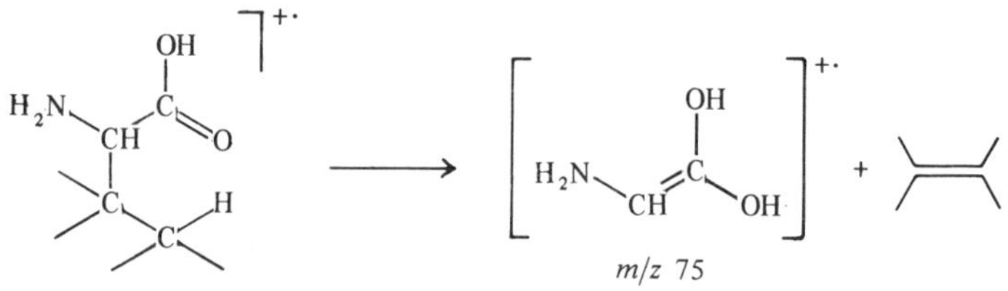

This fragmentation is in accord with the proposed structure, but does not shed any light on the character of the butyl group.

The mass spectra are consistent with the amino acids leucine, isoleucine and norleucine which have the butyl groups $(CH_3)_2CHCH_2$, $C_2H_5(CH_3)CH$ and $CH_3CH_2CH_2CH_2$, respectively. These isomers would be distinguished readily by nuclear magnetic resonance spectroscopy or, if the sample size is limited, by chromatography following derivatization. In fact, compound D is leucine. Note that the analysis would have been much simpler if the amphoteric nature of the compound had been known from the method of isolation.

11.6.   Example E

The mass spectrum (figure 11.5) of an unknown white solid extracted from a plant was acquired and processed by a data system (chapter 3). The spectrum shows the features typical of an aliphatic compound. A long hydrocarbon chain is indicated by the ion series m/z 57, 71, 85, 99 and 113 ($C_nH_{2n+1}^+$) and m/z 55, 69, 83, ..., 209, 233, 237 and 251 ($C_nH_{2n-1}^+$). Since the ions of the two series decrease steadily in abundance, an unbranched chain is indicated (branched chains afford greater ion abundances at branching points; section 10.5). Further information is not readily ascertained from the spectrum; for instance, the molecular ion is difficult to assign. Instead, the computer was programmed to compare the mass spectrum with several thousand reference mass spectra (the library), stored in its memory. The five library entries showing the most resemblance to the unknown spectrum were identified (figure 11.6) and all are long-chain alcohols. The matches are assessed on the basis of 0 for a complete mismatch and 1000 for a perfect match (details of library searching are given in section 3.4). For the first-ranked entry, 1-nonacosanol ($n-C_{29}H_{59}OH$; molecular weight 424), the *fit* figure is high because all the peaks in the reference spectrum occur also in the unknown spectrum with about the same heights, but the *purity* figure is much lower because there are some peaks in the unknown spectrum which are not present in the reference spectrum (see figure 11.7 and section 3.4).

Figure 11.5. Mass spectrum of compound E, as obtained by computer.

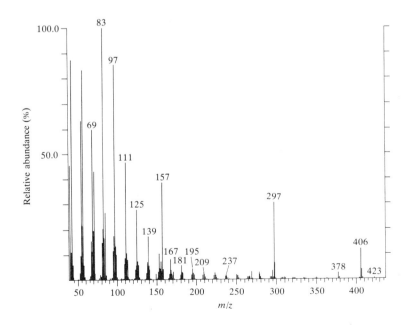

This situation would obtain if the sample were a mixture containing 1-non-acosanol, or if it were a compound with much structural similarity to 1-nonacosanol. As the substance of interest is pure, it must be similar in structure to 1-nonacosanol. It is seen (figure 11.7) that the reference compound, 1-nonacosanol, shows no molecular ion (at m/z 424) but does show ions at m/z 423 (loss of H⋅) and 406 (loss of H_2O). Since the same behaviour is shown by compound E, it implies that it is an isomer of 1-$C_{29}H_{59}OH$.

Since branching of the carbon chain has been excluded and compound E is

Figure 11.6. Computer print-out showing result of library search of spectrum of compound E.

LIBRARY SEARCH OF COMPOUND E
25409 SPECTRA IN LIBRARY SEARCHED FOR MAXIMUM PURITY
  52 MATCHED AT LEAST 7 OF THE 16 LARGEST PEAKS IN THE UNKNOWN

RANK	NAME
1	1-NONACOSANOL
2	1-DOCOSANOL
3	1-HEPTACOSANOL
4	1-HEXACOSANOL
5	1-TETRACOSANOL

RANK	FORMULA	MOL. WT	PURITY	FIT
1	C29.H60.O	424	585	971
2	C22.H46.O	326	554	960
3	C27.H56.O	396	546	971
4	C26.H54.O	382	537	950
5	C24.H50.O	354	523	950

Figure 11.7. Comparison of (*a*) mass spectrum of compound E and (*b*) reference spectrum of 1-nonacosanol from the mass spectra library (bottom).

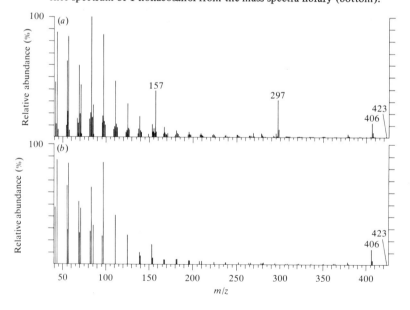

not $1\text{-}C_{29}H_{59}OH$, it is deduced that the unknown substance is a secondary alcohol. This postulate is in agreement with the formation of ions at m/z 423 by loss of H˙ (scheme 11.3; C2 cleavage). A secondary alcohol is expected to undergo C2 cleavage as shown in scheme (11.3) to give ions (1) and (2) which

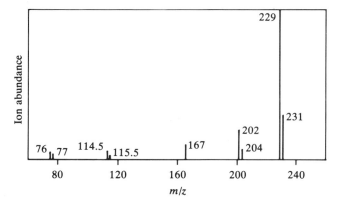

$$(11.3)$$

would not occur in the spectrum of 1-nonacosanol. Inspection of figure 11.7 immediately reveals ions at m/z 157 and 297, not present in the reference spectrum. The ion $C_9H_{19}CH{=}\overset{+}{O}H$ has mass 157 and ion $C_{19}H_{39}CH{=}\overset{+}{O}H$ mass 297. In the 70-eV electron-impact mass spectrum, elimination of the largest radical is favoured (section 10.6) and it is observed that the ion abundances decrease in the order m/z 157 > 297 > 423 as expected. It may be concluded that compound E is 10-nonacosanol, $C_9H_{19}CH(OH)C_{19}H_{39}$, a substance not represented in the mass spectral library.

The structural assignment may be confirmed by nuclear magnetic resonance spectroscopy and by derivatization (e.g. by oxidation to the corresponding ketone) followed by further mass spectrometry.

11.7. Example F

The mass spectrum of compound F (figure 11.8) is typically aromatic in appearance with abundant molecular ions and relatively few fragment ions.

Figure 11.8. Mass spectrum of unknown compound F. Measured accurate masses of ions are 229.0405 and 202.0296. Metastable ions occur at m/z 136.5 and 138.0.

In the molecular ion region there are two peaks at m/z 229 and 231, separated by 2 mass units and with an abundance ratio of 3:1, indicating the presence of chlorine. The odd molecular weight indicates the presence of an odd number of nitrogen atoms and this is confirmed by the accurate mass (229.0405) of the molecular ion since the composition $C_{12}H_8ClN_3$ requires a mass of 229.0408. The error between the measured and required masses is 0.003 in 229.0408, i.e. the error is 1.5 ppm (usually errors of $\pm$ 10 ppm are considered acceptable). Two fragment ions at m/z 202 and 204 in the ratio 3:1 correspond to the loss of 27 mass units from the molecular ions and suggest the expulsion of $C_2H_3^{\cdot}$ or HCN.

The loss of HCN from aromatic amino-compounds and nitrogen heterocyclic compounds is commonly observed and is confirmed by the accurate mass 202.0296 corresponding to the composition $C_{11}H_7ClN_2$ ($C_{12}H_8ClN_3 - C_{11}H_7ClN_2 = $ HCN). Metastable ions at m/z 138 and 136.5 correspond to the losses of ^{35}Cl and ^{37}Cl from m/z 202 and 204, respectively, to yield the ion at m/z 167 containing no chlorine (there is no ion at m/z 169 of one-third the abundance of the m/z 167 ion). The ions at m/z 77 and 76 are usually observed with aromatic compounds containing benzene rings and the relatively abundant doubly charged ions at m/z 114.5 and 115.5 are also indicative of aromatic character. Note that the molecular formula $C_{12}H_8ClN_3$ corresponds to ten double-bond equivalents and signifies a highly unsaturated or condensed ring system.

Thus, the mass spectrum can be used to propose that the compound is aromatic, possibly a condensed polycyclic compound with one or more benzenoid rings and that it contains one chlorine and three nitrogen atoms. The ejection of one nitrogen atom as HCN from the molecular ion before chlorine is ejected strongly suggests that an amino-group is present. Beyond this, further deductions would be speculative unless supported by other experimental information or experience based on similar compounds. For example, the presence of an amino-group could be confirmed by derivatization such as acylation or trimethylsilylation followed by mass spectrometric analysis. Alternatively or in addition, the nuclear magnetic resonance spectrum would yield highly informative data on such a substance. The compound is 1-amino-3-chlorophenazine (F).

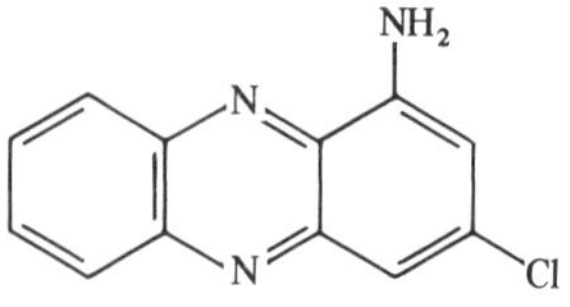

(F)

11.8. Example G

The mass spectrum of compound G (figure 11.9) is typically aliphatic with many abundant fragment ions. The molecular ion at m/z 228 is relatively abundant for an aliphatic compound and suggests the presence of unsaturation or of one or more ring systems. The ion at m/z 210 corresponds to loss of 18 mass units from the molecular ion (metastable ion at m/z 193.4) and must be loss of H_2O; a second expulsion of H_2O is apparent from the presence of ions at m/z 192 and metastable ions at m/z 175.5 ($= 192^2/210$). The two successive losses of H_2O indicate two hydroxyl groups in compound G. The ion at m/z 197 corresponds to ejection of 31 mass units from the molecular ion. This mass loss frequently corresponds to the scission of $CH_3O^\cdot$ from a methyl ester; the accurate masses of the ions at m/z 197 and 228 give a mass difference of 31.0180 mass units (CH_3O requires 31.0184). Confirmation of the ester character is found in the ion at m/z 169 of composition $C_9H_{13}O_3$ (loss of $CH_3OCO^\cdot$ from molecular ion).

$$R-CO-OCH_3^{+\cdot} \to R-CO^+ \to R^+$$

The accurate mass of the molecular ion yields the molecular formula $C_{11}H_{16}O_5$. The fragmentation of the molecular ion has indicated that four of the oxygen atoms reside in two hydroxyl and one methyl ester functions. The molecular formula agrees with the elemental analysis (found: C, 57.9; H, 7.1 per cent; $C_{11}H_{16}O_5$ requires C, 57.9; H, 7.1 per cent), indicating the absence of non-isomeric impurities. Without experience, it would be difficult to proceed further with just the information from the mass spectrum. The compound is actually the aglucone of loganin (G). In a series of closely related compounds with a similar carbon skeleton, the mass spectra show prominent ions at m/z 139 of composition $C_7H_7O_3$. The ion is thought to arise from one particular part of the carbon skeleton, as shown, and may be termed a characteristic ion

Figure 11.9. Mass spectrum of unknown compound G. Metastable ions occur at m/z 193.4 and 175.5. Compositions of ions at m/z 228, 210, 197, 192 and 169 are, respectively, $C_{11}H_{16}O_5$, $C_{11}H_{14}O_4$, $C_{10}H_{13}O_4$, $C_{10}H_{12}O_3$ and $C_9H_{13}O_3$. Elemental analysis: C, 57.9; H, 7.1 per cent.

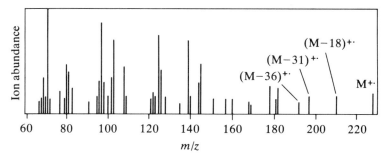

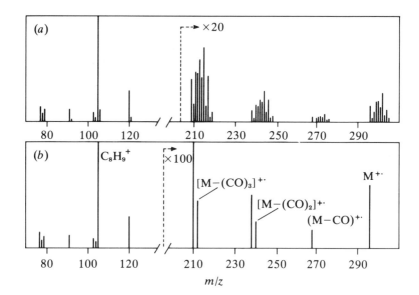

(G)   m/z 139

of the series. It is useful because its abundance varies greatly with the presence and position of a double bond in the five-membered ring and so gives additional structural information.

11.9.   Example H

The mass spectrum of compound H (figure 11.10a) shows groups of peaks at high m/z values but only isolated peaks at low m/z values. There are ten major peaks in the molecular ion region. The last peak at m/z 305 must be a ^{13}C isotope leaving nine other possible isotope peaks in the ratio 1.0 :

Figure 11.10. (a) Mass spectrum of unknown compound H and (b) the same mass spectrum with isotope contributions removed. Metastable ion peaks appear at m/z 236, 208, 101 and 91.8. Ions of very low abundance also occur at m/z 92, 94, 95, 96, 97, 98 and 100. The ion abundances above m/z 200 are drawn to increased scales as shown.

0.2:0.6:1.1:1.2:0.7:1.6:0.3:0.7. An examination of isotope tables suggests molybdenum is present since it has seven isotopes at 92, 94, 95, 96, 97, 98 and 100 in the ratio 1:0.6:1.0:1.0:0.6:1.5 and 0.6 (the peaks at m/z 297 and 303 would then have to be ^{13}C isotope peaks of m/z 296 and 302, respectively). Confirmation of this identification is obtained from the presence of ions at m/z 92, 94, 95, 96, 97, 98 and 100 albeit in low abundance. If the peak at m/z 297 is a ^{13}C isotope contribution, then the abundance of this ion relative to that of m/z 296 enables the ^{13}C isotope contributions to be removed from all the ions between m/z 296 and m/z 305. When this is done, the residual peaks number only seven and are in the ratio 1.0:0.6:1.0:1.1:0.7:1.5:0.7, very close to the ratios required for molybdenum. All isotope contributions other than ^{92}Mo and ^{12}C can then be removed from the spectrum leaving it much simpler to interpret (figure 11.10b).

Successive losses of 28 mass units are observed in the presence of peaks at m/z 296, 268, 240 and 212 (on the evidence of metastable ions, the ions at m/z 238 and 210 arise from m/z 240 and 212, respectively, by loss of H$_2$). Three successive ejections of 28 mass units from an organometallic compound suggest elimination of three CO molecules from a metal complex (confirmed by accurate mass measurement). Thus, the compound contains molybdenum (^{92}Mo, figure 11.10b) and three CO ligands (3 × CO = 84 mass units) and together these account for 92 + 84 = 176 mass units, leaving 296 − 176 = 120 mass units. Abundant ions are found at m/z 120 and these decompose by elimination of 15 mass units (CH$_3$·) to give ions at m/z 105 (metastable ion peak for process at m/z 91.8).

Accurate mass measurement shows that the ion at m/z 105 which also loses H$_2$, has the composition C_8H_9. Apart from molybdenum and the CO groups, the remainder (C_9H_{12}) of compound H must be very unsaturated (four double-bond equivalents) and, together with ions at m/z 77, 78, 79 and 91, this suggests a benzenoid ring, possibly benzylic, with a C_3H_7 side-chain. The prominent loss of methyl from m/z 120 and the relatively small peak at m/z 91 indicate that the side-chain is 2-propyl. In this instance, on the basis of the mass spectrum alone, a tentative total structure can be proposed for compound H, viz. Mo(CO)$_3$ complexed to 2-propylbenzene. The actual structure of (H) is as shown.

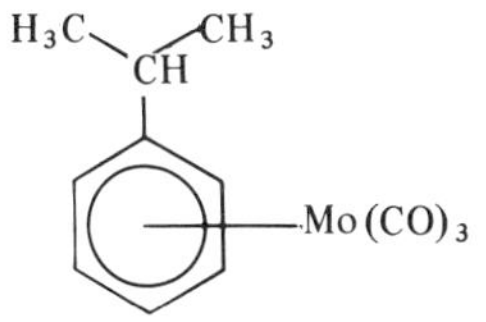

(H)

11.10. Example I

Analysis of a tripeptide showed that the constituent amino acids were glycine, alanine and phenylalanine. The peptide was acetylated at the *N*-terminus and methylated at the carboxyl end:

$$H_2N\text{---}\wedge\wedge\text{---}CO_2H \longrightarrow CH_3CONH\wedge\wedge\wedge CO_2H \longrightarrow CH_3CONH\wedge\wedge\wedge CO_2CH_3$$

The volatility of a peptide can be increased by *N*-methylation which removes amide hydrogen bonding. The acetylated methyl ester was accordingly methylated at each amide nitrogen using strong base and methyl iodide. A partial mass spectrum of the completely methylated peptide is presented in figure 11.11. Table 11.3 represents part of a computer output listing the accurate

Figure 11.11. Partial mass spectrum of peptide (I). See table 11.3 for accurate masses of ions.

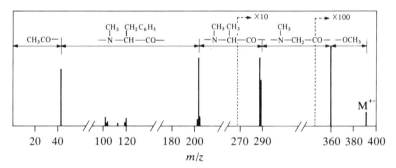

masses of all ions found by using a mass spectrometer/computer data procesing system. From this information, the structure of the peptide can be determined as follows. After methylation, the masses of the amino-acid and acetyl and methoxyl groups are as shown (scheme 11.4). Beginning at one end of

Residue		*Composition*	*Mass*
N-methylglycyl	$-N-CH_2-CO-$ $\quad\vert$ $\quad CH_3$	C_3H_5NO	71.0371
N-methylalanyl	$-N-CH-CO-$ $\quad\vert\quad\ \ \vert$ $\quad CH_3\ \ CH_3$	C_4H_7NO	85.0528
N-methylphenyl- alanyl	$-N-CH-CO-$ $\quad\vert\quad\ \ \vert$ $\quad CH_3\ \ CH_2C_6H_5$	$C_{10}H_{11}NO$	161.0841
Acetyl	CH_3CO-	C_2H_3O	43.0184
Methoxyl	CH_3O-	CH_3O	31.0184

(11.4)

Table 11.3. *Partial output of accurate masses of ions in mass spectrum of peptide (I) from mass spectrometer/computer data acquisition system*

Accurate mass	Ion abundance	Accurate mass	Ion abundance
391.208 40	5	130.050 32	20
361.186 26	4	130.040 34	2
360.188 97	32	129.091 14	12
346.171 89	7	129.055 21	3
300.156 18	101	128.074 52	3
290.156 78	27	128.068 71	7
289.153 30	167	127.148 71	6
288.146 72	237	127.087 69	97
287.143 43	3	127.077 02	5
286.140 76	12	127.062 45	6
276.143 73	7	127.052 97	2
275.138 24	47	126.056 95	10
274.132 22	59	125.132 69	14
262.164 39	14	125.059 18	2
261.161 13	62	123.117 01	10
261.127 04	2	121.101 90	5
260.114 48	5	121.084 17	23
218.115 63	19	121.063 90	10
217.131 95	7	120.081 37	278
217.109 54	64	119.087 03	5
216.102 24	77	119.073 37	188
215.100 53	138	118.087 04	12
213.089 62	6	118.076 48	3
210.071 13	1	118.064 92	40
206.107 62	28	117.079 83	2
205.105 11	331	117.073 71	6
205.085 60	2	117.069 18	7
204.102 26	2414	117.057 26	32
204.082 97	9	116.071 29	47
203.136 98	3	116.063 56	2
203.119 33	3 .	116.048 86	11
203.094 04	208	115.086 86	6
202.085 68	22	115.078 47	3
201.085 31	8	115.054 75	14
200.138 32	3	114.055 27	18
198.095 89	2	113.134 43	5
197.092 47	14	113.130 13	3
192.102 93	6	113.071 91	45
130.088 44	11	113.059 85	4
130.065 29	17	113.048 04	56

the peptide with acetyl, add to its mass (43.0184) the mass of each amino-acid residue in turn; this gives the results presented in scheme (11.5).

$$\text{Acetyl} + N\text{-methylglycyl} = 114.0555$$
$$\text{Acetyl} + N\text{-methylalanyl} = 128.0711$$
$$\text{Acetyl} + N\text{-methylphenylalanyl} = 204.1024 \qquad (11.5)$$

Abundant ions of mass 204.1023 are found in the spectrum (ions of very low abundance occur at m/z 114 and 128 but their accurate masses are not those required) and the first amino acid in the sequence is therefore phenylalanine. Scheme (11.6) shows the masses expected from adding a second residue to phenylalanine.

Acetyl + N-methylphenylalanyl + N-methylglycyl = 275.1396
Acetyl + N-methylphenylalanyl + N-methylalanyl = 289.1552   (11.6)

A prominent peak is found at 289.1533 and alanine must follow phenylalanine in the sequence. The final amino-acid residue must be the remaining glycyl, and to complete the sequence the terminal methoxyl group must be added (11.7).

Acetyl + N-methylphenylalanyl + N-methylalanyl + N-methylglycyl + methoxyl   (11.7)
(360.1923) ◂⋯ (391.2107) ◂⋯

Ions of the expected accurate masses are found at m/z 360 and 391 to complete the sequence and the original must have been a tripeptide, Phe–Ala–Gly (I). A complete structure of the peptide would require knowledge of the D or L nature of the constituent amino-acid residues, but the mass spectrum has yielded the important sequence information. Sequence analysis in this way has now been automated by computerized interpretation of the accurate mass data.

$$H_2NCHCONHCHCONHCH_2CO_2H$$
$$\underset{CH_2C_6H_5}{|} \quad \underset{CH_3}{|}$$
(I)

It should be noted that the consecutive amino-acid residues, Ala–Gly, have the same mass and elemental composition as an asparagine residue after N-methylation (scheme 11.8; R = H). The high-resolution mass spectrometric analysis alone is ambiguous in that these two possibilities are not differentiated.

Gly–Ala = $$-N-CH-CO-N-CH_2-CO-$$
with CR_3 on first N, CR_3 on second N, and CH_3 below the CH.

Asn = $$-N-CH-CO-$$
with CR_3 on N, and $CH_2CON(CR_3)_2$ below the CH.   R = H or D   (11.8)

There are several other coincidences amongst the amino acids (e.g. the masses of Ala–Ala and Gln). The analysis above relies on an amino-acid analysis to eliminate certain possibilities, but amino-acid analysis does not always resolve all ambiguities, as in the case of mixtures of peptides. A simple alternative to amino-acid analysis for removing ambiguities is through the use of isotopically labelled derivatives. If, in the example above (scheme 11.8), iodomethane is replaced with trideuteriated iodomethane, then, after deuteriomethylation, there would be an increase of 6 mass units (two CD_3 groups) in an Ala–Gly grouping and of 9 mass units (three CD_3 groups) in an Asn residue (scheme 11.8; R = D). Thus, differentiation is achieved without the need for high-resolution mass spectrometry. The same technique can be applied successfully to most other ambiguous assignments of amino-acid residues.

12 Further discussion of selected topics

12.1. Ionization and appearance energies

The least energy required to remove an electron from an atom or molecule is the ionization energy. Strictly, this should be the first ionization energy and correspond to the first adiabatic ionizing transition in a molecule. Morse curves for a simple diatomic molecule, M, are illustrated in figure 12.1. Except at high temperatures, most molecules predominantly populate the zeroth vibrational state ($\nu = 0$) and this is the only vibrational state of the molecule considered here. A vertical transition such as v (figure 12.1) on ionization to the ground-state ion, $M^{+\cdot}$, produces the ion in an excited vibrational state ($\nu = 3$); this transition is $0 \to 3$. The shaded area in figure 12.1 defines the range in which vertical transitions are possible from the $\nu = 0$ state of the molecule, M, to different vibrational states of the ion, $M^{+\cdot}$. One of these vertical transitions is called adiabatic (a) since ionization to the ground-state ion, $M^{+\cdot}$, produces the ion in its lowest vibrational state ($\nu = 0$); this is a $0 \to 0$ transition. The $0 \to 0$ transition is that corresponding to the first ionization energy. The $0 \to 1$, $0 \to 2$, etc., transitions are closely separated in energy and are due to vibrational energy levels in the ground state of the ion; these transitions follow each other by a few tenths of an electronvolt. A $0 \to 0$ ionizing transition (a* in figure 12.1) to form the ion ($M^{+\cdot}$) in its first excited state (ionization from the second molecular orbital level) corresponds to the second ionization energy of the molecule (M). Similarly, there are third, fourth and higher ionization energies each of which is accompanied by a series of closely spaced vibrational levels. Rotational energy is frequently neglected in mass spectrometry because the amount of energy involved is slight compared with the total excess of energy in an ion.

The first ionization energy (I) of ethane is represented on the energy level diagram of figure 12.2 which illustrates the energy change in reaction (12.1).

$$C_2H_6 \to C_2H_6^{+\cdot} + e \qquad (12.1)$$

If more energy is gradually imparted in forming the $C_2H_6^{+\cdot}$ ion, a point will be reached at which the ion contains enough vibrational energy for it to fragment as in reaction (12.2).

$$C_2H_6^{+\cdot} \rightarrow CH_3^+ + CH_3^{\cdot} \qquad\qquad (12.2)$$

The energy at which the fragment ion, CH_3^+, first appears is the *appearance energy* (*A*) for that ion (the appearance energy of a molecular ion is its first ionization energy). These ionization and appearance energies can provide fundamental thermochemical information in chemistry, and mass spectrometers may be used to measure them. In the example for ethane given here, the C–C bond dissociation energy may be obtained from the appearance energy of the CH_3^+ ion and the ionization energy of the $CH_3^{\cdot}$ radical. Equations (12.3) illustrate how the dissociation energy D_{C-C} is calculated, assuming there are no excess-of-energy terms.

$$
\begin{aligned}
C_2H_6 &\rightarrow CH_3^+ + CH_3^{\cdot} + e & A \\
CH_3^{\cdot} &\rightarrow CH_3^+ + e & I
\end{aligned}
$$

$$\therefore C_2H_6 \rightarrow CH_3^{\cdot} + CH_3^{\cdot} \qquad\qquad D_{C-C} = A - I \qquad (12.3)$$

Figure 12.1. Morse curves showing ionizing transitions from molecule (M) to ground-state and electronically excited ion ($M^{+\cdot}$). The transition (v) is a 'vertical' one and transitions (a and a*) are adiabatic ones.

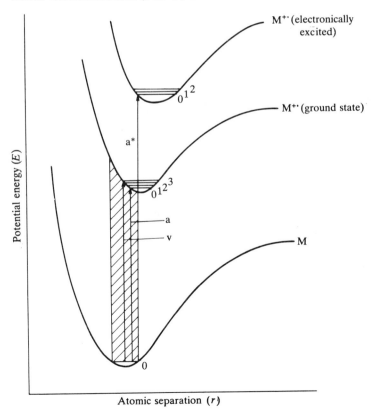

Since a mass spectrometer is a detector of ions, it is only necessary to increase the energy applied to the molecules until molecular or selected fragment ions appear at the detector. The incident energy may be photons from a vacuum monochromator, or electrons. A graph of the ion current versus the incident radiation energy is an *ionization efficiency curve* as illustrated in figure 10.4 for electron-impact ionization. The onset of ionization is the ionization energy, but it can be difficult to assign an accurate value to it, because the position of onset may be somewhat indeterminate. The latter remark applies particularly to values of ionization energies derived from electron impact and there seems little doubt that many of the older values determined in this way will need to be corrected. More recently introduced methods of mathematical smoothing and analysis of raw data promise greater accuracy and consistency. Also, by use of closely defined photon or electron energies and repeated scanning of the ionization efficiency curve, accurate data have become available.

Appearance energies determined by mass spectrometry may have dubious significance. In the example given above (equations 12.3), it was assumed there were no excess-of-energy terms or that these cancelled. These excess-of-energy terms arise in several ways but probably the most important are (i) the molecular and electronic *structures* of the ions produced are *assumed* ones, (ii) the energy of activation for the reverse reaction is assumed to be zero, i.e. $A = I + D_{C-C}$ in equations (12.3) only if the activation energy for the reverse reaction is zero (see figure 9.6), and (iii) for CH_3^+ ions to be detected in the mass spectrometer, the $C_2H_6^{+\cdot}$ ions in equations (12.3) must decompose at a measurable

Figure 12.2. Energy diagram showing relationship between appearance energy of CH_3^+ from C_2H_6 and the ionization energy of C_2H_6. For a ground-state reaction, the C–C bond energy in the ion (D_{C-C}) is $A(CH_3^+) - I(C_2H_6)$.

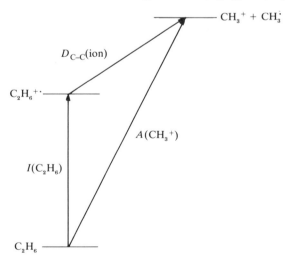

rate and must therefore contain some excess of internal energy. If results determined by mass-spectrometric procedures agree with those found by other methods it may be supposed that the assumptions are satisfactory. Where no check is available, the assumptions must be kept firmly in mind and results regarded as open to question. There are many mass-spectrometric results which do not agree with thermochemical arguments based on other methods of evaluating thermochemical data.

12.2.   Isotope analysis and labelling

The measurement of isotope abundances by mass spectrometry can be made readily to a fair degree of accuracy. If relative abundances are required better than about 3 per cent accuracy, special precautions must be taken and mass spectrometers have been designed specifically for isotope analysis. These spectrometers compare simultaneously two beams of ions for any pair of isotopes and provide very accurate isotope ratios by null methods.

12.2.1. Natural isotopes

A mass spectrometer measures mass-to-charge ratios and the effect of elemental isotopes on the appearance of a mass spectrum has been described in section 1.3. Ions containing carbon (^{12}C) are always accompanied by ions one mass unit greater because of the ^{13}C isotope. The natural abundance of ^{13}C is quite low so that it does not complicate the mass spectrum too much except when there are relatively large numbers of carbon atoms in the compound under investigation. Other elements such as nitrogen and oxygen also contain only low percentages of isotopes other than the main one so that their abundances are of not much significance unless there are large numbers of these atoms in the ion. However, some elements (listed in table 1.4) have abundant isotopes, for example chlorine and bromine, as have many of the metals used in organometallic chemistry. When elements with abundant isotopes are found in mass spectrometry, it is necessary to have some idea of how the appearance of the mass spectrum will depend on their number, i.e. to know what the *isotope pattern* will be.

Ions, M^+, containing one chlorine atom will be accompanied by ions $(M + 2)^+$ of approximately one-third the abundance because the natural relative abundances of ^{35}Cl and ^{37}Cl are in the ratio 3 : 1. When there are two chlorine atoms in the ion, RCl_2^+, the calculation of the relative abundances of the isotope peaks (isotope pattern) becomes more complicated. Three peaks, due to M^+, $(M + 2)^+$ and $(M + 4)^+$ ions, occur corresponding to the possible combinations of ^{35}Cl and ^{37}Cl (formulae 12.4).

$$\underbrace{R\,^{35}Cl\,^{35}Cl}_{M^+} \qquad \underbrace{R\,^{35}Cl\,^{37}Cl}_{(M+2)^+} \qquad \underbrace{R\,^{37}Cl\,^{37}Cl}_{(M+4)^+} \qquad (12.4)$$

Simple calculation of probabilities shows that the ions M^+, $(M + 2)^+$ and $(M + 4)^+$ will have relative abundances of $9:6:1$ (figure 1.6). For more than two chlorine atoms, the calculation becomes more laborious, but expansion of the simple formula (12.5) yields immediately the number and relative abundances of the isotope peaks. In the binomial expansion (12.5), n is the number of atoms to be considered and a and b are relative abundances of the isotopes.

$$(a + b)^n \tag{12.5}$$

For two chlorine atoms in an ion, $n = 2$, $a = 3$ and $b = 1$ and hence $(a + b)^n = (3 + 1)^2 = 9 + 6 + 1$, and these terms are the relative abundances of the chlorine isotope peaks for ions containing two chlorine atoms. For three bromine atoms, $n = 3$, $a = 1$ and $b = 1$ (^{79}Br; $^{81}Br \approx 1:1$; see table 1.4) and so $(a + b)^n = (1 + 1)^3 = 1 + 3 + 3 + 1$. Thus, ions with three bromine atoms would give peaks in the mass spectrum for M^+, $(M + 2)^+$, $(M + 4)^+$ and $(M + 6)^+$ with relative abundances of $1:3:3:1$. The reader familiar with nuclear magnetic resonance spectroscopy will note how the same formula (12.5) is used to obtain spin 'splitting patterns'.

When ions contain two or more different elements with isotopes, the calculation *ab initio* of relative abundances becomes laborious and time consuming and it is better to write a computer program to effect the calculation. In simpler cases, there is a short cut for obtaining the isotope pattern which is best exemplified by two specimen calculations (12.6 and 12.7).

Firstly, consider ions with two chlorine and two bromine atoms.

For Cl_2, $(a + b)^n = (3 + 1)^2 = 9:6:1$ relative
For Br_2, $(a + b)^n = (1 + 1)^2 = 1:2:1$ abundances $\tag{12.6}$

These relative abundances are combined,

$$(961)(121) \therefore (961) \times 1 = 9:\ 6:\ 1$$
$$(961) \times 2 =\ \ 18:12:2$$
$$(961) \times 1 =\ \ \ \ \ \ 9:6:1$$
$$\overline{\text{Total}\ \ 9:24:22:8:1}$$

That is, each of the terms relating to chlorine abundances (961) is multiplied by each term relating to the bromine abundances (121). After each multiplication, the resulting terms are moved one column to the right and finally totalled. Thus, the isotope pattern for Cl_2Br_2 is five ions, M^+, $(M + 2)^+$, $(M + 4)^+$, $(M + 6)^+$ and $(M + 8)^+$, with relative abundances of $9:24:22:8:1$.

Secondly, consider ions having two sulphur atoms and one chlorine:

For S_2, $(a + b)^n = (95 + 5)^2 = 9025:950:25$
($^{32}S:^{34}S \approx 95:5$; see table 1.4)
For Cl, $(a + b)^n = (3 + 1)^1 = 3:1$. (12.7)

Combining these relative abundances (equations 12.7) gives,

$$(9025\ 950\ 25)\ (3\ 1) \therefore\ (9025\ 950\ 25) \times 3 = 27075:\ 2850:\ \ 75$$
$$(9025\ 950\ 25) \times 1 = \qquad\qquad 9025:\ 950:25$$

$$\text{Total}\quad 27075:11875:1025:25$$

$$\text{or}\quad 1083:\ \ 475:\ \ 41:\ 1$$

Therefore, the isotope pattern consists of four peaks for M^+, $(M + 2)^+$, $(M + 4)^+$ and $(M + 6)^+$ in the ratios $1083:475:41:1$. The peak due to $(M + 6)^+$ ions is very much smaller than the other three and might be ignored in the mass spectrum. Notice that the ratio $(2.3:1)$ of the abundances of the M^+ and $(M + 2)^+$ ions is quite different from the $3:1$ it would be if only chlorine and no sulphur were present. Although the natural abundance of ^{34}S is quite modest, its contribution to the spectrum would not be missed by an experienced mass spectrometrist.

Mechanisms of chemical reactions may be investigated by changes in natural isotope abundances as caused by kinetic isotope effects. The technique is best illustrated by reference to an example. The pyrolytic *cis*-elimination of xanthate esters (Chugaev reaction) may proceed by either of two mechanisms (12.8*a,b*). In mechanism (*a*) the bond between the thioether sulphur atom and

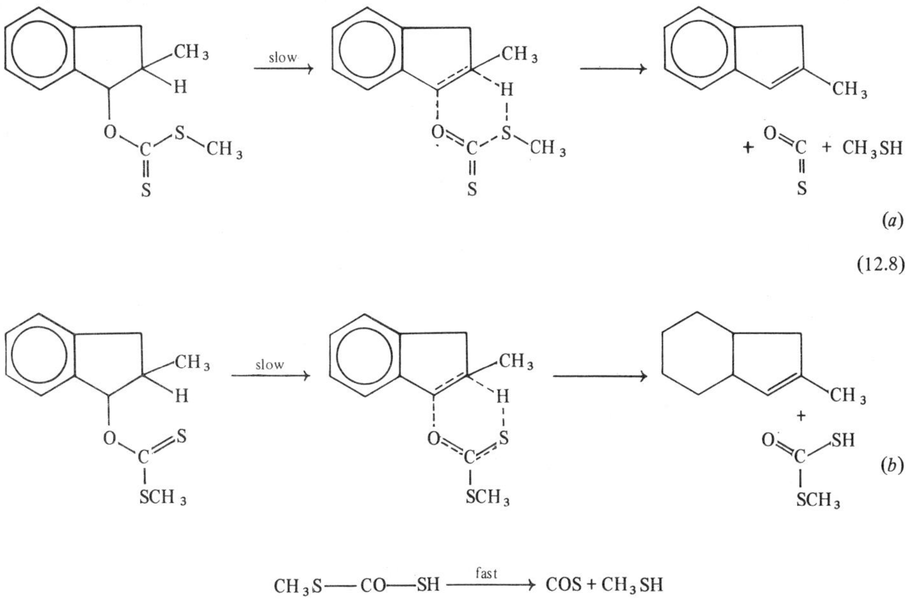

(*a*)

(12.8)

(*b*)

$$CH_3S\!-\!CO\!-\!SH \xrightarrow{\text{fast}} COS + CH_3SH$$

the carbonyl carbon is broken whereas in mechanism (b) the same bond is not significantly altered in the rate-determining step but the double bond between the thione sulphur and carbon atom becomes a single bond. In each case, the rate-determining step is the formation of the cyclic transition state. Therefore, since the $C-^{32}S$ bond is weaker than the $C-^{34}S$ bond, a $^{32}S/^{34}S$ kinetic isotope effect will be observed for either the thioether (a) or the thione sulphur atom (b). Also, a $^{12}C/^{13}C$ isotope effect of the carbonyl carbon atom is anticipated. Theory predicts a relatively large effect for mechanism (a) where dissociation of the $C-S$ bond occurs, and a negligible effect for mechanism (b). An isotope effect manifests itself as a change in the natural isotope content for the particular atom as the reaction progresses from starting material to products with different relative rates for the different isotopes. The change is monitored readily by mass spectrometry once the carbonyl carbon atom and the thione and thioether sulphur atoms are separated. The gaseous products COS and CH_3SH were swept out of the reaction vessel, chemically separated and subjected to degradative processes such that the sulphur atoms were converted to SO_2 and the carbonyl carbon atom incorporated into CO_2. No significant change of the natural $^{12}C/^{13}C$ ratio was observed and there was only a small change of the natural isotopic composition of the thioether sulphur atom, indicating that the bond between them is not cleaved in the rate-determining step. On the other hand, a large change in the isotope ratio for the thione sulphur atom, and hence a large isotope effect, was observed, proving that the bonding at this atom is altered during formation of the transition state. It was concluded that mechanism $(12.8b)$ is operating in the Chugaev reaction. This and other applications of isotope effects have been discussed (Shiner & Buddenbaum, 1975).

Measurement of the natural isotopic content of substances has other, diverse applications. For example, archaeologists have long used the number of tree-rings as a method of dating (dendrochronology) and the thickness of each ring as an indication of the climate of that year. Mass spectrometry has been used to quantify climatic changes by measuring extremely accurately the changes in the $^{12}C/^{13}C$ ratio in tree-rings. The measurements can differentiate differences in the average temperature for each year to about 0.2 °C. The variations in stable isotope ratios are attributed to either a temperature-dependent isotope effect during photosynthesis in the tree or to variation in the composition of the atmosphere. Insight into thermal history and chronology may also be obtained by measurement of the $^{16}O/^{18}O$ ratio in carbonates in limestone. As this rock once formed the sea-bed, the ratio of ^{16}O to ^{18}O varies in successive layers, depending on the temperature of the sea at the time the deposits were formed.

In fields more familiar to the chemist, metastable ion analysis can be used in conjunction with natural isotopes to aid structural elucidation. Consider an unknown substance which shows a metastable ion for the loss of 87 mass units from the molecular ion at m/z 282 and an isotope pattern in the molecular ion region indicating two sulphur atoms and about 15 carbon atoms. This would suggest an elemental composition of either $C_{15}H_{22}OS_2$ or $C_{15}H_6O_2S_2$. To a good approximation, the $(M + 1)^{+\cdot}$ ions at m/z 283 are derived from molecules with one ^{13}C atom randomly distributed throughout the structure. Using one of the techniques which detects selectively product ions of a given precursor ion (chapter 8), it is possible to focus on $(M + 1)^{+\cdot}$ ions and to observe losses of 87 mass units (^{13}C retained in ion) and 88 mass units (^{13}C in neutral particle ejected) without interference from ions derived from other sources. The ratio of the resulting metastable ion peaks for loss of 87 and 88 mass units is $2:1$, so that out of every three $(M + 1)^{+\cdot}$ ions undergoing this fragmentation only one, on average, ejects its ^{13}C atom. Therefore, the number of carbon atoms in the neutral species must be a third of that in the precursor ion. This fact necessitates that the number of C atoms in the original molecule is a multiple of three and is consistent with a compound with 15 carbon atoms. It follows that the neutral species contains five carbon atoms (60 of the 87 mass units) and that its composition is $C_5H_{11}O$, there being no nitrogen atoms in the molecule. Therefore, the original molecule cannot be $C_{15}H_6O_2S_2$ but must be $C_{15}H_{22}OS_2$ (Beynon, Morgan & Brenton, 1979). This low-resolution mass-spectrometric analysis contributed greatly to the total structural determination of the substance under consideration as compound (1). Note that, normally, this type of information is only gained

(1)

through the use of high-resolution mass spectrometry (accurate mass measurement). The same reasoning *cannot* be applied to normal ions because of interfering fragmentation of molecular ions.

12.2.2. Labelling with stable isotopes

One of the simplest uses of isotope labelling is the determination of the number of active hydrogen atoms in a molecule. Treatment of a ketone, $R_2CHCOCH_2R$, that has three enolizable hydrogens with either D_2O or CH_3OD containing a slight amount of alkali will give the deuteriated ketone, $R_2CDCOCD_2R$, so that the molecular weight increases by 3 mass units. The increased molecular weight is readily detectable by mass spectrometry and it is unnecessary for there to be complete incorporation of label. Indeed, even if incorporation of deuterium were complete, it may happen that, on putting the labelled compound into the mass spectrometer, exchange with protons from water adsorbed on the surface inside the machine would occur; the incorporation of the isotopic label would then seem incomplete in the first place. The use of stable isotopes as an aid to structural elucidation, by which many ambiguities may be resolved, has been described in section 5.3 and exemplified in chapter 11 (see example I).

Labelling of compounds specifically for mass-spectrometric work is usually done either to clarify a fragmentation sequence or to investigate the mechanisms of mass-spectrometric reactions. The 'shift' technique can be used to help unravel fragmentation pathways. For example, the mass spectrum of a methyl ester may be compared with that of the trideuteriomethyl ester; all peaks which shift by 3 mass units must incorporate the methyl group.

Most labelling has been done to gain some insight into mass-spectrometric fragmentation reactions and may be illustrated for the case of toluene. The fragmentation of toluene in the mass spectrometer has been investigated extensively; the major features of the fragmentation are shown in scheme (12.9). The elimination of a hydrogen atom from the molecular ion of toluene was at first believed to occur from the side-chain methyl group to give the benzyl cation, $C_7H_7^+$, at m/z 91. However, various deuterium labelling experiments showed that the ejected hydrogen was almost randomly derived from any of the original eight hydrogen atoms in the toluene molecule. Further, ^{13}C-labelling of toluene in the side-chain or in the ring showed that the ejection of C_2H_2 from the fragment ion, $C_7H_7^+$, was again a random process in

$$\left[\begin{array}{c} CH_3 \\ \bigcirc \end{array}\right]^{+\cdot} \xrightarrow{-H^\cdot} C_7H_7^+ \xrightarrow{-C_2H_2} C_5H_5^+ \xrightarrow{-C_2H_2} C_3H_3^+ \qquad (12.9)$$

m/z 92 $\qquad$ m/z 91 $\left(\begin{array}{c}\bigcirc\!\!\!+\end{array}\right)$ $\qquad$ m/z 65 $\qquad$ m/z 39

(2)

that the labelled carbon atom was not specifically retained in or ejected from the decomposing ion, $C_7H_7^+$. These observations, together with measurements of the heat of formation of the ion, $C_7H_7^+$, suggested that a tropylium structure (2; scheme 12.9) was plausible for the ion. If the toluene molecular ion at m/z 92 first rearranged to a hydrotropylium ion, the random loss of hydrogen atom to give the tropylium ion at m/z 91 could be explained and also the random loss of ^{13}C-label as C_2H_2 from this ion was understandable (scheme 12.9).

Experiments with the doubly ^{13}C-labelled toluene (3; scheme 12.10) have shown that, although the two ^{13}C-labels are adjacent in the molecule before ionization, when the ion at m/z 93 fragments, the loss of ^{13}C-label in the ejected C_2H_2 corresponds to complete randomization of all the carbon atoms as well as hydrogen in the ion at m/z 93 (scheme 12.10).

$$
\text{(3)} \quad
\begin{array}{c}
^{13}CH_3 \\
\text{[benzene ring with }^{13}\text{]}
\end{array}
\xrightarrow{-e}
\;\; ^{13}C_2\,^{12}C_5H_8^{+\cdot} \;\;
\xrightarrow{-H^\cdot}
$$

$$
m/z\ 94
$$

$$
^{13}C_2\,^{12}C_5H_7^+ \xrightarrow{\text{random}}
\left\{
\begin{array}{c}
-\,^{12}C_2H_2 \\
-\,^{13}C\,^{12}CH_2 \\
-\,^{13}C_2H_2
\end{array}
\right\}
\longrightarrow
\left\{
\begin{array}{c}
^{13}C_2\,^{12}C_3H_5 \\
^{13}C\,^{12}C_4H_5 \\
^{12}C_5H_5
\end{array}
\right\}^+
\qquad (12.10)
$$

$$
m/z\ 93 \qquad\qquad\qquad\qquad\qquad\qquad\qquad m/z\ 65\text{-}67
$$

The above results show that the randomization of the carbon atoms may occur in the molecular ion which would not therefore have the toluene structure; alternatively, it could occur in the $C_7H_7^+$ ion before it fragments or it could be the result of the $C_7H_7^+$ ion not having a tropylium structure. More recent experiments with techniques such as field ionization kinetics, collisional activation and ion cyclotron resonance (section 12.4), which allow ions with specific energies and lifetimes to be studied, reveal a complex picture. Carbon and hydrogen scrambling in the molecular ion of toluene $(C_7H_8^{+\cdot})$ occurs by at least four mechanisms and there may be a dynamic equilibrium between the toluene and cycloheptatriene structures. Low-energy $C_7H_7^+$ ions from toluene may have either of two structures, presumably tropylium and benzyl ions. The two can be distinguished because the former is unreactive in ion/molecule reactions with neutral toluene whereas the latter is reactive. In the absence of ion/molecule reactions, the benzyl ion, $C_6H_5CH_2^+$, is long lived and stable. However, substantial rearrangement by several mechanisms precedes fragmentation of high-energy $C_7H_7^+$ ions. Isotope labelling for determination of fragmentation mechanisms has been reviewed (Holmes, 1975).

To be unequivocal, labelling experiments in mass spectrometry need to give either 100 per cent positive or 100 per cent negative results. Intermediate values may be interpreted as being due to scrambling of the label, the operation of alternative mechanisms or the occurrence of rearrangement. Hydrogen is particularly prone to scrambling in the mass spectrometer, and deuterium labelling experiments are frequently invalidated by this effect.

For good quantitative results and less ambiguity in isotope studies, it is necessary to have as great an enrichment of the isotopic label as possible. The methods used to introduce labels are legion, but often some ingenuity is required in the synthesis of labelled compounds from expensive starting materials, the available range of which is limited. Labelled acetic acid, $CH_3\,^{13}CO_2H$, enriched to 65 per cent with ^{13}C is expensive but is in fact one of the cheapest ^{13}C-labelled compounds commercially available. By working on a small scale, the cost of the labelling experiment may be reduced considerably. The following synthesis of ^{13}C-labelled 1,4-diphenylbutadiene required a total of about 100 mg of acetic acid (scheme 12.11).

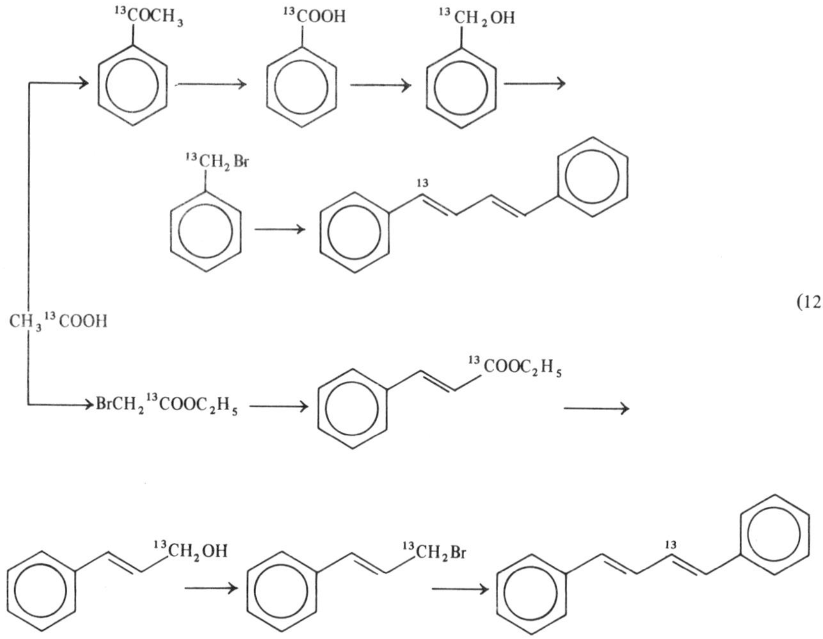

$^{13}COCH_3$ $^{13}COOH$ $^{13}CH_2OH$

$^{13}CH_2Br$

13

(12.11)

$CH_3{}^{13}COOH$

$\rightarrow BrCH_2{}^{13}COOC_2H_5 \longrightarrow$ $^{13}COOC_2H_5$ $\longrightarrow$

$^{13}CH_2OH$ $^{13}CH_2Br$ 13

The incorporation of deuterium labels is usually easier and the required chemicals cheaper than those for ^{13}C-labels. Thus, D_2O or CH_3OD may be used to exchange deuterium for hydrogen in amines, amides, alcohols and enolizable carbonyl groups. Similarly, lithium aluminium deuteride is readily available and can be used to reduce many compounds. The use of an ^{18}O isotope is relatively expensive and fewer ^{18}O-labelled compounds are available compared with deuterium and ^{13}C.

In the fields of biochemistry, medicine and toxicology, before the advent of mass spectrometry, the fate of a substance in biological tissue was often investigated by employing radioactive substrates. Mass spectrometry has largely superseded such studies because it does not require that the isotope be radioactive, thus removing radiation hazards to both experimental organism and investigator. The metabolism of exogenous compounds such as drugs and pollutants as well as endogenous substances may be studied. A substance is 'labelled' with a stable isotope at a site which does not affect and is not affected by the natural processes under investigation and, after incubation, any products of the labelled compound will have the same isotopes in the same ratio as the substrate, providing unequivocal evidence of origin. For example, a sample of progesterone, containing some unlabelled, some mono-, di- and mainly tri-deuteriated molecules (4), was incubated with frog ovaries. After appropriate work up, several compounds with the same isotope pattern in their molecular ion regions were identified by gas chromatography/mass spectrometry. Figure 12.3 shows the high-mass regions of the mass spectra of the

Figure 12.3. Partial mass spectra of (*a*) deuteriated progesterone (4) used for incubation with biological tissue, (*b*) a product of the incubation and (*c*) natural pregnanolone (5). Note that the sizes of peaks in the molecular ion regions of spectra (*a*) and (*b*) are the same and that the major peaks of spectrum (*b*) are the same as those in (*c*) but displaced to higher mass by 3 mass units.

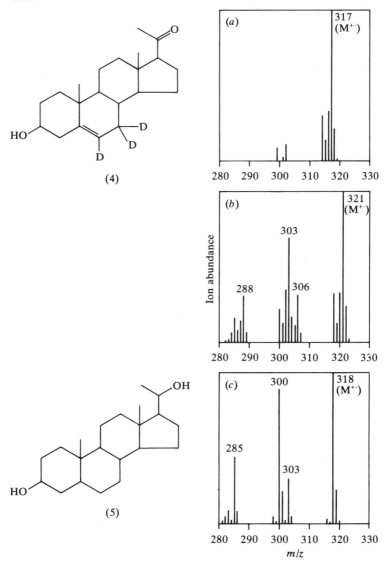

labelled progesterone substrate (*a*) and one extracted product (*b*) together with a reference spectrum of pregnanolone (*c*; structure 5). The extracted product was identified as pregnanolone by its retention time and mass spectrum, its high-mass ions being 3 mass units greater than those of natural pregnanolone (figure 12.3*b*,*c*). Note also that the extracted pregnanolone must have been a product of the progesterone because the mass spectra of the two compounds show the same isotope pattern for deuterium content in the molecular ion region (figure 12.3*a*,*b*). The proportion of unlabelled product is not taken into account for this will change unpredictably if the organism has 'natural pools' of substrate and/or products.

The above example illustrates a principle which is very widely applied for tracing the metabolic fate of compounds (see, for example, Baillie, 1978; Klein & Klein, 1979*a*; Caprioli & Bier, 1980).

12.3. 'Involatile' substances and examination of surfaces

The technique of field desorption was developed for analysis of thermally unstable compounds which are not sufficiently volatile to give appreciable vapour pressure and allow ionization in the gaseous phase, as required for electron impact, and chemical and field ionization. Field desorption has been discussed in section 7.6 and a method of ionizing complex compounds of low volatility by intense heating, applicable to combined liquid chromatography/mass spectrometry, is covered in section 4.3. Here, some other methods designed mainly for 'involatile' compounds are outlined.

12.3.1. *Modified electron impact and chemical ionization*

The effect of temperature on a mass spectrum was discussed in section 10.3. It was pointed out that, especially for thermally labile substances, the rate of heating rather than the absolute temperature seems to be the deciding factor in the competition between evaporation (breaking of intermolecular bonds) and decomposition (breaking of intramolecular bonds). Very rapid heating favours evaporation because there is less time for the molecules to transfer energy into dissociative vibrational states and because the molecules spend less time in the region of relatively high gas density around the sample probe where bimolecular interactions may lead to decomposition. Rapid heating, sometimes called *flash desorption*, can be used in conjunction with electron impact or chemical ionization. Rapid rates of heating vary between $10°/s$ and $5000°/s$ and mass spectra of involatile organic salts such as acetylcholine chloride (6) and thermally unstable compounds like steroid glycosides can be recorded. When EI ionization is employed, the typical fragmentations of normal EI mass spectra are observed as well as enhanced abundances of molecular ions. By this technique, the amino-acid sequence of an underivatized heptapeptide amide was determined (Daves, 1979).

$$\text{CH}_3\text{COOCH}_2\text{CH}_2 - \overset{\overset{\displaystyle \text{CH}_3}{|}}{\overset{+}{\text{N}}} - \text{CH}_3 \quad \text{Cl}^-$$
$$\underset{|}{\text{CH}_3}$$

(6)

In many mass spectrometers with direct insertion probes, once molecules of the sample are volatilized from the probe, they must travel some distance before they enter the beam of electrons where ionization occurs. During flight time, unimolecular decomposition of thermally labile compounds is possible. In an ion source used for CI, the volatilized molecules travel to the region rich in reactant gas ions (the *ion plasma*) and equivalent to the electron beam in EI ion sources. Because of the higher pressures used in chemical ionization, uni- or bimolecular decomposition may occur before ionization. Decomposition can be minimized by using long direct probes such that the molecules of sample are volatilized directly into the electron beam or ion plasma. Several new terms have been coined for this technique, the common ones being *in-beam ionization, direct EI* or *direct CI, desorption EI* or *desorption CI* and *plasma desorption*, but in fact the method is an old one. For many years, users of older mass spectrometers with direct probes of adjustable length have found that the best mass spectra are obtained with the longest probes and with the sample coated on the outside rather than the inside of the cup at the tip of the probe. This arrangement ensures greater exposure of the sample to the electron beam. The recent 'rediscovery' of the method has brought some refinements (Horning, Mitchell *et al.*, 1979). Instead of using simple probe tips of glass, samples have been coated on carbon, tungsten, platinum, rhenium, copper, gold, Vespel, Teflon and quartz in an effort to reduce irreversible adsorption of the sample on the probe. Probe tips have even been coated with a dimethylsiloxane film to hold but not absorb polar compounds. The rapid heating technique can be combined with the in-beam method when the sample is loaded onto the wire of a heating element which can be placed in close proximity to the zone of ionization. The precise mechanism of ionization is not fully understood. Ionization could occur (*a*) in the solid phase with subsequent desorption of ions, (*b*) in the gas/solid or gas/liquid interface or (*c*) in the gaseous phase after desorption of intact neutral molecules. Current experiments favour simultaneous desorption and ionization (method (*b*)) or ionization after desorption (method (*c*)). The success of the technique is due to three factors: minimal distance between sample and electron beam or ion plasma, evaporation from an inert surface and rapid heating. An example of the in-beam technique is shown in figure 12.4 which compares the conventional CI and in-beam CI mass spectra of cholesterol glucoside (7) with ammonia as reactant gas. Note that quasi-molecular ions $(M + NH_4)^+$ at *m/z* 566 are observed only with the in-beam method. Ions at *m/z* 385 and 369 are

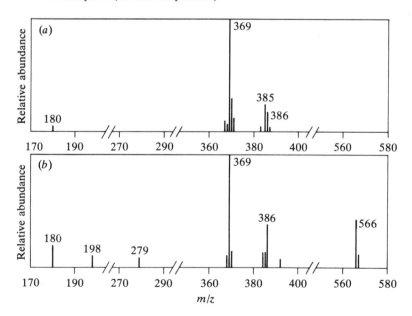

(7) mol. wt = 548

due to the steroid moiety with or without the glucosidic oxygen atom, respectively. The 'cluster' of ions near these two values arises from hydrogen transfers. The glucose unit is also represented at m/z 198 $(C_6H_{12}O_6 + NH_4)^+$ and m/z 180 $(C_6H_{10}O_5 + NH_4)^+$, these peaks being clearer in the in-beam spectrum. The technique may also be used in the negative-ion mode: in-beam negative ion CI yields useful mass spectra for polyhydroxylated compounds like (7) with OH^- as reactant gas ions, and underivatized di-, tri- and tetra-

Figure 12.4. Partial mass spectra of cholesterol glucoside (7) obtained by chemical ionization with ammonia, (*a*) using a conventional direct insertion probe (conventional CI spectrum) and (*b*) a longer, rapidly heated, direct insertion probe (in-beam CI spectrum).

saccharides with Cl⁻ as reactant gas ions give abundant $(M + Cl)^-$ ions and some diagnostic fragment ions due to elimination of one or two sugar units.

12.3.2. *Bombardment of surfaces*

When a surface is bombarded with particles or high-energy radiation, ions may be formed at that surface. Bombardment of the target surface with ions, 'hot' atoms or photons causes ionization by a variety of mechanisms, but in each case the principle of rapid heating is employed, as discussed below.

For *laser desorption*, polar non-volatile molecules (including salts) are coated onto a metal surface to give a thin layer which is subsequently subjected to a short but intense laser pulse which causes thermal desorption of alkali ions present as impurities in the metallic probe. The alkali metal ions become attached to molecules (M) to give cationized species, $(M + Na)^+$, $(M + K)^+$, etc. Alternatively, upon laser heating, alkali metal ions of salts intentionally added to the sample undergo ion/molecule reactions to give quasi-molecular ions $(M + Na)^+$, $(M + K)^+$ and so on. The cationized particles are desorbed from the probe surface under the influence of rapid laser heating, accelerated out of the ion source and analysed in a conventional mass spectrometer. The method may be considered as a rapid heating (flash desorption) technique since the wavelength of the laser beam does not seem to affect the spectra markedly. If the laser radiation were adding directly to the vibrational energy of the molecules or ions, laser wavelengths corresponding to absorption bands in their infrared spectra would generate more ions and induce more fragmentation than those in regions where the sample and its ions do not absorb, and this effect is not observed (Posthumus, Kistemaker, Meuzelaar & ten Noever de Brauw, 1978). As with field desorption, little fragmentation is observed and useful mass spectra are obtained from steroid conjugates, amino acids and peptides, nucleotides and oligosaccharides. For example, sucrose affords abundant quasi-molecular ions $(C_{12}H_{22}O_{11} + Na)^+$ and $(C_{12}H_{22}O_{11} + K)^+$ at *m/z* 365 and 381, respectively, along with cationized fragments corresponding to cleavage at the glycosidic linkage with concomitant hydrogen transfer. By laser desorption, digitoxin, a steroid conjugated to a pentasaccharide, affords a very impressive mass spectrum with abundant $(M + Na)^+$ ions at *m/z* 1251 as well as characteristic fragment ions.

Alternatively, a laser beam of low intensity can be trained continuously on a sample to generate gaseous cationized ions of thermally labile compounds. Positive and negative ions are formed and the analyser part of the mass spectrometer can be set to detect either. Contrary to the high-intensity pulse method, the wavelength of the laser beam is important inasmuch as ion yield and degree of fragmentation depend strongly on the infrared absorption characteristics of the sample (Stoll & Röllgen, 1979). The organic ammonium salt $(C_4H_9)_4N^+I^-$ yields only $(C_4H_9)N^+$ ions by this technique although ions from thermal decomposition products of the salt may be induced by increasing somewhat the power of the laser beam.

A substantial review of the uses of lasers in mass spectrometry has been published (Conzemius & Capellen, 1980).

Another approach utilizing the principle of rapid heating is *radionuclide ionization* (*^{252}Cf plasma desorption*). When the radionuclide ^{252}Cf decays, it produces two fission products (e.g. $^{142}Ba^{18+}$ and $^{106}Tc^{22+}$) which move apart in opposite directions, there being in total about 40 different pairs of product ions. One of the nuclear fission fragments travels towards a sample which is coated on a very thin nickel foil. Each highly energetic ion passes through the foil and, on impact with the sample, causes a localized 'hot spot' (about 10 000 °K). The sudden deposition of such a large amount of energy in the sample results in rapid volatilization before decomposition can occur. Ionization occurs through ion-pair formation and ion/molecule reactions, generation of positive and negative ions being equally likely. For instance, some organic molecules, M, form a desorbed and loosely bound dimer $(M)_2$, which affords quasi-molecular or molecular ions with or without proton transfer (scheme 12.12).

$$(M)_2 \nearrow^{M^{+\cdot} + M^{-\cdot}}_{\searrow (M + H)^+ + (M - H)^-} \qquad (12.12)$$

In the positive ion mode, most substances give molecular or quasi-molecular ions and generally some fragment ions are observed. The type of fragmentation is similar to that caused by chemical ionization with isobutane (section 7.2). In the negative-ion mode, fragmentation is usually less favourable.

The ions are drawn from the region of ionization and into the analyser, a time-of-flight mass spectrometer. The nuclear fission fragment which moves off in the opposite direction to that which causes ionization is detected and serves as a time marker for the fission event and hence the onset of ionization. Measurement of the difference in time between ionization and detection of an ion leads to mass assignment, the time taken to traverse the flight tube being proportional to the square root of the mass of the ion (section 2.4). A mass spectrum is developed over a period of minutes to hours by monitoring the ions formed from each individual fission event. The time-of-flight spectrometer is used because, unlike all other analysers, it has no mass range limitations, i.e. ions of several thousand mass units can be detected.

The ^{252}Cf plasma desorption method (Macfarlane & Torgerson, 1976b; Macfarlane, 1980) requires samples in the nanogramme range, but relatively few molecules are used during the analysis and the remainder can be recovered from the nickel foil afterwards. It is particularly useful for natural products such as toxins, antibiotics, amino acids, peptides and oligonucleotides. The natural analgesic peptide β-endorphin gave $(M + Na)^+$ ions at m/z 3487, the protein insulin afforded molecular ions (m/z 5790) and the sequence of decanucleotides has been determined.

A substance to be analysed by *secondary-ion mass spectrometry* (*ion bombardment*) is coated on a metallic surface and bombarded with *primary ions* (e.g. $Ar^{+\cdot}$) of high kinetic energy. At or just above the surface of the metal, complex charge-exchange processes and ion/molecule reactions occur, yielding *secondary ions*. The secondary ions of interest to the mass spectrometrist are molecular ($M^{+\cdot}$ or $M^{-\cdot}$) or quasi-molecular ions [mainly $(M + H)^+$, $(M - H)^-$ or $(M + Met)^+$ where Met is a metallic element] generated from molecules (M), together with any fragment ions. When the dipeptide, phenylalanylglycine (8), is deposited on silver and bombarded with argon ions, $C_6H_5^+$, $C_6H_5CH_2^+$, $C_6H_5CH_2CH=\overset{+}{N}H_2$ and $(M + H)^+$ ions, along with Ag^+ ions, are

$$H_2N-\underset{\underset{CH_2C_6H_5}{|}}{CH}-CONHCH_2COOH$$

(8)

observed in the positive-ion mode whilst, in the negative-ion mode, $(M - H)^-$ ions are accompanied by $AgCl_2^-$ and low-mass ions such as H^-, C^- and Cl^- (Benninghoven & Sichtermann, 1978). The surface on which the sample is adsorbed is usually, but not always, metallic. For example, the primary-ion beam can be scanned along a developed paper chromatogram and, wherever a 'spot' of substance occurs, the component is identified by its secondary-ion mass spectrum. Natural products in some biological tissues (e.g. a fungus) may be determined directly by focussing the primary-ion beam on the tissue sample.

Many reviews of secondary-ion mass spectrometry are to be found (see, for example, Benninghoven & Bispinck, 1979).

Despite the usefulness of the secondary-ion technique, there are difficulties in its application due to charging effects at the surface being bombarded. These difficulties are avoided if the surface is bombarded by 'hot' (fast) atoms instead of ions. Fast atom beams can be obtained by charge exchange of an ion beam with neutral gas molecules (e.g. $Ar + Ar^{+\cdot}$ (fast) $\rightarrow Ar^{+\cdot} + Ar$ (fast)). As a result of *fast atom bombardment* (FAB), organic substances coated on the target surface are desorbed as positive and negative ions. Currently, the technique is being developed vigorously, particularly for use with involatile samples. Usually, abundant molecular or quasi-molecular ions are obtained with sufficient fragment ions to enable structure determination. Mass spectra of peptides and glycosides can be obtained without derivatization in both positive and negative ion modes. Unlike field-ionization and field-desorption methods, fast atom bombardment is relatively simple and, unlike radionuclide ionization, does not pose radiation hazards.

12.3.3. The nature of surfaces

It is outside the scope of this book to discuss in any detail the examination of metallic and non-metallic surfaces, but here it is noted that bombardment by photons (laser desorption or *laser-microprobe mass analysis*) or charged particles (secondary-ion or *ion-probe mass spectrometry*) can give some insight into the composition of the outer few nanometers of such surfaces. Where the laser- or ion-beam bombardment occurs, a microscopic crater is formed so that, by focussing on the same spot, the change in composition of the surface layers with depth may be probed. For example, the distribution of oxygen, hydrogen or deuterium with depth in germanium and silicon films can be determined by secondary-ion mass spectrometry (primary ions, Cs^+). Likewise, the laser-microprobe technique has been used to study the distribution of hydrogen in non-metallic inclusions in steel. The techniques are useful for studying diffusion through solids (e.g. diffusion of copper in crystals of silver), corrosion at metallic surfaces, traces of impurities in semi-conductors and polymers, reactions at surfaces, and the interfaces (seams) between different materials or dislocations in crystals. A Si/SiO_2 interface, when probed by ion bombardment, gave ions Si^+, Si^-, SiO^+ and SiO_2^-. When the abundances of these ions were plotted against ion bombardment time (i.e. depth), the width of the transition region could be estimated. Much literature is devoted to modern methods of analysing films and surfaces (see, for instance, Blattner & Evans, 1980; Werner, 1980).

12.3.4. Spark-source mass spectrometry

This is another means of obtaining mass spectra of involatile, inorganic compounds. The powdered sample is added to a relatively large amount of graphite and the resulting mixture is compressed into rods. Two rods are mounted close together in the ion source of a mass spectrometer and an electrical potential (> 10 kV) is applied between them. An electric discharge (spark) occurs in the gap between the rods (electrodes) and ions of the electrode material are formed there. The ions are accelerated out of the ionization region, analysed in the mass spectrometer and detected by the photographic plate method (section 2.5). This method of detection is used because all ions are detected simultaneously thereby compensating for erratic formation of ions.

Spark-source mass spectrometry has little application in organic chemistry. Covalent bonds are broken rather easily under the conditions of ionization, resulting in too large a number of fragment ions. Its greatest potential lies in the elemental analysis of involatile inorganic substances. A large number of elements (X) can be detected simultaneously as X_a^{b+} where a and b are small integers. The measurements are quantitative since the abundances of ions are related to the degree of darkening produced on the photographic plate. Spark-source mass spectrometry has been used to determine the nature of archaeological samples, traces of impurities in semi-conductors at the level of 1 part in 10^9 and the amounts of toxic metallic elements in biological samples such as teeth, tumour tissue and body organs.

12.4.   Ion cyclotron resonance spectroscopy

This method of mass spectroscopy has features which are different from the usual methods and merits some extra discussion. Two major differences between ion cyclotron resonance and the more commonly met methods of mass spectroscopy lie in the production of long-lived ions and the ability to examine ion/molecule collisions much more readily.

Under the influence of crossed magnetic and electric fields, ions describe a cycloidal path of frequency, $\omega_c = zE/m$, and have a drift velocity, $v = E/B$, where E is the electric field, B the magnetic field, m the mass of the ion and z its charge (figure 12.5). Both positive and negative ions obey the same laws and are distinguished only by the direction of rotation. In ion cyclotron resonance spectroscopy, ions are formed in an 'open' ion source (having no exit slits) and then subjected to the crossed electric and magnetic fields so that they travel through an analyser to the detector (figure 12.6). Space-charge effects would lead to the ions drifting to the walls of the analyser, and a 'trapping' potential is used to prevent this. In the region of the detector there is no trapping potential and the ions striking the walls there are recorded as the total ion current. Because the ions follow a long cycloidal path, the actual time taken to traverse the analyser is 5–10 ms although the analyser itself is only a few centimeters long. The time may be compared with the microsecond time of flight in many other mass spectrometers. The relatively long flight time in ion cyclotron spectroscopy leads to a high probability that ion/molecule collisions will occur before the ions are collected and it is possible therefore to investigate ion/molecule reactions (Lehman & Bursey, 1976).

Since all ions proceed through the analyser at the same drift velocity, individual ion species are examined by applying a radio-frequency electro-

Figure 12.5. Side view of path of ion beam in ion cyclotron resonance spectroscopy (see figure 12.6 for top view). The ion cyclotron frequency (ω_c) and drift velocity (v) are shown. The electric (E) and magnetic field (B) are crossed (E is in the plane of the paper; B is vertical to plane of paper).

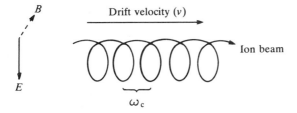

magnetic field to the ion beam in the analyser. Although all ions have the same drift velocity, their individual cyclotron frequencies (ω_c) are proportional to their m/z values. When the frequency of the applied radio-frequency field equals that of the ion-cyclotron frequency, absorption of energy occurs from the radio-frequency field; this resonance condition of energy absorption is measured to assign m/z values.

It is outside the scope of this book to discuss ion/molecule reactions (Knewstubb, 1969) but it may be mentioned that, apart from their intrinsic interest, they can give valuable thermochemical data such as proton and electron affinities through estimation or comparison of equilibrium constants. For example, from a knowledge of the pressures of the pyridines in the ion cyclotron resonance cell and measurement of the abundances of the protonated pyridines (structures 9 and 10; reaction 12.13), the free energy changes in the proton-transfer equilibria and the relative basicities of the pyridines can be measured in the gas phase (Taagepera, *et al.*, 1972).

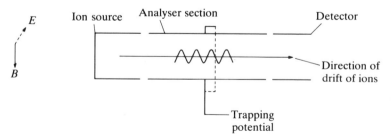

$$(12.13)$$

(9) (10)

Many other relative basicities or acidities have been obtained from similar gas-phase equilibria and compared with pK_a values determined in solution. Similarly, rate constants for gas-phase reactions can be measured (Lehman & Bursey, 1976, pp. 126-69) and compared with collision rate constants calculated from 'Langevin' equations (for leading references, see Johnstone, 1979). For systems in which reaction occurs each time an ion collides with a neutral

Figure 12.6. Top view of path of ion beam in ion cyclotron resonance spectroscopy showing side-to-side oscillatory motion of ions impressed on their cycloidal path (see figure 12.5). The ions produced in the ion source pass through an analyser section to a detector. A trapping potential prevents the ions striking the sides of the analyser. The electric field (E) is perpendicular to, and the magnetic field (B) is in the plane of the paper.

molecule, the measured rate constant and the collision (momentum-transfer) rate constants are equal.

The investigation of ion/molecule reactions by ion cyclotron resonance spectroscopy can yield considerable information on ion structures. The use of the method in this way may be illustrated by a single example. The ion resulting from six-centre rearrangement of 2-hexanone is considered to be the enol of acetone (11; scheme 12.14); the ion resulting from simple fragmentation

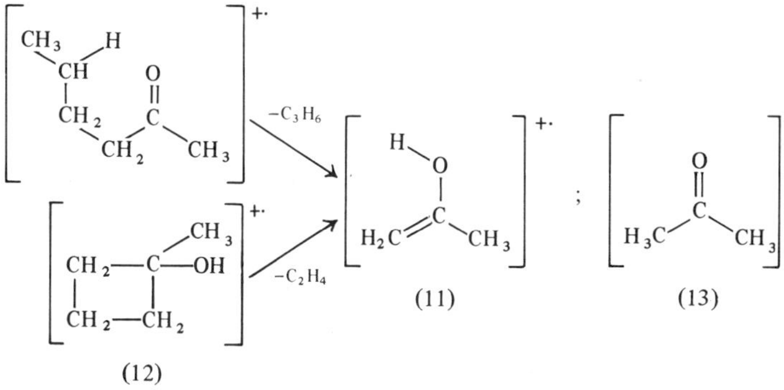

(12.14)

(11)

(12)

(13)

of the methyl cyclobutanol (12) is also thought to possess the structure (11). These fragment ions were allowed to collide with molecules in the analyser section of the ion cyclotron resonance spectrometer and subsequent ion/molecule reactions were investigated. No difference was found in the behaviour of the two ions (11) at m/z 58 from 2-hexanone or the methyl cyclobutanol (12), from which it was concluded the ions had the same structure. One typical ion/molecule reaction of the ion at m/z 58 is shown in scheme (12.15). The molecular ion at m/z 86 of the cyclobutanol (12) fragments to

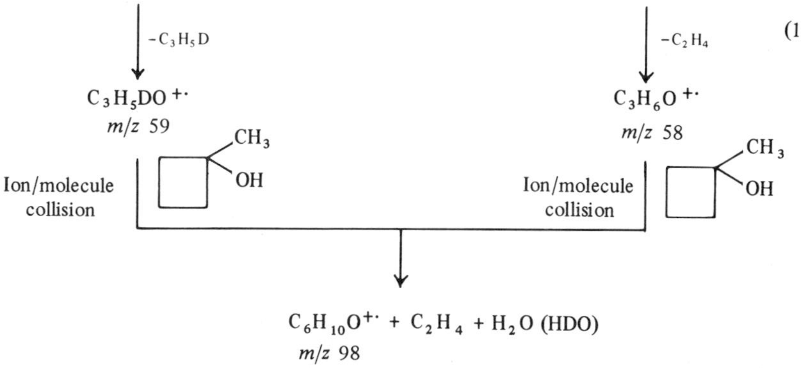

(14) m/z 102

m/z 86

(12.15)

$-C_3H_5D$

$-C_2H_4$

$C_3H_5DO^{+\cdot}$
m/z 59

$C_3H_6O^{+\cdot}$
m/z 58

Ion/molecule collision

Ion/molecule collision

$C_6H_{10}O^{+\cdot} + C_2H_4 + H_2O$ (HDO)
m/z 98

give the ion (11) at m/z 58 ($C_3H_6O^{+\cdot}$) which, by ion/molecule collision with a neutral molecule of the cyclobutanol (12), then affords an ion, $C_6H_{10}O^{+\cdot}$, at m/z 98. To differentiate the ion, $C_3H_6O^{+\cdot}$, at m/z 58 produced by the cyclobutanol (12) from the ion of the same composition produced by fragmentation of 2-hexanone, the latter was labelled with deuterium in the 5-position. The 5,5-d_2-hexanone (14) fragmented like 2-hexanone to give the ion (11) but now of composition C_3H_5DO at m/z 59. In the ion cyclotron resonance spectrometer, this ion at m/z 59 also underwent an ion/molecule reaction with neutral methyl cyclobutanol (12) to give the ion, $C_6H_{10}O^{+\cdot}$, at m/z 98 (scheme 12.15). Thus, the ions, $C_3H_6O^{+\cdot}$, from methyl cyclobutanol and C_3H_5DO from the hexanone (14), when produced together in the spectrometer, were both found to react with neutral methyl cyclobutanol molecules to give the ion, $C_6H_{10}O^{+\cdot}$, at m/z 98. The molecular ion of acetone (13; scheme 12.14) was investigated in the same way but its ion/molecule reactions were found to be different from those of the fragment ion (11). It may be concluded that the fragment ions of composition C_3H_6O derived from both the methyl cyclobutanone (12) and 2-hexanone (scheme 12.14) have the same structure which is different from the structure of the molecular ion of acetone (13). Although structures of ions may be compared in this way by ion cyclotron spectroscopy, it should be borne in mind that the actual structures may not be as postulated since rearrangement may occur in an ion/molecule reaction.

Negative-ion spectra may be measured as readily as positive ones by ion cyclotron resonance spectroscopy. An interesting application of this facility is the measurement of electronic excitations in molecules resulting from impact with low energy electrons. In the reaction sequence (12.16), electrons of energy E_1 are captured by SF_6 and the abundance of SF_6^- ions determined by ion cyclotron resonance. In the presence of other gas molecules, energy may be transferred from the electrons with energy E_1 due to electronic excitation of the molecules; in this process the electrons after impact have less

$$M + e(E_1) \rightarrow M^* + e(E_2); (E_2 < E_1)$$
(neutral molecule) (electronically excited)

$$SF_6 + e(E_1 \text{ or } E_2) \rightarrow SF_6^- \qquad (12.16)$$

energy (E_2). The electronic excitation energy of the molecule is then $E_1 - E_2$. The change in the impacting electron energy leads to a change in the abundance of SF_6^- ions. Thus, by monitoring the ion current due to SF_6^- ions as the energy of the impacting electrons is gradually increased, electronic excitation processes in the other gas molecules can be investigated.

It is also possible to carry out a technique of double resonance spectroscopy by irradiating one ion at its ion cyclotron frequency whilst observing the effect of this on the abundance of another ion at a different frequency.

If an ion/molecule reaction (12.17) is suspected, ion A^+ can be irradiated in the ion source at its cyclotron frequency, ω_A, and ion C^+ observed at its reso-

$$A^+ + B \rightarrow C^+ + D \qquad (12.17)$$

nance frequency, ω_C. If C^+ ions are derived from A^+ ions, then changing the translational energy of A^+ ions by irradiation at ω_A changes the size of the peak due to C^+ and the interrelationship between A^+ and C^+ can be demonstrated. For example, the ion at m/z 59 resulting from a double McLafferty rearrangement in the molecular ion of 4-nonanone-1,1,1-d_3 has been considered to have either structure (15) or (16) in reaction scheme (12.18).

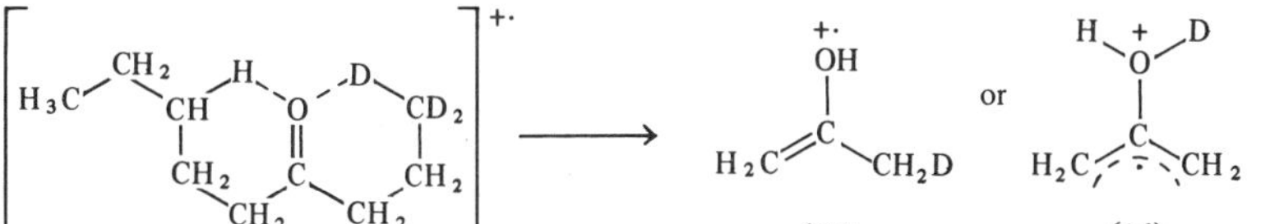

$$(C_3H_5DO^{+\cdot}; m/z\ 59) \qquad (12.18)$$

The molecular ion of a ketone such as 4-nonanone is readily protonated by collision with radical-ions that have hydrogen or an oxygen atom (reaction 12.19). Therefore, collision of the ion at m/z 59 with 4-nonanone-1,1,1-d_3 (molecular weight 145) could lead to transfer of H^+ or D^+ with formation of protonated 4-nonanone at m/z 146 $[(M + H)^+]$ or m/z 147 $[(M + D)^+]$ depending on the structure of the ion at m/z 59 (structure 15 or 16 in reaction 12.18). The ion at m/z 147 was observed at its cyclotron frequency and the ion at m/z 59 irradiated at its cyclotron frequency. No change was found in the abundance of the ion at m/z 147 and therefore it could not have been

$$CH_3(CH_2)_4\overset{\overset{\displaystyle O}{\|}}{C}(CH_2)_2CD_3 + C_3H_5DO^{+\cdot} \rightarrow$$
$$m/z\ 59$$

$$CH_3(CH_2)_4\overset{\overset{\displaystyle \overset{+}{O}-H(D)}{\|}}{C}(CH_2)_2CH_3 + CD_3 + C_3H_4D(H)O$$
$$[(M + H)^+]\ m/z\ 146$$
$$[(M + D)^+]\ m/z\ 147 \qquad (12.19)$$

formed by collision with the ion at m/z 59. On the other hand, the abundance of the ion at m/z 146 was strongly affected when the ion at m/z 59 was irradiated. It was concluded from this *double resonance* experiment that the ion at m/z 59 could not have had the structure (16) in scheme (12.18) because transfer of deuterium from the ion to 4-nonanone was not detectable.

As well as recording a mass spectrum by measuring the energy loss of the irradiating field at each ion cyclotron resonance frequency, a mass spectrum can be recorded by monitoring the total ion current at high field strengths. Under these conditions, when resonance occurs, an ion is ejected onto the sides of the cell before reaching the detector and therefore there is a decrease in the total ion current. These changes in the total ion current at each m/z value yield the mass spectrum.

It is possible to assess the relative contributions of different ion/molecule reactions leading to a common ion. Consider the two ion/molecule reactions (12.20).

$$A^+ + B \rightarrow C^+ + D$$
$$X^+ + B \rightarrow C^+ + Z \tag{12.20}$$

If the mass spectrum could be recorded without ion A^+ (or X^+), the relative importance of the two reactions leading to the formation of C^+ could be assessed. In ion cyclotron resonance spectroscopy this assessment can be made by 'ejecting' A^+ (or X^+) from the cell by application of an oscillating electric field to the trapping plates (figure 12.6). Since the ions are oscillating between the trapping plates as well as following a cycloidal path, then at resonance the oscillations increase and the ions are ejected to the walls. Ejection of A^+ would reduce the abundance of C^+ ions to that due solely to their formation from X^+ ions. The technique can be used with any number of reactant ions so long as they can be ejected selectively.

12.5. Literature on mass spectrometry

To close this introductory text, pointers to more specialized informtion are provided to help interested readers further their knowledge of mass spectrometry. There are four major categories of literature: books, periodicals, journals and compilations of mass spectra, which are discussed in turn. Many books on general and specialist mass spectrometry have been referred to at the appropriate points in the text. Other than these, several books may be recommended for the following topics: ion/molecule reactions (Franklin, 1979), collision processes (Cooks, 1978), computers (Carrick, 1979), mass spectrometry of inorganic and organometallic compounds (Litzow & Spalding, 1973), biomedical, environmental and forensic applications (Watson, 1976), and biochemical applications of mass spectrometry (Waller, 1972; Waller & Dermer, 1980), analysis of drugs and metabolites by gas chromatography/ mass spectrometry in several volumes (see Gudzinowicz & Gudzinowicz, 1978, 1979), analysis of organic pollutants in water (Keith, 1976), vacuum technology (Roth, 1976) and practical mass spectrometry (Middleditch, 1979).

There are too many periodicals on mass spectrometry for complete coverage here; only the most important ones are mentioned. The two most useful sources of comprehensive information are the biennial Specialist Periodical Reports on *Mass Spectrometry* (published by the Royal Society of Chemistry, formerly the Chemical Society) and reviews in *Analytical Chemistry* (American

Chemical Society). The former publication contains chapters on the main fields of mass spectrometry, extending from theory and energetics to methods and results, and occasional chapters critically appraising specialized areas such as organic geochemistry, food science and environmental aspects. The reviews in Analytical Chemistry (for instance, Burlingame, Baillie, Derrick & Chizhov, 1980) cover all areas of mass spectrometry in a highly condensed fashion but form a useful summary of the primary literature. Bibliographies of biomedical and environmental applications of stable isotopes have been compiled by Klein & Klein (1978, 1979*b*), these being up-dated biennially from 1979. Most other periodicals are the published proceedings of conferences and symposia. All aspects of mass spectrometry are covered at the International Conferences on Mass Spectrometry, the proceedings of which are published triennially as *Advances in Mass Spectrometry* (Heyden), and the *Annual Conferences on Mass Spectrometry and Allied Topics* (American Chemical Society). The former (see, for example, Quayle, 1980) present complete and contemporary reports of mass spectrometry. Two other worthwhile synopses of conferences, with explicit titles, are *Recent Developments in Mass Spectrometry in Biochemistry and Medicine* (Plenum Press) and *Quantitative Mass Spectrometry in Life Sciences* (Elsevier), both of which are published annually. Many published proceedings of symposia are useful for assessing the state of a particular subject, but are not necessarily up-to-date because there may be a long delay between the conference and publication resulting from it. It is recommended that the reader take notice of the date of the conference as well as that of the publication. Many of the research papers presented at such conferences appear in greater detail at a later date in the primary journals, and not infrequently before the conference has begun!

Three excellent journals in the English language are devoted solely to mass spectrometry. The *International Journal of Mass Spectrometry and Ion Physics* (Elsevier) concentrates on theory, energetics, instrumentation and methodology whilst the main themes of *Organic Mass Spectrometry* (Heyden) and *Biomedical Mass Spectrometry* (Heyden) are self-evident. A glance at the bibliography at the end of this book will show that many references have been made to these journals. Some journals containing substantial numbers of research papers on mass spectrometry are *Journal of Chemical Physics* (American Institute of Physics) concerned with theoretical aspects, *Analytical Chemistry* (American Chemical Society) concentrating on methodology and computers, *Analytical Biochemistry* (Academic Press), and *Journal of Chromatography* (Elsevier) and *Journal of Chromatographic Science* (Preston Technical Abstracts Company), both with many reports of gas and liquid chromatography/ mass spectrometry. Frequently, these and other journals include highly useful reviews of specialized areas within mass spectrometry. Such articles are far too numerous to mention individually here, but many have been referred to in the foregoing text.

Two abstracting journals are devoted to mass spectrometry. Published monthly by the Mass Spectrometry Data Centre (managed by the United Kingdom Chemical Information Service, UKCIS), the *Mass Spectrometry Bulletin* contains a comprehensive list of references, selected from current research literature, that are relevant to mass spectrometry. Each abstract in the list contains key words defining the subject area of the original paper. The key words form the basis of a subject index, making it easy for the reader to find abstracts of interest to him; a useful author index is also included. A more specialized abstracting service is provided by *Gas Chromatography-Mass Spectrometry Abstracts* (published bimonthly by PRM Science and Technology Agency) and gives a valuable aid to literature searching in the combined analytical techniques. Selected entries in the well known and popular journal *Chemical Abstracts* (American Chemical Society) which are concerned with all aspects of mass spectrometry are published separately and biweekly through UKCIS as *CA Selects: Mass Spectrometry*. Similarly, the *Scientific Documentation Centre* (UK) prints reference cards with the title and authorship of most mass spectrometric research papers. It should be borne in mind that abstracting journals, whilst being very convenient, are necessarily retrospective.

One route towards interpretation of an 'unknown' mass spectrum is to compare it with many reference spectra of known compounds for a possible match. Many data systems contain mass spectral libraries for comparison with mass spectra of unknown compounds (section 3.4). Such libraries, supplied by the computer manufacturers, may run to several tens of thousands of reference spectra. If a computer is not available, there are many different sources of published compilations of mass spectra for manual comparisons. Usually these collections contain abridged rather than whole spectra for the sake of brevity. One of the best known collections is the *Eight-Peak Index of Mass Spectra* from the Mass Spectrometry Data Centre. For each compound in the collection, the largest eight peaks and origin of the mass spectrum, the molecular weight, elemental formula and name are listed. The compilation of over 30 000 spectra is arranged in several different ways for easy reference. A similar strategy is used for the *Compilation of Mass Spectral Data* (Cornu and Massot, 1975). Containing over 7000 spectra, the ten largest peaks of each spectrum are recorded. The *Atlas of Mass Spectral Data* (Stenhagen, Abrahamsson & McLafferty, 1969) consists of the tabulated mass spectra of over 6000 compounds in order of their molecular weights and over 18 000 line diagrams of mass spectra are presented in the *Registry of Mass Spectral Data* (Stenhagen, Abrahamsson & McLafferty, 1974). The very large, computerized mass-spectral library of the US Environmental Protection Agency and National Institutes of Health is available in book form as the *EPA/NIH Mass Spectral Data Base* (US Government Printing Office). All of the above compilations are general in the sense that there is no bias towards any particular type of compound. Smaller, specialized libraries containing only the mass spectra of drugs, steroids, pesticides, etc. are also available from various sources. These and other collections, as well as abstracting journals, are discussed further in the book by Middleditch (1979, pp. 179-217).

Finally, the reader is referred to the International Union of Pure and Applied Chemistry (IUPAC) recommendations on nomenclature for mass spectrometry (Zerbi & Beynon, 1977). Unfortunately, in the past, practitioners of mass spectrometry, unrestrained by such standardization, produced much unnecessary symbolism and jargon to confound new-comers and experts alike. Happily, as the expertise of an individual in mass spectrometry increases, his ability to decipher the jargon increases. We hope this book will enable the reader to appreciate better the valuable techniques hiding behind such obfuscations.

BIBLIOGRAPHY

Adams, R. P., Granat, M., Hogge, L. R. & von Rudloff, E. (1979). *Journal of Chromatographic Science*, **17**, 75-81.

Arpino, P. (1975). *Recherches (Paris)*, **6**, 769-72.

Baillie, T. A. (1978). *Stable Isotopes. Applications in Pharmacology, Toxicology and Clinical Research*. London: Macmillan.

Barber, M. & Elliott, R. M. (1964). 12th ASTM E-14 Meeting on Mass Spectrometry, Montreal.

Barber, M., Bordoli, R. S., Sedgwick, R. D. & Tyler, A. N. (1981). *Nature*, **293**, 270-5.

Beckey, H. D. (1961). *Zeitschrift für Naturforschung*, **16a**, 505-10.

Beckey, H. D. (1977). *Principles of Field Ionisation and Field Desorption Mass Spectrometry*. Oxford: Pergamon.

Beckey, H. D. (1979). *Organic Mass Spectrometry*, **14**, 292.

Beckey, H. D. & Röllgen, F. W. (1979). *Organic Mass Spectrometry*, **14**, 188-90.

Beckey, H. D. & Schulten, H.-R. (1975). *Angewandte Chemie, International Edition*, **14**, 403-15.

Benninghoven, A. & Bispinck, H. (1979). *Modern Physics in Chemistry*, **2**, 391-421.

Benninghoven, A. & Sichtermann, W. K. (1978). *Analytical Chemistry*, **50**, 1180-4.

Bentley, T. W. (1979). In *Mass Spectrometry*, vol. 5, ed. R. A. W. Johnstone, pp. 71-74. London: The Chemical Society.

Beynon, J. H. (1960). *Mass Spectrometry and its Applications to Organic Chemistry*. Amsterdam: Elsevier.

Beynon, J. H. & Caprioli, R. M. (1980). In *Biochemical Applications of Mass Spectrometry, First Supplementary Volume*, ed. G. R. Waller & O. C. Dermer, pp. 89-102. New York: Wiley-Interscience.

Beynon, J. H., Caprioli, R. M., Baitinger, W. E. & Amy, J. W. (1970). *Organic Mass Spectrometry*, **3**, 455-77.

Beynon, J. H. & Cooks, R. G. (1975). In *MTP International Review of Science, Physical Chemistry Series Two*, vol. 5, ed. A. Maccoll, pp. 159-205. London: Butterworths.

Beynon, J. H., Morgan, R. P. & Brenton, A. G. (1979). *Philosophical Transactions of the Royal Society of London*, Series *A*, **293**, 157-66.

Beynon, J. H., Saunders, R. A. & Williams, A. E. (1968). *The Mass Spectra of Organic Molecules*. Amsterdam: Elsevier.

Beynon, J. H. & Williams, A. E. (1963). *Mass and Abundance Tables for Use in Mass Spectrometry*. Amsterdam: Elsevier.

Biemann, K. (1962). *Mass Spectrometry, Organic Chemical Applications*. New York: McGraw-Hill.

Biemann, K. (1979). *Journal of Molecular Evolution*, **14**, 65-70.

Biller, J. E. & Biemann, K. (1974). *Analytical Letters*, **7**, 515-28.

Bjoerkhem, I. (1979). *CRC Critical Reviews in Clinical Laboratory Sciences*, **11**, 53-105.

Blakley, C. R., Carmody, J. J. & Vestal, M. L. (1980). *Journal of the American Chemical Society*, **102**, 5931-3.

Blattner, R. J. & Evans, C. A., Jr (1980). *Journal of Educational Modules for Materials Science and Engineering*, **2**, 1-43.

Blau, K. & King, G. (1978). *Handbook of Derivatives for Chromatography*. London: Heyden.

Bowie, J. H. (1971). In *Mass Spectrometry*, vol. 1, ed. D. H. Williams, pp. 91-138. London: The Chemical Society.

Bowie, J. H. (1973). In *Mass Spectrometry*, vol. 2, ed. D. H. Williams, pp. 90-142. London: The Chemical Society.

Bowie, J. H. (1975). In *Mass Spectrometry*, vol. 3, ed. R. A. W. Johnstone, pp. 288-95. London: The Chemical Society.

Bowie, J. H. (1977). In *Mass Spectrometry*, vol. 4, ed. R. A. W. Johnstone, pp. 237-41. London: The Chemical Society.

Bowie, J. H. (1979). In *Mass Spectrometry*, vol. 5, ed. R. A. W. Johnstone, pp. 279-84. London: The Chemical Society.

Bowie, J. H. & Williams, B. D. (1975). In *MTP International Review of Science, Physical Chemistry Series Two*, vol. 5, ed. A. Maccoll, pp. 89-127. London: Butterworths.

Boyd, R. K. & Beynon, J. H. (1977). *Organic Mass Spectrometry*, 12, 163-5.

Brooks, C. J. W. & Edmonds, C. G. (1979). In *Practical Mass Spectrometry*, ed. B.S. Middleditch, pp. 57-126. New York: Plenum Press.

Brooks, C. J. W., Edmonds, C. G., Gaskell, S. J. & Smith, A. G. (1978). *Chemistry and Physics of Lipids*, 21, 403-16.

Brooks, C. J. W. & Middleditch, B. S. (1973). In *Mass Spectrometry*, vol. 2, ed. D. H. Williams, pp. 302-335. London: The Chemical Society.

Brooks, C. J. W. & Middleditch, B. S. (1975). In *Mass Spectrometry*, vol. 3, ed. R. A. W. Johnstone, pp. 296-338. London: The Chemical Society.

Brooks, C. J. W. & Middleditch, B. S. (1977). In *Mass Spectrometry*, vol. 4, ed. R. A. W. Johnstone, pp. 146-85. London: The Chemical Society.

Brooks, C. J. W. & Middleditch, B. S. (1979). In *Mass Spectrometry*, vol. 5, ed. R. A. W. Johnstone, pp. 142-85. London: The Chemical Society.

Brown, P. & Djerassi, C. (1967). *Angewandte Chemie, International Edition*, 6, 477-96.

Bruins, A. P., Jennings, K. R. & Evans, S. (1978). *International Journal of Mass Spectrometry and Ion Physics*, 26, 395-404.

Budzikiewicz, H., Djerassi, C. & Williams, D. H. (1964a). *Interpretation of Mass Spectra of Organic Compounds*. San Francisco: Holden-Day.

Budzikiewicz, H., Djerassi, C. & Williams, D. H. (1964b). *Structure Elucidation of Natural Products*, vols. 1 and 2. San Francisco: Holden-Day.

Budzikiewicz, H., Djerassi, C. & Williams, D. H. (1967). *Mass Spectrometry of Organic Compounds*. San Francisco: Holden-Day.

Burlingame, A. L., Baillie, T. A., Derrick, P. J. & Chizhov, O. S. (1980). *Analytical Chemistry*, 52, 214R-58R.

Campana, J. E. (1980). *International Journal of Mass Spectrometry and Ion Physics*, 33, 101-17.

Caprioli, R. M. & Bier, D. M. (1980). In *Biochemical Applications of Mass Spectrometry, First Supplementary Volume*, eds. G. R. Waller and O. C. Dermer, pp. 895-925. New York: Wiley-Interscience.

Carrick, A. (1979). *Computers and Instrumentation*. London: Heyden.

Carroll, D. I., Dzidic, I., Stillwell, R. N., Haegele, K. D. & Horning, E.C. (1975). *Analytical Chemistry*, 47, 1308-12.

Chapman, J. R. (1978). *Computers in Mass Spectrometry*. London: Academic Press.

Chen, E. C. M. (1979). In *Practical Mass Spectrometry*, ed. B. S. Middleditch, pp. 127-49. New York: Plenum Press.

Chesnavich, W. J. & Bowers, M. T. (1978). *Journal of Chemical Physics*, 68, 901-10.

Claeys, M., Muscettola, G. & Markey, S. P. (1976). *Biomedical Mass Spectrometry*, 3, 110-6.

Conzemius, R. J. & Capellen, J. M. (1980). *International Journal of Mass Spectrometry and Ion Physics*, 34, 197-271.

Cooks, R. G. (1978). *Collision Spectroscopy*. New York: Plenum Press.

Cooks, R. G., Beynon, J. H., Caprioli, R. M. & Lester, G. R. (1973). *Metastable ions*. Amsterdam: Elsevier.

Cornu, A. & Massot, R. (1975). *Compilation of Mass Spectral Data*, 2nd edn. London: Heyden.

Cottrell, T. L. (1965). *Dynamic Aspects of Molecular Energy States*, pp. 32-44. Edinburgh: Oliver & Boyd.

Daves, G. D., Jr (1979). *Accounts of Chemical Research*, 12, 359-65.

Dawson, P. H. (1976). *Quadruple Mass Spectrometry and its Applications*. Amsterdam: Elsevier.

de Leenheer, A. P. & Cruyl, A. A. (1980). In *Biochemical Applications of Mass Spectrometry, First Supplementary Volume*, ed. G. R. Waller and O. C. Dermer, pp. 1169-207. New York: Wiley-Interscience.

de Leenheer, A. P. & Roncucci, R. R. (1977). *Quantitative Mass Spectrometry in Life Sciences*, vol. 1. Amsterdam: Elsevier.

de Leenheer, A. P., Roncucci, R. R. & van Petegham, C. (1978). *Quantitative Mass Spectrometry in Life Sciences*, vol. 2. Amsterdam: Elsevier.

Derrick, P. J. (1977). In *Mass Spectrometry*, vol. 4, ed. R. A. W. Johnstone, pp. 132-45. London: The Chemical Society.

Dewar, M. J. S. & Worley, S. D. (1969). *Journal of Chemical Physics*, 50, 654-67.

Dromey, R. G., Buchanan, B. G., Smith, D. H., Lederberg, J. & Djerassi, C. (1975). *Journal of Organic Chemistry*, 40, 770-4.

Dromey, R. G. & Foyster, G. T. (1980). *Analytical Chemistry*, 52, 394-8.

Dromey, R. G., Stefik, M. J., Rindfleisch, T. C. and Duffield, A. M. (1976). *Analytical Chemistry*, 48, 1368-75.

Dubrin, J. & Henchman, M. J. (1972). In *MTP International Review of Science, Physical Chemistry Series One*, vol. 9, ed. J. C. Polanyi, pp. 213-45. London: Butterworths.

Fales, H. M., Milne, G. W. A., Winkler, H. U., Beckey, H. D., Damico, J. N. & Barron, R. (1975). *Analytical Chemistry*, 47, 207-19.

Field, F. H. & Franklin, J. L. (1957). *Electron Impact Phenomena*. New York: Academic Press.

Franklin, J. L. (1979). *Ion-Molecule Reactions, Parts I and II*. Stroudsburg: Hutchinson & Ross.

Franzen, J., Küper, H. & Riepe, W. (1974). *Analytical Chemistry*, 46, 1683-90.

Futrell, J. H., Ryan, K. & Siek, L. W. (1965). *Journal of Chemical Physics*, 43, 1832-3.

Games, D. E. (1979). In *Mass Spectrometry*, vol. 5, ed. R. A. W. Johnstone, pp. 287-92. London: The Chemical Society.

Garland, W. A. & Min, B. H. (1979). *Journal of Chromatography*, 172, 279-86.

Gaskell, S. J., Finney, R. W. & Harper, M. E. (1979). *Biomedical Mass Spectrometry*, 6, 113-6.

Gaskell, S. J., Pike, A. W. & Millington, D. S. (1979). *Biomedical Mass Spectrometry*, 6, 78-81.

Grönneberg, T. O., Gray, N. A. B. & Eglinton, G. (1975). *Analytical Chemistry*, 47, 415-9.

Gudzinowicz, B. J. & Gudzinowicz, M. J. (1978). *Analysis of Drugs and Metabolites by Gas Chromatography-Mass Spectrometry*, vol. 1: Respiratory Gases, Ethyl Alcohol, and Related Toxicological Materials; vol. 2: Hypnotics, Anticonvulsants, and Sedatives; vol. 3: Antipsychotic, Antiemetic, and Antidepressant Drugs; vol. 4: Central Nervous System Stimulants; vol. 5: Analgesics, Local Anaesthetics, and Antibiotics. New York: Dekker.

Gudzinowicz, B. J. & Gudzinowicz, M. J. (1979). *Analysis of Drugs and Metabolites by Gas Chromatography-Mass Spectrometry*, vol. 6: Cardiovascular, Antihypertensive, Hypoglycemic, and Thyroid-Related Agents. New York: Dekker.

Hammond, G. S. (1955). *Journal of the American Chemical Society*, 77, 334-8.

Hertz, H. S., Hites, R. & Biemann, K. (1971). *Analytical Chemistry*, 43, 681-91.

Hill, H. C. (1972). *Introduction to Mass Spectrometry*. London: Heyden.

Holland, J. F. (1979). *Organic Mass Spectrometry*, 14, 291.

Holland, J. F., Soltmann, B. & Sweeley, C. C. (1976). *Biomedical Mass Spectrometry*, 3, 340-5.

Holmes, J. L. (1975). In *MTP International Review of Science, Physical Chemistry Series Two*, vol. 5, ed. A. Maccoll, pp. 207-87. London: Butterworths.

Holmes, J. L. & Terlouw, J. K. (1980). *Organic Mass Spectrometry*, 15, 383-96.

Horning, E. C., Carroll, D. I., Dzidic, I., Haegele, K. D., Horning, M. D. & Stillwell, R. N. ←
(1974). *Journal of Chromatography*, 99, 13-21.
Horning, E. C., Mitchell, J. R., Horning, M. G., Stillwell, W. G., Stillwell, R. N., Nowlin,
J. G. & Carroll, D. I. (1979). *Trends in Pharmacological Science*, 1, 76-81.
Horning, M. G., Stillwell, W. G., Nowlin, J., Lertratanangkoon, K., Carroll, D., Dzidic, I.,
Stillwell, R. N., Horning, E. C. & Hill, R. M. (1974). *Journal of Chromatography*, 91,
413-23.
Howe, I. (1973). In *Mass Spectrometry*, vol. 2, ed. D. H. Williams, pp. 34-42. London:
The Chemical Society.
Hunt, D. F. & Crow, F. W. (1978). *Analytical Chemistry*, 50, 1781-4.
Hunt, D. F., Stafford, Jr, G. C., Crow, F. W. & Russell, J. W. (1976). *Analytical Chemistry*, 48, 2098-105.
Hunt, D. F., Stafford, G. C., Shabanowitz, J. & Crow, F. W. (1977). *Analytical Chemistry*, 49, 1884.
Isenhour, T. L., Kowalski, B. R. & Jurs, P. C. (1974). *CRC Critical Reviews in Analytical Chemistry*, 4, 1-44.
Jennings, K. R. (1965). *Journal of Chemical Physics*, 43, 4176-7.
Jennings, K. R. (1971). In *Mass Spectrometry: Techniques and Applications*, ed. G. W. A.
Milne, pp. 419-58. New York: Wiley-Interscience.
Jennings, K. R. (1977). In *Mass Spectrometry*, vol. 4, ed. R. A. W. Johnstone, pp. 203-
216. London: The Chemical Society.
Jennings, K. R. (1979). *Philosophical Transactions of the Royal Society of London*,
Series A, 293, 125-33.
Jennings, W. (1978). *Gas Chromatography with Glass Capillary Columns*. New York:
Academic Press.
Johnstone, R. A. W. (1979). In *Mass Spectrometry*, vol. 5, ed. R. A. W. Johnstone,
pp. 1-63. London: The Chemical Society.
Johnstone, R. A. W., Povall, T. J., Baty, J. D., Pousset, J.-L., Charpentier, C. & Lemonnier,
A. (1974). *Clinica Chimica Acta*, 52, 137-42.
Jolly, W. L. & Gin, C. (1977). *International Journal of Mass Spectrometry and Ion Physics*, 25, 27-37.
Justice, J. B. & Isenhour, T. L. (1974). *Analytical Chemistry*, 46, 223-6.
Karni, M. & Mandelbaum, A. (1980). *Organic Mass Spectrometry*, 15, 53-64.
Keith, L. H. (1976). *Identification and Analysis of Organic Pollutants in Water*. Ann
Arbor: Ann Arbor Science.
Kenyon, C. N., Melera, A. & Erni, F. (1980). *Journal of Chromatographic Science*, 18,
103-4.
Kimble, B. J. (1978). In *High Performance Mass Spectrometry*, ed. M. L. Gross, pp.
120-49. Washington: American Chemical Society.
Kiser, R. W. (1965). *Introduction to Mass Spectrometry and its Applications*. New Jersey:
Prentice-Hall.
Klein, E. R. & Klein, P. D. (1978). *Biomedical Mass Spectrometry*, 5, 91-111, 321-30,
373-9 and 425-32.
Klein, E. R. & Klein, P. D. (1979a). *Stable Isotopes - Proceedings of the Third International Conference*. New York: Academic Press.
Klein, E. R. & Klein, P. D. (1979b). *Biomedical Mass Spectrometry*, 6, 515-45.
Klots, C. E. (1976). *Journal of Chemical Physics*, 64, 4269-75.
Knapp, D. R. (1979). *Handbook of Analytical Derivatization Reactions*. New York:
Wiley & Sons.
Knewstubb, P. F. (1969). *Mass Spectrometry and Ion-Molecule Reactions*. London:
Cambridge University Press.
Kwok, K.-S., Venkataraghavan, R. & McLafferty, F. W. (1973). *Journal of the American Chemical Society*, 95, 4185-94.
Lacey, M. J. & Macdonald, C. G. (1979). *Analytical Chemistry*, 51, 691-5.
Lawson, G. & Todd, J. F. J. (1972). *Chemistry in Britain*, 8, 373-80.
Lederberg, J. (1964). *Compilation of Molecular Formulas for Mass Spectrometry*.
San Francisco: Holden-Day.

Lehman, T. A. & Bursey, M. M. (1976). *Ion Cyclotron Resonance Spectroscopy.* New York: Wiley.

Lehmann, W. D. & Schulten, H.-R. (1978). *Angewandte Chemie, International Edition,* 17, 221-38.

Light, J. C. (1967). *Discussions of the Faraday Society,* No. 44, 12-29.

Litzow, M. R. & Spalding, T. R. (1973). *Mass Spectrometry of Inorganic and Organometallic Compounds.* Amsterdam: Elsevier.

McCormick, A. (1977). In *Mass Spectrometry,* vol. 4, ed. R. A. W. Johnstone, pp. 85-101. London: The Chemical Society.

McCormick, A. (1979). In *Mass Spectrometry,* vol. 5, ed. R. A. W. Johnstone, pp. 121-41. London: The Chemical Society.

McFadden, W. H. (1973). *Techniques of Combined Gas Chromatography/Mass Spectrometry.* New York: Wiley-Interscience.

McFadden, W. H. (1979). *Journal of Chromatographic Science,* 17, 2-16.

McFadden, W. H. *et al.* (1980). *Journal of Chromatographic Science,* 18, 97-115.

Macfarlane, R. D. (1980). In *Biochemical Applications of Mass Spectrometry, First Supplementary Volume,* ed. G. R. Waller and O. C. Dermer, pp. 1209-18. New York: Wiley-Interscience.

Macfarlane, R. D. & Torgerson, D. F. (1976*a*). *International Journal of Mass Spectrometry and Ion Physics,* 21, 81-92.

Macfarlane, R. D. & Torgerson, D. F. (1976*b*). *Science,* 191, 920-5.

McKelvey, J. M., Alexandratos, S., Streitwieser, A. Abboud, J. L. M. & Hehre, W. J. (1976). *Journal of the American Chemical Society,* 98, 244-6.

McLafferty, F. W. (1963). *Mass Spectral Correlations.* Washington: American Chemical Society.

McLafferty, F. W. (1966). *Interpretation of Mass Spectra.* New York: Benjamin.

McLafferty, F. W. (1977). *Analytical Chemistry,* 49, 1441-3.

McLafferty, F. W. (1980). *Accounts of Chemical Research,* 13, 33-9.

Mahan, B. H. (1976). In *MTP International Review of Science, Physical Chemistry Series Two,* vol. 9, ed. A. D. Buckingham, pp. 25-65. London: Butterworths.

Majer, J. R. & Boulton, A. A. (1970). *Nature,* 225, 658-60.

Marcus, R. A. (1952). *Journal of Chemical Physics,* 20, 359-64.

Marcus, R. A. (1975). *Journal of Chemical Physics,* 62, 1372-84.

Massey, H. S. W. (1976). *Negative ions.* London: Cambridge University Press.

Mather, R. E. & Todd, J. F. J. (1979). *International Journal of Mass Spectrometry and Ion Physics,* 30, 1-37.

Meili, J., Walls, F. C., McPherron, R. & Burlingame, A. L. (1979). *Journal of Chromatographic Science,* 17, 29-42.

Meisel, W. S. (1972). *Computer Oriented Approaches to Pattern Recognition.* New York: Academic Press.

Mellon, F. A. (1975). In *Mass Spectrometry,* vol. 3, ed. R. A. W. Johnstone, pp. 117-42. London: The Chemical Society.

Mellon, F. A. (1977). In *Mass Spectrometry,* vol. 4, ed. R. A. W. Johnstone, pp. 59-84. London: The Chemical Society.

Mellon, F. A. (1979). In *Mass Spectrometry,* vol. 5, ed. R. A. W. Johnstone, pp. 100-20. London: The Chemical Society.

Middleditch, B. S. (1979). *Practical Mass Spectrometry.* New York: Plenum Press.

Milberg, R. M. & Cook, J. C., Jr (1979). *Journal of Chromatographic Science,* 17, 17-23.

Millard, B. J. (1978*a*). *Quantitative Mass Spectrometry.* London: Heyden.

Millard, B. J. (1978*b*). In *Quantitative Mass Spectrometry in Life Sciences,* vol. 2, ed. A. P. de Leenheer, R. R. Roncucci & C. van Peteghem, pp. 83-102. Amsterdam: Elsevier.

Millington, D. S. & Smith, J. A. (1977). *Organic Mass Spectrometry,* 12, 264-5.

Milne, G. W. A. & Lacey, M. J. (1974). *CRC Critical Reviews in Analytical Chemistry,* 4, 45-104.

Minnikin, D. E. (1978). *Chemistry and Physics of Lipids,* 21, 313-47.

Munson, M. S. B. & Field, F. H. (1966). *Journal of the American Chemical Society*, 88, 2621-30.
Pesyna, G. M., Venkataraghavan, R., Dayringer, H. E. & McLafferty, F. W. (1976). *Analytical Chemistry*, 48, 1362-8.
Pierce, A. E. (1968). *Silylation of organic compounds*. Illinois: Pierce Chemical Company.
Posthumus, M. A., Kistemaker, P. G., Meuzelaar, H. L. & ten Noever de Brauw, M. C. (1978). *Analytical Chemistry*, 50, 985-91.
Quayle, A. (1980). *Advances in Mass Spectrometry*, vol. 8. London: Heyden.
Reed, R. I. (1962). *Ion production by Electron Impact*. New York: Academic Press.
Reed, R. I. (1966). *Applications of Mass Spectrometry to Organic Chemistry*. New York: Academic Press.
Rose, M. E. (1981). *Organic Mass Spectrometry*, 16, 323-4.
Rose, M. E. (1982). In *Carotenoids 6. Proceedings of the Sixth International Symposium of Carotenoids*, ed. G. Britton & T. W. Goodwin, pp. 167-74. Oxford: Pergamon Press.
Rosenstock, H. M., Draxl, K., Steiner, B. W. & Herron, J. J. (1977). *Journal of Physical and Chemical Reference Data*, vol. 6, Supplement no. 1, 786 pp.
Rosenstock, H. M., Wallenstein, M. B., Wahrhaftig, A. L. & Eyring, H. (1952). *Proceedings of the National Academy of Science*, U.S., 38, 667-78.
Roth, A. (1976). *Vacuum Technology*. Amsterdam : North-Holland.
Russell, D. H., McBay, E. H. & Mueller, T. R. (1980). *International Laboratory*, 49-61.
Safron, S. A., Weinstein, N. D., Herschblach, D. R. & Tully, J. C. (1972). *Chemical Physics Letters*, 12, 564-8.
Schulten, H.-R. (1977). *Methods of Biochemical Analysis*, 24, 313-448.
Schulten, H.-R. (1979). *International Journal of Mass Spectrometry and Ion Physics*, 32, 97-283.
Sen, N. P., Miles, W. F., Seaman, S. & Lawrence, J. F. (1976). *Journal of Chromatography*, 128, 169-73.
Shiner, V. J., Jr & Buddenbaum, W. E. (1975). In *MTP International Review of Science, Physical Chemistry Series Two*, vol. 5, ed. A. Maccoll, pp. 129-58. London: Butterworths.
Simmons, D. S., Colby, B. N. & Evans, C. A. Jr (1974). *International Journal of Mass Spectrometry and Ion Physics*, 15, 291-302.
Slagle, J. R. (1971). *Artificial Intelligence, the Heuristic Programming Approach*. New York: McGraw-Hill.
Smith, D. H., Olsen, R. W., Walls, E. C. & Burlingame, A. L. (1971). *Analytical Chemistry*, 43, 1796-1806.
Snedden, W. & Parker, R. B. (1976). *Biomedical Mass Spectrometry*, 3, 295-8.
Spalding, T. R. (1979). In *Mass Spectrometry*, vol. 5, ed. R. A. W. Johnstone, pp. 312-46. London: The Chemical Society.
Stenhagen, E., Abrahamsson, S. & McLafferty, F. W. (1969). *Atlas of Mass Spectral Data*. New York: Interscience.
Stenhagen, E., Abrahamsson, S. & McLafferty, F. W. (1974). *Registry of Mass Spectral Data*. New York: Wiley & Son.
Stimpson, B. P. & Evans, C. A., Jr (1978). *Biomedical Mass Spectrometry*, 5, 52-63.
Stoll, R. & Röllgen, F. W. (1979). *Organic Mass Spectrometry*, 14, 642-5.
Sweeley, C. C., Elliot, W. A., Fries, I. & Ryhage, R. (1966). *Analytical Chemistry*, 38, 1549-53.
Taagepera, M., Henderson, W. G., Brownlee, R. T. C., Beauchamp, J. L., Holtz, D. & Taft, R. W. (1972). *Journal of the American Chemical Society*, 94, 1369-70.
ten Noever de Brauw, M. C. (1979). *Journal of Chromatography*, 165, 207-33.
Todd, J. F. J. & Lawson, G. (1975). In *MTP International Review of Science, Physical Chemistry Series Two*, vol. 5, ed. A. Maccoll, pp. 289-348. London: Butterworths.
van Marlen, G. & Dijkstra, A. (1976). *Analytical Chemistry*, 48, 595-8.
Vestal, M., Wahrhaftig, A. L. & Johnston, W. H. (1962). *Theoretical Studies in Basic Radiation Chemistry*, Aeronautical Research Laboratory Report, pp. 62-426.
von Ardenne, M., Steinfelder, K. & Tümmler, R. (1971). *Electronenanlagerungs-Massenspektrographic organischer Substanzen*. Berlin: Springer Verlag.

Waller, G. R. (1972). *Biochemical Applications of Mass Spectrometry*. New York: Wiley-Interscience.

Waller, G. R. & Dermer, O. C. (1980). *Biochemical Applications of Mass Spectrometry, First Supplementary Volume*. New York: Wiley-Interscience.

Wangen, L. E., Woodward, W. S. & Isenhour, T. L. (1971). *Analytical Chemistry*, **43**, 1605-14.

Ward, S. D. (1971). In *Mass Spectrometry*, vol. 1, ed. D. H. Williams, pp. 253-87. London: The Chemical Society.

Ward, S. D. (1973). In *Mass Spectrometry*, vol. 2, ed. D. H. Williams, pp. 264-301. London: The Chemical Society.

Watson, J. T. (1976). *Introduction to Mass Spectrometry. Biomedical, Environmental and Forensic Applications*. New York: Raven Press.

Werner, H. W. (1980). *Materials Science and Engineering*, **42**, 1-12.

Weston, A. F., Jennings, K. R., Evans, S. & Elliott, R. M. (1976). *International Journal of Mass Spectrometry and Ion Physics*, **20**, 317-27.

Wexler, S. & Parks, E. K. (1979). *Annual Review of Physical Chemistry*, **30**, 179-213.

Wilson, J. M. (1971). In *Mass Spectrometry*, vol. 1, ed. D. H. Williams, pp. 1-30. London: The Chemical Society.

Wilson, J. M. (1973). In *Mass Spectrometry*, vol. 2, ed. D. H. Williams, pp. 1-32. London: The Chemical Society.

Wilson, J. M. (1975). In *Mass Spectrometry*, vol. 3, ed. R. A. W. Johnstone, pp. 86-116. London: The Chemical Society.

Wilson, J. M. (1977). In *Mass Spectrometry*, vol. 4, ed. R. A. W. Johnstone, pp. 102-31. London: The Chemical Society.

Wolkoff, P., van der Greef, J. & Nibbering, N. M. M. (1978). *Journal of the American Chemical Society*, **100**, 541-5.

Zerbi, G. & Beynon, J. H. (1977). *Organic Mass Spectrometry*, **12**, 115-8.

Zerilli, L. F. (1979). *Chromatography Symposia Series*, **1**, 59-71.

INDEX